rary of Congress Cataloging-in-Publication Data

strength testing in soils : field and laboratory studies/
. Richards, ed.
. — (STP: 1014)
om the International Symposium on Laboratory and Field
r Strength Testing, held at Tampa, Fla., 22-23 January
sored by ASTM Committee D-18 on Soil and Rock.
publication code number (PCN) 04-010140-38."
bibliograhies and index.
-8031-1188-6
strength of soils—Testing—Congresses. 2. Vane shear
ngresses. I. Richards, Adrian F. II. International
m on Laboratory and Field Vane Shear Strength Testing
mpa, Fla.) III. ASTM Committee D-18 on Soil and Rock.
: ASTM special technical publication; 1014.
V36 1988
6—dc19 88-7684
CIP

NOTE

The Society is not responsible, as a body,
for the statements and opinions
advanced in this publication.

Peer Review Policy

blished in this volume was evaluated by three peer reviewers. The authors the reviewers' comments to the satisfaction of both the technical editor(s) Committee on Publications.

f the papers in this publication reflects not only the obvious efforts of the technical editor(s), but also the work of these peer reviewers. The ASTM Publications acknowledges with appreciation their dedication and contri- and effort on behalf of ASTM.

Printed in Baltimore, MD
December 1988

Vane Shear Stre
in Soils:
Field and Labora

Adrian F. Ric

STP 1014

ASTM
1916 Race Street
Philadelphia, PA 19103

Lil

Vane shea
Adrian
p. c
Papers
Vane She
1987, spo
"ASTM
Include
ISBN 0
1. Shea
tests—C
Symposi
(1987: T
IV. Serie
TA710.5
624.1'51

Each paper p
addressed all o
and the ASTM
The quality
authors and th
Committee on
bution of time

Foreword

The International Symposium on Laboratory and Field Vane Shear Strength Testing was held at Tampa, FL, on 22–23 January 1987. ASTM Committee D-18 on Soil and Rock sponsored the symposium. Adrian F. Richards, Adrian Richards Company, and Michael Perlow, Jr., Valley Foundation Consultants, Inc., served as chairmen of the symposium. Adrian F. Richards is also editor of the resulting publication.

Contents

Overview

This overview summarizes the results of the International Symposium on Laboratory and Field Vane Shear Strength Testing that was held in Tampa, FL, in Jan. 1987. The objectives of the symposium were to review the state of knowledge of the vane shear test (VST) and to provide the latest information on test theory, methods, and interpretation for the purpose of improved standardization of the field and laboratory vane tests. The need for a symposium at this time was based on the fact that the brief published results of the previous ASTM vane symposium appeared over two decades ago [*1*]. The vane literature since then has been extensive, including a short Australian overview of field vane testing and standardization by Walker [*2*] and an extensive overview by Aas et al. [*3*].

The field vane test was standardized by the ASTM for land testing in 1972 (ASTM Method for Field Vane Shear Test in Cohesive Soil [D 2573]), the laboratory vane test was standardized in 1987 (D 4648), and the offshore vane test has not yet been standardized by ASTM. Consequently, the time appeared auspicious to overview the entire subject of vane testing by holding an international symposium. It was also intended to help provide guidance to the ASTM Committee D-18 on soil and rock subcommittees concerned with the different problems of standardization using the various vane tests.

This Special Technical Publication (STP), presenting 22 papers from the symposium, has been organized into seven parts for simplicity of use as follows: Part I provides state-of-the-art reviews of the vane test on land and offshore. Part II is concerned with field vane theory and interpretation, while Part III covers the same topics for the laboratory vane. Part IV provides information on new laboratory test methods. Part V compares field vane testing to laboratory testing and other methods of in-situ testing. Part VI presents papers on the practice of vane testing on land, and Part VII does the same for vane testing offshore.

The 22 papers are intended for both theoreticians and practitioners involved with vane shear strength testing. They also should be useful to geotechnical engineers, geologists, and others concerned with laboratory and in-situ testing of soft soil, or sediment, and the application of test results to foundations, problems of soil instability, calibration of in-situ testing equipment, and other purposes.

The symposium was organized to include consideration of a number of problems in vane shear strength testing. These can be expressed as a series of questions: What is the range of soils suitable for vane testing? What are the principal advantages and disadvantages of the vane test? Does the field vane test yield data of the same or different reliability compared to other in-situ tests? In addition, What is the engineering significance of peak and residual (post-peak) strength? What corrections, if any, are appropriate for field vane testing? Does the method of soil remolding significantly influence the calculation of sensitivity? Most of these problems have been dealt with in a number of the symposium papers.

Recommendations from the symposium papers for standardization improvement are grouped and summarized for the ASTM D-18 subcommittees and others responsible for writing standards on vane testing. Suggestions for future work, based on information presented in many of the papers, are summarized following the recommendations. These summaries are intended to assist the reader interested in vane standardization or in the possibilities for future research in vane shear strength testing.

Symposium Contributions

In the following summary of some of the important points contained in the papers included in this book, the use of symbols follows the use given by the authors of the papers.

Chandler, in his state-of-the-art paper on the use of the field vane on clays, reported international acceptance of vane dimensions and test procedures resulting in what he called a "standard test." The standard test should result in undrained vane strengths in almost all uniform clays using a recommended relationship of $c_{uv} = 0.91\ M/\pi D^3$, where M is the maximum recorded torque. An approximate ratio of the field vane strength to CK_0UC triaxial strength, $V_r = 0.55 + 0.008\ I_p$ was stated to be only marginally dependent on the overconsolidation ratio (OCR) of the clay. Finally, the field vane strength, with an accuracy of about $\pm 25\%$ of the measured value, may be given for "normal" clays ($m \approx 0.95$) by the relationship of $c_u/\sigma_v' = S_1\ (\mathrm{OCR})^m$, where S_1 is the undrained strength ratio at OCR = 1.

Young, McClelland, and Quiros, in their state-of-the-art paper, summarized the results of an international survey they conducted on the practice of offshore vane testing. They found that the predominant use of the vane test was with the offshore petroleum industry, where measurements have been reliably and efficiently made in water depths exceeding 1000 m and with penetration depths as great as 440-m subseabed. In normally consolidated clay deposits (s_u/σ_{vo}' from 0.2 to 0.3) recommended adjustment values of the undrained shear strength are 0.7 to 0.8 for the design of axially loaded piles, 1.0 for development of *p-y* curves for laterally loaded piles, and 0.8 to 0.9 for bearing capacity and slope stability problems. The field vane was particularly useful to determine the strength of marine deposits having gas (low fluid saturation) or of high sensitivity. In these soils, the undrained shear strength may be significantly understated if it is obtained from tests on samples collected using high-quality samplers. The authors concluded that the vane shear test should become a standard test for offshore geotechnical investigations.

Becker, Crooks, and Been, in their interpretation of the field vane test in clays, presented evidence that strength ratios normalized using vertical preconsolidation pressures do not provide a good basis for comparison because they are not sufficiently refined. They suggest that a more rational comparison results from using strength ratios based on horizontal yield stresses, although they are not yet prepared to recommend that s_u/σ_{hy}' be established as the basis for developing alternative vane correction factors.

Silvestri and Aubertin discussed anisotropy and field vane testing. In sensitive clay deposits, the degree of strength anisotropy, or s_{uh}/s_{uv}, was found to vary between 1.14 and 1.41. They corroborated the Davis and Christian elliptical failure criterion.

Ortigão and Collet found that the Aas and others field vane corrections for the Rio de Janeiro clay were too conservative. Other cases of embankment failures in highly sensitive clays in which the Aas et al. corrections were found to be too conservative are also reported. The authors state that a correction factor was not necessary to obtain a safety factor near unity. They had no explanation to account for the differences between the results of their investigations and those reported by Aas et al.

Roy and LeBlanc performed field and laboratory vane tests in clay. They found that vane insertion produced disturbance and generated pore pressures that resulted in a reduced shear strength and that time effects led to an increased strength. A better designed vane blade, they suggested, would reduce disturbance and result in a 5 to 10% increase in the measured strength. They further recommended that not more than 1 min elapse between the time of vane insertion and the commencement of testing.

Karube, Shibuya, Baba, and Kotera used a cylinder shear test apparatus for the purpose of making vane test analyses in the laboratory. Reconstituted clays were tested. They found that the shear strength mobilized on the horizontal vane shear surface was larger than on the vertical shear surface. The difference was attributed to the different magnitudes of the effective normal stress acting on the two shear surfaces.

De Alencar, Chan, and Morgenstern performed a finite-element analysis of the laboratory vane test using an elasto-plastic constitutive relationship with strain-softening behavior. They showed that the peak torque is dependent upon the peak strength, the residual strength, and also on the rate of post-peak softening of the soil. The effect was reported to be particularly strong in very strain sensitive soils. The authors concluded that progressive failure has to be empirically corrected in practice, which requires a knowledge of the complete stress-strain curve obtained from laboratory testing to interpret the vane test and other in-situ tests involving shear failure.

Chaney and Richardson examined residual and remolded vane shear strength measured in the laboratory. They believed that the residual strength was reached after a 180° revolution of the vane. The remolded strength was dependent upon the method of remolding. The authors recommended a minimum of three vane revolutions to remold using the field vane, which yields a higher strength than either laboratory vane or hand remolding. Vane remolding was shown to be related to anisotropy, while hand remolding did not show this relationship.

Veneman and Edil, in the last symposium paper on the theory and interpretation of laboratory testing, studied shear structures developed during vane rotation using optical thin-section techniques. In soft and very soft clays of low plasticity, they found that the failure surface was a shear zone about equal to the vane diameter. The authors concluded that calculations of the undrained strength based on a fully developed cylindrical surface tend to underestimate the actual soil strength, and that the type of soil determines the magnitude of the deviation.

Pamukcu and Suhayda used a triaxial vane device equipped with a computer-aided data acquisition system for the detection of low strain shear deformations. They reported that the ratio of maximum static shear modulus to maximum dynamic shear modulus was about 0.85 in artificially prepared soft kaolinite specimens.

Almeida and Parry performed miniature vane and cone penetration tests in a bed of Gault clay overlying kaolin, which had been consolidated from slurry, in a centrifuge operating at 100 *g*. They found that the vane strengths compared well to theoretical strengths and that curves of point resistance with depth were similar to curves of vane strength with depth.

Tsutsumi, Y. Tanaka, and T. Tanaka used the laboratory vane test in a novel way to study the hardening characteristics of cements intended for soil stabilization in Japan. If the cement hardens too rapidly the blades used for mixing soil and cement cannot be extracted from the mixture. The authors found that the laboratory vane test could be successfully used to detect slight differences in early age hardening of the treated soils.

Lefebvre, Ladd, and Paré compared field vane strength with the undrained shear strength measured by triaxial and simple shear methods in the laboratory on marine clay specimens cut from block samples. The authors found very good agreement between the field vane strength and the laboratory test results for two clay deposits of low plasticity, high sensitivity, and medium to low OCR. Average correction factors were found to be unity, in agreement with the Bjerrum correction and in disagreement with the correction factors proposed more recently by Aas et al. The authors conclude that the field vane appears to be a reliable tool for profiling the undrained strength of low plasticity and sensitive clays for embankment stability, and they corroborate the use of the Bjerrum correction factors for these soils.

Greig, Campanella, and Robertson compared field vane test results at sites composed of soft organic clays, clayey silts, and sensitive clays to test results using the dilatometer, pressuremeter, piezocone, and screw plate. All of the in-situ test results were in reasonable agreement, despite the differences in failure mechanisms. At three of the sites the dilatometer results were different in value, compared to the results from other test methods, but similar in profile. A conclusion of the authors was that the piezocone and dilatometer provide continuous profiles of data that are equivalent to the field vane data if locally evaluated empirical correction factors are applied appropriately to the dilatometer and piezocone data.

Garga, in the first of the symposium papers on vane testing on land, investigated soft clays having variable amounts of sand, silt, and organic matter. At the site of a new port near Rio de Janeiro, field vane strengths were found to be similar to unconfined compression and UU triaxial tests. The Aas method to determine the anisotropy ratio S_h/S_v was not corroborated, and the strength anisotropy at the site could not be determined using vanes having different height to diameter ratios. Garga reported that the vane strength increase at the site was not the same as the effective stress increase underneath the test fills, and that the vane test could not be reliably used to monitor the consolidation progress of the soil.

Nagarkar, Rode, Shurpal, and Dixit used the field vane in soft, sensitive normally consolidated clays near Bombay to obtain strength profiles for the design of embankments. Conventional sampling and laboratory tests on samples from the same site also were undertaken. Laboratory strengths were found to be 40 to 60% lower than the field vane test results. The use of a vane guard resulted in strength values about 12% higher than when a guard was not used and the soil was more disturbed. Calculated and observed settlements were essentially the same, leading the authors to conclude that the field vane test, especially with the vane guard, is a highly reliable method for soft soil investigations.

Johnson, Hamilton, Ebelhar, Mueller, and Pelletier, in the first of the symposium papers on field vane testing offshore, compared results from the vane and cone penetrometer tests to results from laboratory vane, UU triaxial, and CU triaxial tests. The purpose of their study was to obtain a stress history and normalized soil engineering properties (SHANSEP) shear strength for the Gulf of Mexico deepwater clays, as part of the American Petroleum Institute's recommended practice to use normalized shear strength for static pile foundation design. The authors reported that UU triaxial strengths deviate significantly from SHANSEP and the in-situ shear strength at deeper depths, probably caused by stress relief during sampling. They concluded that in-situ strength and SHANSEP provide a means of interpreting shear strength for pile design, at least for the Gulf of Mexico deepwater clays.

Quiros and Young repeated approximately the same approach for the slightly overconsolidated Pleistocene clays of the Santa Barbara Channel, CA. They found that laboratory vane tests on specimens from pushed wireline samples had strengths about 30% less than field vane tests, and that the UU triaxial tests on specimens from pushed samples were about the same as the field vane $\pm 10\%$. The UU triaxial and field vane data were found to be in good agreement with the SHANSEP profile.

Geise, Ten Hoope, and May reported on the design, construction, and use of the Fugro field vane for wireline and seabed operations. They concluded with a number of recommendations relative to the ASTM field vane standard (D 2573): (1) For soils having shear strengths of less than 20 kPa, conventional heave compensation is inadequate and a hard-tie system should be used if the recommended vane test 1 m below the drillbit is to be followed; otherwise, vane tests should be performed at a minimum distance of 1.5 m, or five times the borehole diameter, for these soft soils. (2) Two or three vane rotations at 1.0°/s are adequate for remolded tests offshore. (3) ASTM Type A1 (rectangular and tapered) and G1 (rectangular) vane blades should not be used offshore because of their excessively high area ratios. The blades need to be redesigned to have an area ratio of 12% or less. (4) The ASTM D 2573 field vane standard is considered to be satisfactory for offshore testing without further modification other than the above recommendations.

Kolk, Ten Hoope, and Ims compared field vane tests on normally consolidated silts and clays, which had various amounts of carbonate, with the field piezocone and UU triaxial tests, laboratory vane, and Torvane tests. For soils from the North Sea and off the west coast of India, the field vane strength without correction factors compared well to the UU triaxial test results. The remolded UU triaxial strength compared well to the residual field vane strength (called the "post-peak" strength in the paper) at one site; data were unavailable from two other sites. The laboratory vane and the Torvane strength results were generally lower than results from the field vane; however, when piston sampling using the hard-tie heave compensation system, there was less divergence of data than when using push sampling without the hard-tie. The authors believed that K_0 could be estimated from field vane strength data together with triaxial results.

Silva and Wyland, in the final symposium paper, discussed their latest results using a remotely operated seabed field vane to obtain shear strength profiles in water depths of 6000 m and to a penetration depth of 1.5 m. The authors reported that the in-situ strengths were considerably greater than the strengths obtained from laboratory vane tests on core samples collected nearby. Field and laboratory vane strength profiles were found to be similar.

Recommendations for Vane Test Standardization

Three papers contained recommendations relevant to the various ASTM vane shear test standards. These are briefly listed in Table 1.

Midway through the symposium there was a panel discussion on standardization of the VST. Before the meeting, a panel steering committee put together a series of suggestions for VST standards (Table 2). There appeared to be no disagreement to these suggestions following their presentation to the participants.

Directions for Future Research and Development

A number of papers contained suggestions for future work. These are given in Table 3. The research and development suggested is that determined by the present author from information presented in the papers. He assumes responsibility in the event that he has misinterpreted the original authors, who are listed for the purpose of assisting a reader desiring to obtain additional information.

Summary and Conclusions

The standardization panel, referred to previously, posed some questions for discussion (Table 4) at the end of the first day that were considered to be important to the VST. At the end of the symposium most of these questions were put to the participants together

TABLE 1—*Recommendations for vane test standardization.*

Authors	L = Land O = Offshore	Recommendation
Roy and Leblanc	L	A maximum of 1 min should elapse from the time of insertion of the vane until the beginning of a test. Vane blades redesigned to cause less soil disturbance would be preferable over those specified.
Young et al.	O	Use their stated correction factors for normally consolidated Gulf of Mexico or similar offshore soils for piles, *p-y* curves, and bearing capacity and slope stability problems. Monitor or control the following for quality testing: vane blade geometry, vane rotation rate, bottomhole test penetration, drilling fluid weight and pressure, and torque calibration.
Geise et al.	O	A hard-tie heave compensation system should be used for quality testing when the soil shear strength is 20 kPa or less. Only in this case can the recommended test depth of 1 m below the drillbit be considered; without the hard-tie, the minimum test depth should be not less than 1.5 m below the drillbit. Two or three vane revolutions, made at a rotation speed of 1.0°/s, are considered adequate for the remolded strength test. Some of the vane blades specified by the ASTM need redesign to reduce the area ratio to 12% or less, while at the same time maintaining structural integrity for use in very stiff soils. The field vane standard is satisfactory if these recommendations are included.

TABLE 2—*Panel suggestions for field vane test standards.*

Hardware details
- dimensions: rectangular, with $H/D = 2$
  - $H = 125 \pm 25$ mm
  - $D = 62.5 \pm 12.5$ mm
  - blade thickness $\approx$ 2 mm (area ratio $\approx$ 12%)
- sleeved rods to avoid friction on rods
- geared drive

Procedural details
- depth below base of borehole: $\geq 5 \times$ hole diameter
- rest period before vane rotation: ≤ 5 min
- rate of rotation: 6 to 12°/min
- interpretation: $c_u = 0.86\ M/\pi D^3$ (M = max torque, and D = vane blade diameter)
- sensitivity should be reported, procedure requires standardization
- for offshore testing, specification is also required for mud pressure, motion compensation, and so forth
- the above procedural details apply to "uniform" soils having $S_t \leq$ 10 to 15 and $c_d \leq 100$ m^2/year; thus, vane test data must be accompanied by geological data

TABLE 3—*Suggested future VST research and development.*

Authors	R & D Suggested from Paper
Young et al.	Investigate how to be able to determine K_0, G, and E as part of the offshore VST.
Becker et al.	Further study the stress-strain characteristics of the soil for better interpretation of the VST. Refine the $S_u/\sigma'p - I_p$ approach for comparing vane strength and the strength operational in field failures by knowledge of horizontal stress components. Evaluate the assumptions needed to obtain σ'_{hy}.
Silvestri and Aubertin	Investigate uncertainties restraining the anisotropic analysis presented. Evaluate vane test stress relief effects. Determine the stress path generated by the VST for all stress directions.
Roy and Leblanc	Continue investigations of the failure mode in the VST.
De Alencar et al.	Obtain stress-strain curve, and relate it to the VST for the interpretation of results. Further investigate the relationship of the peak strength and progressive failure in high sensitivity soils.
Chaney and Richardson	Evaluate relationships of peak vane strength, remolded laboratory vane strength, and remolded field vane strength with vane orientation. Determine the relationship among the hand remolded strength, the laboratory vane remolded strength, and the I_L of the soil, especially with regard to the calculation of sensitivity.
Ortigão and Collet; also, Lefebvre et al.	Investigate further the problems in applying the Bjerrum and the Aas et al. VST corrections related to embankment failures with respect to resolving disagreement between the two methods.
Greig et al.	Extend comparisons among the different in-situ tests and the standard laboratory tests to the full range of applicable soil types in engineering practice.
Garga	Propose a universally correct method for analyzing the VST.
Johnson et al.	Extend procedures relating to SHANSEP, the VST, and laboratory tests for the same types of soils for confirmation, as well as to other soil types and geographic regions.
Geise et al.	Evaluate the extent of disturbance below the drillbit in different soil types and under different operating conditions relative to quality testing and sampling offshore.
Kolk et al.	Continue studies into obtaining K_0 reliably from the VST and triaxial strength data.
Silva and Wyland	Further evaluate stress relief effects, particularly when soil samples are raised from great water depths.

TABLE 4—*Panel proposed questions for discussion by participants.*

1. Do standards need to differ onshore and offshore?
2. What rotation is required to define the remolded shear strength for the calculation of sensitivity? What rotation rate should be used?
3. What "back-up" geological data are required? (For instance, plasticity index, natural water content, overconsolidation ratio, and soil profiles.)
4. When should or should not use be made of the field vane?
5. What is the field vane shear strength to be used for?

with answers that arose out of the symposium papers and discussions to learn if there was any disagreement. There appeared to be none; consequently, the following summary may be considered to represent the conclusions of the symposium participants. It has been slightly modified in this paper to take into account information contained in the revised symposium papers.

Standardization of the vane shear test is well established in both Europe and North America. With regard to the ASTM VST standardization, it is suggested that the existing D 2573 field vane standard be augmented by adding to it the differing test methods required for offshore use. In addition, the D 2573 standard needs further elaboration and clarification, and suggestions for this have been given in Tables 1 and 2. There appears to be no need at this time to have a separate field vane standard for offshore testing.

All types of field vane test methods appear to yield data of about the same degree of reliability and repeatability compared to other types of in-situ tests. The vane is particularly useful because it can be used to obtain the sensitivity of soil in situ. However, sensitivities calculated from field vane and laboratory measurements may be different, and a relationship between the two has not been standardized. The definitions of the post-peak or residual shear strength, the different types of sensitivity calculations, and perhaps other terms, need to be standardized.

Obtaining vane data is easy, but what to do with the data is more difficult. It was expressed during the symposium that the field vane shear strength is useful as a benchmark. In particular, the vane strength may be used to determine the cone factor for the cone penetrometer and piezocone tests. It is more appropriate to calibrate the various cones using shear strength data from in-situ measurements than from strength measurements made in the laboratory.

The suitability of the field vane to the type of soil appeared to be partly related to the user's experience with regard to the soil type, familiarity with the test method, and comprehension of the particular problem under investigation. While there was general consensus for the use of the VST in normally consolidated clays, opinion was divided on the suitability of the VST for other soil types. After so many decades of vane testing, this uncertainty still has not been resolved.

There was a difference of opinion among the symposium participants whether or not to correct field vane test results and, if corrections are to be applied, what the magnitude of the correction should be. In this regard, there was a concern expressed regarding comparison of VST results with routine and sophisticated laboratory results because of the sampling problem and attendant disturbance that usually affects the quality of strength data from the laboratory tests. Clearly, this is an area needing further investigation.

Bjerrum established a linkage between vane strength and embankment failures. A comparable linkage to other types of foundations appears not yet to be attained, except for some offshore applications. Several papers presented at the symposium pointed out disagreement between the Bjerrum and the Aas et al. methods of relating VST results to embankment failures. This area, also, would appear to be appropriate for further study and analysis.

There was general agreement that the laboratory vane test provided a simple and convenient index of shear strength. Everyone also agreed that the field vane test provided data that can be related to foundation design.

Acknowledgments

M. Perlow, Jr., assisted with the initial preparation of the symposium; unfortunately, illness prevented his attendance at the meeting and subsequent interaction. R. C. Chaney provided substantial assistance in many ways, including identifying the names of reviewers for the symposium papers. My employer at the time of the symposium, Fugro Geotechnical Engineers B.V., provided considerable support; N. Withers, H. Zuidberg, and others discussed a number of aspects and made useful suggestions. A number of colleagues graciously contributed time to meet on several occasions at the symposium to help plan for the panel presentations, which were made by E. Brylewski, R. Chandler, B. Ims, and A. Young, and to prepare the information given in Tables 2 and 4. The following persons are thanked for their constructive and helpful reviews of this paper: R. Chandler, J. Geise, J. ten Hoppe, J. Nieuwenhuis, E. Richards, J. de Ruiter, and H. Zuidberg.

References

[1] ASTM, *Vane Shear and Cone Penetration Resistance Testing of In-Situ Soils, STP 399,* American Society for Testing and Materials, Philadelphia, 1966, 47 pp.

[2] Walker, B. F., "Vane Shear Strength Testing," *In-Situ Testing for Geotechnical Investigations,* M. C. Ervin, Ed., A. A. Balkema, Rotterdam, The Netherlands, 1983, pp. 65–72.

[3] Aas, G., Lacasse, S., Lunne, T., and Høeg, K., "Use of In Situ Tests for Foundation Design on Clay," *Use of In Situ Tests in Geotechnical Engineering,* S. P. Clemence, Ed., American Society of Civil Engineers, New York, 1986, pp. 1–30.

Adrian F. Richards
Adrian Richards Company,
Aaslmeer, The Netherlands;
symposium chairman and editor.

Part I: State-of-the-Art Reviews

Richard J. Chandler[1]

The In-Situ Measurement of the Undrained Shear Strength of Clays Using the Field Vane

REFERENCE: Chandler, R. J., **"The In-Situ Measurement of the Undrained Shear Strength of Clays Using the Field Vane,"** *Vane Shear Strength Testing in Soils: Field and Laboratory Studies, ASTM STP 1014,* A. F. Richards, Ed., American Society for Testing and Materials, Philadelphia, 1988, pp. 13–44.

ABSTRACT: A brief review of the standard dimensions, insertion procedures, and rates of rotation of the field vane is followed by consideration of those factors that influence the undrained shear strength measured by the field vane. These factors include the disturbance effects of vane insertion, the consequences of a "rest" period before shear, strain-rate effects, the rotation rate necessary to ensure undrained conditions during shear, and the method of calculation of c_u. Following a discussion of the relative magnitude of c_u measured by the field vane, it is concluded that there is a consistent relationship with other measures of c_u, depending on the plasticity index, provided the other strength measurements are made on specimens reconsolidated to the in-situ stresses. After consideration of a number of case records reported in the literature, it is suggested that for "standard" clays the approximate expression, $c_u/\sigma_v' = S_1\,(\text{OCR})^{0.95}$, may be useful as a check on the measured field vane strength, or as an estimate of OCR. In this equation the strength ratio $S_1 = (c_u/\sigma_v')$ is that given by Bjerrum for "Young" clays, for which OCR $\cong$ 1.0.

KEY WORDS: field vane shear tests, in-situ undrained shear strength, soil mechanics, state-of-the-art review, strain-rate effects, undrained strength relationships, plasticity index relationships

The field vane is the most widely used method for the in-situ determination of the undrained strength of soft clays. Originally used in Sweden in 1919, it has been employed extensively on a worldwide basis since the late 1940s, following pioneering development work by, among others, Carlsson [*1*], Skempton [*2*], and Cadling and Odenstad [*3*]. A detailed history of the use of the field vane is given by Flodin and Broms [*4*].

It is the purpose of this paper to (1) review the current understanding of the use of the field vane test to measure the in-situ undrained strength of soft clays and (2) to show how the field vane strength relates to other measurements of undrained strength. In order to do this, international understanding of the "standard" elements of the field vane test is reviewed, and the influences of the various stages of the test on the measured undrained strength are considered in some detail. These influences include the effects of vane insertion, of varying periods of delay following insertion, and of rate of rotation. The interpretation of the test to obtain the undrained strength from the maximum torque is also considered.

The second half of the paper discusses relationships between the field vane undrained strength and other measurements of undrained strength, and such factors as plasticity

[1] Reader in soil mechanics, Department of Civil Engineering, Imperial College, London SW7 2BU, United Kingdom.

index and overconsolidation ratio (OCR). Such relationships are widely used for comparison of strength data from different sites and from different clay types, and also for the selection of design parameters. A number of case records, which give both field vane and triaxial strength measurements, have been reviewed. These form a "data base," which is presented in the Appendix, and is used to examine the various undrained strength relationships.

The discussion is limited to the field vane test per se, and little attempt is made to extend the interpretations to the strength mobilized in full-scale field failures. The reason for this is that the considerable complications that arise from the back-analysis of field failures (for example, three-dimensional effects, strength anisotropy, strain-rate effects, and progressive failure, and so forth) put such a discussion well beyond the scope of the present paper. Moreover, the subject has recently been reviewed [5].

The "Standard" Field Vane Test

The first topic that must be addressed is the extent of international agreement on the basic experimental details of the field vane test. While the writer has not made a comprehensive review of the many national standards specified for the performance of the field vane test, even a cursory perusal of the literature reveals that there is in practice considerable international agreement concerning most of the essential elements of the test. These elements are summarized in Fig. 1. Throughout this paper reference to a "standard" field vane test is to be taken as one complying with most, if not all, these criteria. There seems to be universal agreement that the vane consists of four blades set at right-angles, with a

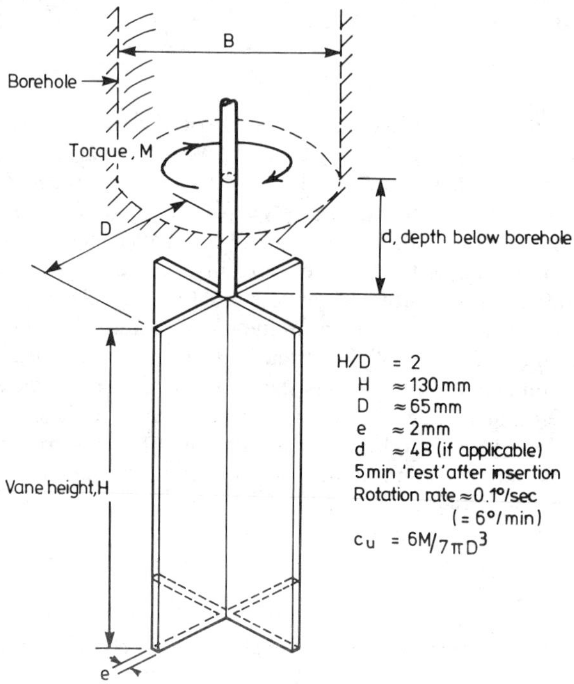

FIG. 1—*Summary of the most commonly used dimensions and other details of the field vane test.*

height:width (or diameter) ratio H/D of 2:1; the main variation is that of the size of the field vane.

Vanes of different dimensions are sometimes specified to allow larger vanes to be used in the softer clays, so that torque measuring systems can, for greater accuracy, operate over comparatively limited ranges. However, as will be seen later, strain-rate effects are important in the vane test, and, since these effects vary with the vane diameter, there probably is benefit to be gained by restricting the use of the field vane to one particular size. Since the most widely used dimensions appear to be $H = 130$ mm and $D = 65$ mm, these would seem to be the most appropriate for standardization.

Other dimensions on which there is wide agreement are the vane blade thickness ($\cong$2 mm) and the area ratio (the ratio of the volume of soil displaced by the vane to the soil volume swept by the rotated vane), which is almost universally specified to be less than 12%.

Rod friction should be accounted for with the use of sleeved rods or a slip coupling.

A major area of potential difference in the various national standards arises where field vane tests are carried out below the base of a borehole, in contrast to the more usual use of a self-contained vane device. Where a borehole is used it is clearly necessary to insert the vane to a depth below the base of the boring that is sufficient to ensure that the vane is in truly undisturbed soil (probably at least four borehole diameters), and that the in-situ stresses are not affected by the presence of the borehole. The use of drilling fluid in the borehole may be of considerable value in this case, particularly in a sensitive clay.

The usual "rest-period" following vane insertion seems to be 5 min, and the rate of rotation is almost invariably specified as either 6° or 12°/min. In practice this typically results in failure occurring at about 1 min. These two factors significantly influence the measured undrained strength and are reviewed in detail in the following sections.

The final major factor is the conventional interpretation that

$$c_u = 6M/7\pi D^3 \tag{1}$$

where c_u is the undrained shear strength, and M the maximum recorded torque. This interpretation is also discussed later; it appears likely to yield a conservative value of c_u, though perhaps only marginally so.

All the field vane test results that are quoted in this paper are presumed to comply with these various criteria. Such tests are referred to as "standard" tests. In particular, it is assumed throughout that c_u has been computed using Eq 1.

It will become apparent later that the sensitivity of the clay S_t may be a significant factor influencing the measured undrained strength. Where possible this should always be measured, and for convenience and consistency it seems most appropriate to do this with the field vane. An internationally agreed procedure for making this measurement is urgently required. It must be appreciated, however, that other test methods may well yield different values of sensitivity [*6*].

Shear Stress Distribution Around the Vane

The problem of the distribution of shear stress around the periphery of the rotated vane, which is central to the interpretation of the field vane test, has recently been reviewed by Wroth [*6*]. He cites the results of two investigations: an elastic, three-dimensional finite-element analysis reported by Donald et al. [*7*], and complementary experimental results by Menzies and Merrifield [*8*]. In the former case the shear stresses were determined on a section midway between the tips of the vanes, while in the latter case the stresses were measured at the tips of an instrumented vane.

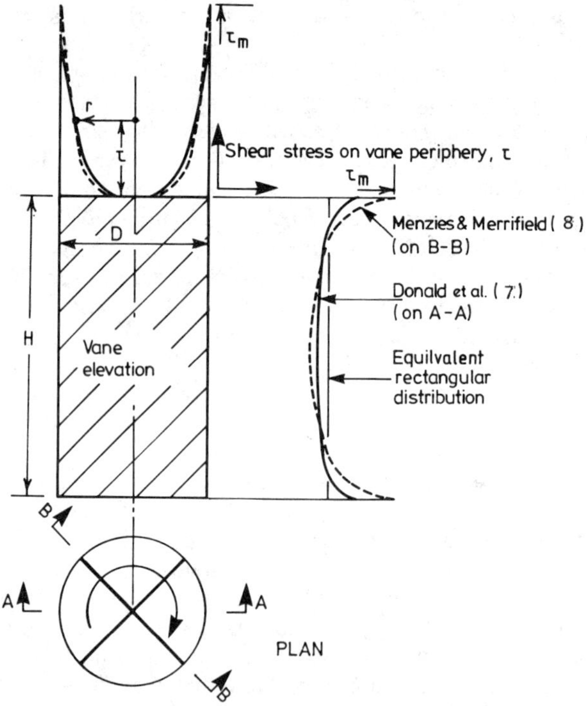

FIG. 2—*Shear stress distribution on the cylindrical surface described by the rotating vane. The shear stress distribution on the horizontal surfaces is given by* π/π_m = $(r/D/2)^n$.

As can be seen in Fig. 2 the results of both these studies show that the shear stresses on the vertical sides of the cylindrical potential failure surface are reasonably close to the conventional assumption of a uniform shear stress distribution, though with small peaks at the top and bottom corners. In contrast, the shear stress distribution on the horizontal portions of the vane is highly nonuniform, with dramatic peaks at the corners. The stress distributions determined in the two different investigations are, however, encouragingly similar.

From these results Wroth suggested that the shear stress distribution at the top and bottom surfaces of the vane periphery could be given by a polynominal of the form

$$\tau/\tau_m = \left(\frac{r}{D/2}\right)^n \tag{2}$$

where τ_m is the maximum shear stress, assumed to occur simultaneously around the entire vertical perimeter; r is defined in Fig. 2.

Thus, assuming the soil strength to be isotropic, the portion of the total torque M acting on the two horizontal surfaces is

$$\begin{aligned} M_h &= 2\int_0^{D/2} 2\,\pi r^2 \tau\, dr \\ &= \frac{\pi\, D^3 \tau_m}{2(n+3)} \end{aligned} \tag{3}$$

Since the portion of total torque provided around the vertical surface (of height H) is, from the conventional analysis

$$M_v = (\pi D^2 H \tau_m)/2$$

the ratio of the torques is

$$M_h/M_v = D/H(n + 3) \qquad (4)$$

If the conventional interpretation of uniform shear stress across the top of the vane is followed, $n = 0$ in Eq 2, and Eq 3 becomes

$$M_h = (\pi D^3 \tau_m)/6$$

The ratio of the torques is then, taking $H = 2D$

$$M_h/M_v = 1/6 \qquad (5)$$

Noting that Menzies and Merrifield's data for London clay give $n \cong 5$ [6], the observed polynominal shear stress distribution on the horizontal surfaces yields

$$M_h/M_v = 1/16 \qquad (6)$$

As Wroth [6] points out, this result has important consequences. First, the strongly non-uniform shear stress distribution on the horizontal surfaces casts doubt on the method of determining strength anisotropy using a range of vanes with differing H/D ratios. Second, the vertical surfaces contribute 94% of the resistance to the total torque, not 86% as suggested by the conventional interpretation; and the inferred shear strength will thus be dominantly that exhibited by the vertical planes. Providing the soil is isotropic with respect to shear strength the conventional interpretation will thus underestimate c_u.

The undrained strength behavior of the soil is, however, likely to be anisotropic, most markedly so with normally consolidated, low plasticity clays. Higher plasticity normally consolidated clays and all lightly overconsolidated clays will be more nearly isotropic with respect to undrained strength [9]. There is little guidance as to the likely ratio of the strength of vertical to horizontal planes c_{uv}/c_{uh}. Assuming, however, that the minimum ratio is likely to be about 0.6, the influence of this degree of anisotropy on the magnitude of c_{uv} may be examined. From Table 1 it is seen that, with a polynomial distribution of

TABLE 1—*Comparison of c_u obtained from different interpretations of the "standard" vane test (with* H = 2D), *assuming uniform shear stress distribution on vertical surfaces.*

Parameters	Conventional Interpretation	After WROTH [6]
Shear stress distribution on horizontal surfaces	uniform; $\tau/\tau_m = 1.0$	nonlinear; $\tau/\tau_m = (2r/D)^5$
Isotropic undrained strength ($c_{uv} = c_{uh}$)	$c_u = 0.86M/\pi D^3$	$c_u = 0.94M/\pi D^3$
Anisotropic undrained strength ($c_{uv} = 0.6c_{uh}$)	$c_{uv} = 0.78M/\pi D^3$	$c_{uv} = 0.91M/\pi D^3$

NOTE: M = maximum torque; τ = shear stress; τ_m = maximum shear stress; $c_{uv} = c_u$ on vertical planes; and $c_{uh} = c_u$ on horizontal planes.

shear stress on the horizontal surfaces, the isotropic case yields a value of c_{uv} somewhat higher than the (probably extreme) degree of anisotropy chosen for comparison, a multiplying factor of 0.94 compared to 0.91.

Thus it appears that it is perhaps unwise to use a factor of 0.94 in arriving at an estimate of c_u for design purposes, though the figure of 0.91 may not be unreasonable. Many, however, will be content to use the conventional, but apparently conservative value of 0.86, with the undrained strength being obtained from

$$c_u = 0.86M/\pi D^3 \tag{7}$$

Vane Insertion Effects

It is to be expected that the insertion of a field vane into the ground will result both in disturbance of the soil fabric and displacement of the soil particles. This disturbance will modify to some extent the undrained strength of the soil.

The two main consequences of the soil displacement caused by vane insertion are as follows:

(1) a local destruction of interparticle bonds within the soil, which particularly with the more sensitive clays may reduce significantly the available undrained strength in the vicinity of the vane; and

(2) a local displacement of soil particles together with a corresponding increase in the pore pressure around the vane, which with dissipation will result in an increase in effective stress.

There will be a corresponding increase in the available undrained strength. To some extent these two effects will be compensating, but the relative magnitudes of each may be expected to depend at least in part on the sensitivity of the clay involved. Field evidence, discussed later, suggests that both effects are particularly important when sensitivity >15.

Fabric Disturbance

A number of authors have discussed the nature of the disturbance caused by vane insertion, and Fig. 3 shows Cadling and Odenstad's [*3*] diagrammatic representation of this. This figure also shows the concept of the "perimeter ratio," an alternative to the area ratio, which expresses the vane blade thickness as a ratio of the vane peripheral distance, and was used by La Rochelle et al. [*10*] as a measure of potential fabric disturbance.

The actual disturbance caused by the insertion of a laboratory vane into a specimen of reconstituted kaolin is shown in Fig. 4. The soil specimen containing the vane was impregnated with Breox Peg 8000 to harden it before cutting and grinding to prepare a thin section (see Ref *11* for further details of this technique). When the thin-section is viewed between crossed polarizing filters the region where the clay fabric has been displaced by vane insertion shows as a pale area around the vane. The area of disturbance is indeed comparable to that postulated by Cadling and Odenstad.

La Rochelle et al. [*10*], investigated the consequence of insertion disturbance with field vane tests carried out in four adjacent borings. They used vanes of overall dimensions $H = 95$ mm, $D = 47.5$ mm, but with four different blade thicknesses ranging from 1.6 to 4.7 mm. In every other respect the tests were "standard." The results of this investigation were presented as undrained strength plotted against the perimeter ratio α (Fig. 5). There is a clear relationship, with the strength increasing as the perimeter ratio is reduced. Extrapolation to $\alpha = 0$ provides an estimate of the undrained strength were it possible to insert the field vane without causing any soil disturbance. The resulting "undisturbed" strength at this particular site (where the clay had a sensitivity $\cong 12$), exceeded that measured in

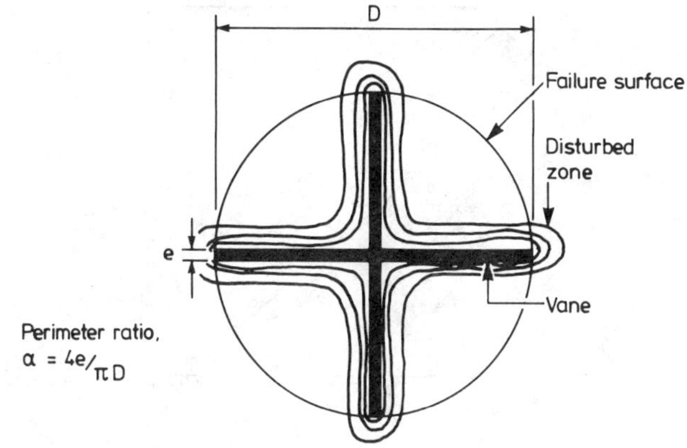

FIG. 3—*Vane insertion disturbance* [3, 10].

the "standard" test (blade thickness 1.95 mm) by 15%. The difference would presumably have been less if the larger "standard" vane, though with a similar blade thickness, had been used. La Rochelle et al. also carried out a similar investigation at a second site where the sensitivity was less (≅7), finding that similar extrapolation of the strength data to the hypothetical case of zero blade thickness indicated an "undisturbed" strength 11% higher than measured with a 1.95-mm-thick field vane. These results indicate that sensitivity is a factor in controlling the effect of vane insertion disturbance on the measured undrained strength.

FIG. 4—*Vane insertion disturbance (white area) around a laboratory vane inserted in reconstituted kaolin. Scale is in mm.*

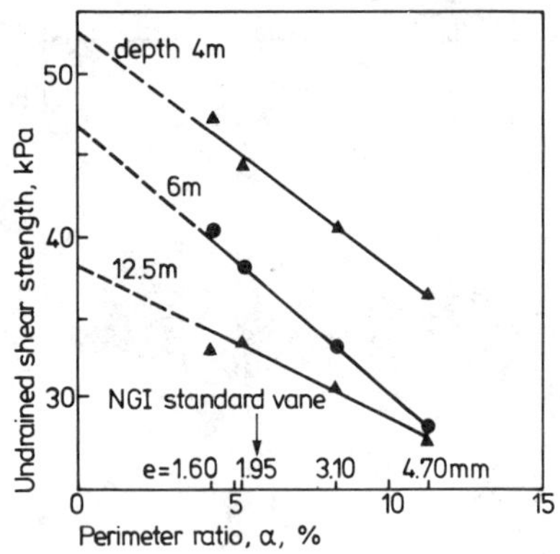

FIG. 5—*Effect of vane blade thickness* e *on measured undrained strength* [10].

Consolidation Following Vane Insertion

The only observations of pore-pressure measurements following vane insertion of which the writer is aware relate to laboratory tests. Kimura and Saitoh [*12*], using reconstituted clays, measured pore-pressure increases of 75% of the vertical effective consolidation pressure, and quoted Matsui and Abe [*13*] as predicting, by numerical methods, pore pressures equal to 50% of the vertical effective stress.

A number of authors [*14–16*][2] have reported that increases in undrained strength (compared with the "standard" test) occur if a "rest" period is allowed between insertion and rotation of the field vane. Moreover, the increase in strength becomes greater as the rest period becomes longer. This is attributable to the dissipation of the pore pressure set up by vane insertion, resulting in an increase in the lateral effective stresses. Such tests are sometimes referred to as "consolidated undrained" tests. The effect is illustrated in Fig. 6 with data from two sites in Sweden [*16*]. Both are highly plastic, sensitive clays.

The results shown in Fig. 6 are "standard" tests except that rest periods of up to seven days following vane insertion were allowed for consolidation. The strengths are normalized with respect to the "standard" undrained strength, that is, that measured after a delay of 5 min. It is seen that 90% of the seven-day strength is reached after periods of 10–20 h. That the more plastic of the two clays (Askim) apparently consolidates the more quickly suggests that in these two cases the sensitivity of the less plastic Bäckebol clay may be the more important factor, with a large zone of disturbance giving rise to longer drainage paths, and hence resulting in a lower dissipation rate.

Other similar results are summarized in Table 2. The data relating to Canadian clays presented by Roy and Leblanc[2] show much more rapid consolidation than do Torstensson's [*16*] data. Here the difference can be attributed to the high coefficient of consolidation c_v of these much less plastic clays, and this is discussed further in the context of strain-rate effects.

It is of interest to note that except for the two low plasticity, highly sensitive Norwegian clays reported by Aas [*15*] the eventual gain in strength caused by consolidation is very similar (≅20%) despite the considerable variation in placticity. Though Aas does not give information on the time period that was required for consolidation, beyond implying that it was complete within a rest period of one to three days before testing, he does give the

[2] Roy, M. and Leblanc, A., in this publication, pp. 117–128.

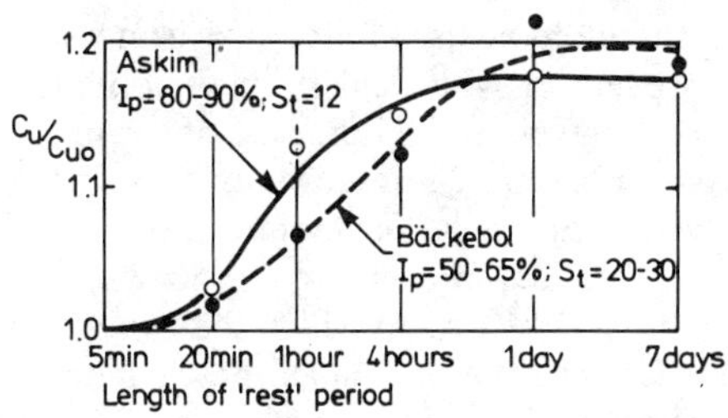

FIG. 6—*Effect of delay between insertion and rotation of field vane on measured undrained strength* [16].

"consolidated" strengths, which proved to be 40 to 50% in excess of those recorded in "standard" tests. This much greater increase in strength compared with the other clays listed in Table 2 strongly suggests that a significantly higher degree of insertion disturbance occurred with these two extremely sensitive clays.

It is concluded, therefore, that vane insertion causes disturbance to the clay fabric that will result in an underestimate of the available undrained strength. Based on the findings of La Rochelle et al. [*10*], the underestimation will be up to 15%, depending on sensitivity. It is inferred from Aas's [*15*] data, however, that the effect is almost certainly greater with extremely sensitive clays. Assuming that, typically, vane insertion causes a reduction of 10% on the true undisturbed strength, and that there is a subsequent 20% strength increase if consolidation is allowed before test, then the "consolidated" undrained strength is about 10% above the true undrained strength. If a similar consolidated undrained strength is assumed to apply to Aas's data, then vane insertion must have reduced the undisturbed undrained strength of these two highly sensitive clays by about 25%.

Effect of Rate of Vane Rotation

At the standard rate of rotation, 0.1°/s or 6°/min, failure with the field vane is typically attained in 30 to 60 s. This is a much shorter time than in the laboratory undrained triaxial compression (UU) test, where times to failure are about 5 to 15 min and is very much faster than the times taken for full-scale field failures to occur. In the subsequent discussion the latter will be assumed to take about seven days (10^4 min).

TABLE 2—*Consolidation results following vane insertion.*

Site	I_p	S_t	Rest Period Required for 90% Increase in Strength, h	% Increase in c_u in 24 h	Reference
Lierstranda, Norway	6	50 to 150	. . .	40	15
Manglerud, Norway	8	40 to 170	. . .	52	15
Unspecified, Norway	. . .	. . .	4	19	14
Saint-Alban, Canada	6 to 18	4 to 14[a]	5	18	Footnote 2
Saint-Louis de Bonsecours, Canada	13 to 19	8[a]	1	22	Footnote 2
Bäckebol, Sweden	50 to 65	20 to 30	20	20	16
Askim, Sweden	80 to 90	12	10	18	16

[a] With laboratory vane.

Thus if field vane strengths are to be used for comparison with the results of other measurements of undrained strength or with field strengths obtained from back-analyses of failures, consideration must be given to the influence of strain-rate on the measured vane strength. Two aspects have to be taken into account. One is the viscous, or rheological effect, which leads to the development or mobilization of increased strength under undrained conditions as the rate of vane rotation is increased. The other effect occurs if the coefficient of consolidation of the clay in which the vane is inserted is not sufficiently low with respect to the rate of vane rotation. Under these circumstances consolidation may occur as the rotation rate is reduced, leading to a change in the observed strength.

Viscous rate effects may be expected to be related to shear strain rates. Though these are at present unknown in the vane test, it seems reasonable to assume that the relevant strain rates in the soil are related to the vane tip velocity [*17*]. Thus tests carried out using vanes of different diameter should be performed at rotation rates chosen to give similar tip velocities. That is, at rates inversely proportional to the vane diameter. In contrast, consolidation effects will depend, inter alia, on the length of the relevant drainage path. This distance, discussed in the following section, must be quite short in the vane test, being the distance from the region of the vane periphery into the surrounding mass of clay to where the pore pressures are unaffected by the vane rotation. If it is presumed that the drainage-path length is proportional to (vane diameter)^2, then, for equal degrees of consolidation, tests using different vanes should be carried out at rates of rotation inversely proportional to the square of the vane diameter.

These two conflicting effects complicate comparison of data from laboratory-scale vane tests with those from the field vane. For this reason discussion of the effects of different vane rotation rates will be restricted to a review of work carried out with the field vane.

Studies of the effects of field vane rotation rates are comparatively few. Among the more important are those of Torstensson [*16*] and Wiesel [*18*] who report results from plastic Swedish clays, and Roy and Leblanc[2] who tested two low plasticity Canadian clays.

The Swedish tests, the results of which are shown in Fig. 7, were carried out with vane rotation rates between 200 and 2×10^{-4} °/min. Each test was performed 24 h after vane insertion, and the measured strengths were normalized with respect to "standard" tests. In the case of Wiesel's tests the results were normalized in relation to undrained strengths measured following a 24-h rest after insertion. The data shown are the average obtained over a range of depths. For convenience, in the absence of knowledge of shear-strain rates, all the results have been expressed in terms of time to failure. The corresponding approximate vane rotation rates are also shown. The use of time to failure provides a means of comparison with other measures of undrained strength, and also enables the data to be used to check a simplified theory of consolidation that is discussed later.

All the strain-rate effect results shown in Fig. 7 indicate a reduction in strength with increasing times to failure. In spite of this observation, it seems probable that drainage must have occurred during the very slow tests; this point is discussed further in the following section.

In contrast to the behavior of the Swedish plastic clays, the low plasticity Canadian clays tested by Roy and Leblanc show a very different trend (Fig. 8). Again the data have been plotted against time to failure, normalized to the strength obtained from "standard" tests, on which basis the two clays tested show very similar results. They are therefore considered as one data set. The vane rotation rates used were no slower than about 1°/min, with corresponding times to failure of about 6 to 7 min. Though there is a reduction in strength, albeit small, at high rates of rotation, once the rates of rotation fall below the "standard" rate the strengths begin to increase again markedly. As Roy and Leblanc observe, this presumably indicates progressively higher degrees of consolidation. If the trend of strength-

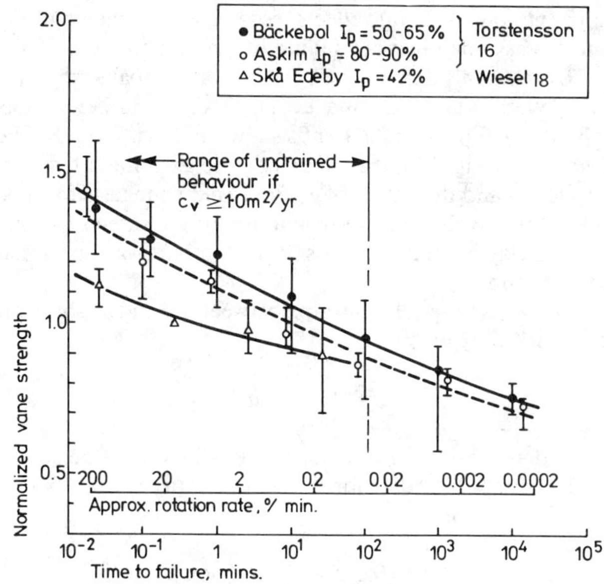

FIG. 7—*Effect of rate of rotation on field vane strength, high* I_p *clays.*

loss observed at high rotation rates (when conditions are presumably undrained) is extrapolated to lower rotation rates as shown in Fig. 8, then consolidation during slow rotation is seen to result in a strength increase of about 20%.

It appears from comparison of Figs. 7 and 8 that the viscous rate effects vary with the plasticity index I_p of the clay. Such an effect was postulated by Bjerrum [*19*], who gave a multiplying factor μ, which related field vane strength to the strength obtained from the back-analysis of full-scale field failures. The factor $\mu = \mu_A\mu_R$, the two latter terms corre-

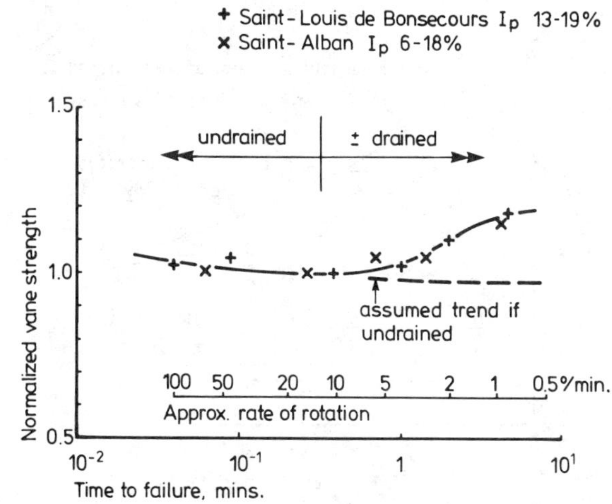

FIG. 8—*Effect of rate of rotation on field vane strength, low* I_p *clays.*

sponding to anisotropic and strain-rate effects respectively, as shown in Fig. 9. If the "standard" field vane test is assumed to reach failure in 1 min, and if field failures occur with $t_f = 10^4$ min, then Torstensson's data from Fig. 7 can be compared directly with Bjerrum's relationship for μ_R. Wiesel's and Roy and Leblanc's data can be extrapolated (the latter only very approximately), following the trends shown in Figs. 7 and 8. The results of these comparisons are shown in Fig. 10*a*; the apparent close agreement between Bjerrum's estimated relationship for μ_R and the values of μ_R from field vane tests must be to some degree fortuitous, but none-the-less the relative magnitude of the rate-effect and its dependence on the plasticity of the clay is confirmed. A similar comparison can be made (Fig. 10*b*) for times to failure of 100 min.

For all values of $I_p > 5\%$, the relationships between μ_R and I_p shown in Fig. 9 may be expressed by the empirical expression

$$\mu_R = 1.05 - b(I_p)^{1/2} \quad (8)$$

where the value of b depends on the time to failure t_f for which the correction factor μ_R is required. For $t_f = 100$ min, $b = 0.030$; for $t_f = 10\,000$ min, $b = 0.045$. Or, more generally, for 10 min $< t_f <$ 10 000 min

$$b = 0.015 - 0.0075 \log t_f \quad (9)$$

Rate of Vane Rotation to Ensure Undrained Conditions

Blight [*20*] has developed an approximate theory by which the rate of vane rotation required to ensure undrained conditions at failure may be checked. The basic assumptions made are (1) that vane insertion pore pressures have dissipated; (2) that excess pore pressures caused by vane rotation are uniform within a sphere of radius *a* (sphere of influence) centered on the mid-point of the vane (Fig. 11); (3) that torque and excess pore pressures both increase linearly with time up to failure, but that at failure the torque remains constant, while pore pressures fall; (4) that the surface of the sphere of influence is effectively a drainage boundary, with the pore pressures maintained at hydrostatic values; and (5) that

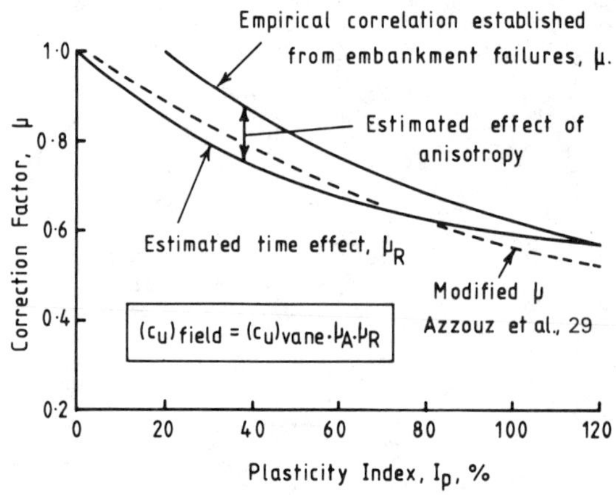

FIG. 9—*Factors relating field vane and field failure strengths* [19].

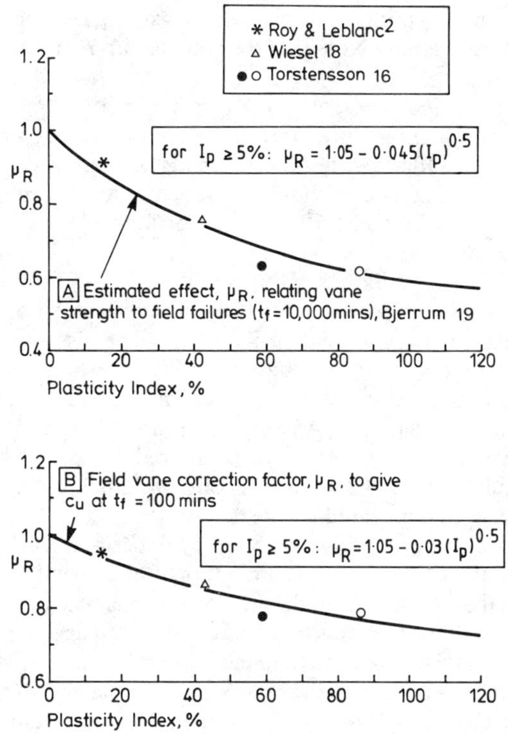

FIG. 10—*Factor μ_R to correct field vane strength for strain-rate effects.*

the pore pressures at the vane periphery may be represented by those at a radius R from the center of the sphere of influence.

It is assumed that the maximum torque M is related to the degree of consolidation U at the vane periphery and hence at radius R, such that

$$U = (M - M_0)/(M_1 - M_0) \tag{10}$$

where M_0 is the torque with zero drainage and M_1 the torque measured in very slow tests in which U = 100%. It is then possible to compare the theoretical solution, relating U to

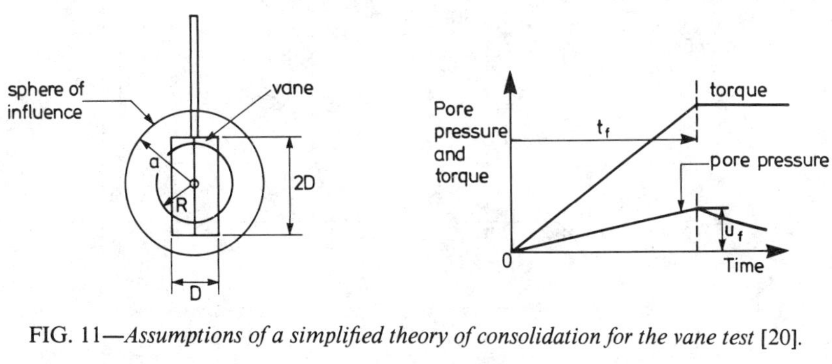

FIG. 11—*Assumptions of a simplified theory of consolidation for the vane test* [20].

the time factor T, with the results of vane tests of varying duration, and therefore with varying degrees of consolidation. Note that the time factor $T = c_v t_f / D^2$, in which c_v is the coefficient of consolidation.

Blight used vanes of various sizes, all with $H = 2D$, in a silt ($k = 1 \times 10^{-7}$ m/s; c_v = 370 m^2/year), and these gave measured torques in slow tests that were 50% or so greater than in rapid, presumed undrained tests. The results of these tests, which are plotted in Fig. 12, show acceptable agreement with the theoretical solution provided that $a = D$, and $R = D/2$. (Note that in Blight's Fig. 5 the lower portions of Curve 1 and the "empirical curve" should be interchanged.)

Matsui and Abe [*13*], who used reconstituted kaolin to carry out laboratory vane tests with measured pore pressures, concluded that their experimental data supported Blight's theory. Roy and Leblanc's[2] vane test data from Saint-Alban also support the theory. Their results, already shown in Fig. 8, are also plotted in Fig. 12 where agreement with the theoretical relationship is obtained if c_v is taken as 250 m^2/year. Leroueil et al. [*21*] reported values of c_v at in-situ stresses at Saint-Alban in the range 440 to 915 m^2/year (average 700 m^2/year). Thus the value of c_v of 250 m^2/year required to give a "fit" to the theoretical relationship seems quite reasonable when it is considered that disturbance and consolidation following vane insertion must reduce the magnitude of c_v in the zone around the vane.

Blight's approximate theory is thus supported by experimental data and can therefore be regarded as a reasonable basis for the examination of the rate of vane rotation that must be used to ensure that substantially undrained conditions apply during the vane test. It also follows, as inferred in the previous section, that the drainage path length is quite short, and is approximately equal to the vane diameter D.

Undrained conditions can, for practical purposes, be assumed to apply if the degree of consolidation $U < 10\%$. From Fig. 12 the corresponding time factor $T < 0.05$. In "standard" field vane tests, the time to failure is typically 1 min, and D = 65 mm. Thus for $U < 10\%$, c_v must be equal to or less than 110 m^2/year (3.5×10^{-2} cm^2/s). Since most soft clays do indeed show values of c_v less than this, the unintentional measurement of drained or partly drained strength with the "standard" field vane test in uniform clays seems unlikely.

However, it is also evident from this analysis that the longer term tests reported by Tor-

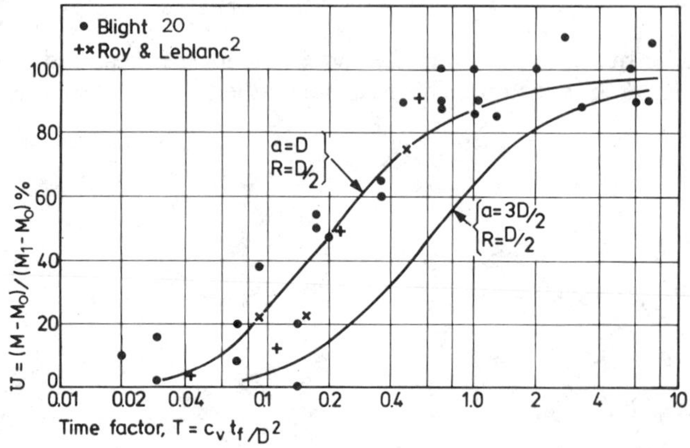

FIG. 12—*Theoretical drainage curves for the vane test* [20] *compared with experimental results.*

stensson [*16*], shown in Fig. 7, with times to failure of up to seven days, are unlikely to be undrained. This cannot be checked precisely as the relevant magnitude of c_v is not known, but values lower than 1 m^2/year (3×10^{-4} cm^2/s) would seem to be unlikely. If this lower bound value is taken to apply to the two clays tested by Torstensson, it is found that less than 10% consolidation (indicating undrained conditions) will only occur with t_f less than 110 min.

The reason for the continued reduction in the measured strengths with increasing times to failure reported by Torstensson, which contrasts so markedly with the Canadian experience[2] that the strength increases with increasing t_f, is a matter for speculation. One explanation is that the strongly structured, low plasticity Canadian clays tend to dilate when sheared by the field vane, resulting in an increase in the effective stresses. In contrast, the highly plastic Swedish clays are perhaps contractant. Thus, when rotation rates are slow enough to allow consolidation in these clays there will be arching in the soil around the vane, resulting in reducing effective stresses. This will lead to a corresponding reduction in the measured strength. Thus the strain rate effect factor μ_R (Fig. 10*a*), which is in part based on Torstensson's results, may lead to an underestimate of the true undrained strength at high values of t_f.

The Undrained Strength Measured with the Field Vane

Any discussion of the (undrained) strength measured by the field vane must include consideration of the stress system applied by the vane and of the undrained strength anisotropy exhibited by soft clays. In this section of the paper these two factors are discussed, leading on to consideration of the relationships between the strengths measured by the field vane and by other test methods. Finally, empirical relationships between the undrained strength ratio c_u/σ'_v and plasticity index are explored.

Stress Conditions at Failure in the Vane Test

It has previously been shown that the shearing resistance around the vane is dominated by the stresses on the vertical planes. For all practical purposes, therefore, the mode of failure that must be considered is that around the vertical periphery of the vane. This, Wroth [*6*] has suggested, is that of direct shear. Recent work at Imperial College (M. Mahmoud, unpublished) appears to confirm that the stress conditions at failure are, in fact, either direct shear or the closely analogous simple shear. Mahmoud's study of the vane test may be compared with that of the laboratory direct shear test carried out by Morgenstern and Tchalenko [*11*], who concluded that the shear structures that develop at peak shear stress in the shear box result from simple shear conditions. It is worth noting that this conclusion has recently been confirmed by an analytical study of the direct shear test [*22*].

Mahmoud, working with similar optical thin-section techniques to those used by Morgenstern and Tchalenko, has examined the shear structures that develop around the periphery of the laboratory vane. The successive development of these structures at three different stages of the vane test is shown in Figs. 14 to 16; Fig. 13 indicates the relative positions of each stage on the torque-rotation curve.

The similarities of the structures observed in the vane test when compared to those described by Morgenstern and Tchalenko for the direct shear test are striking. This is to be seen in Fig. 17, where the structures at maximum shear stress in the two test types are compared. It is difficult to escape from the conclusion that simple shear conditions also probably apply at peak shear stress in the vane test; a direct shear mode develops as the stresses fall post-peak with the formation of a continuous peripheral shear surface.

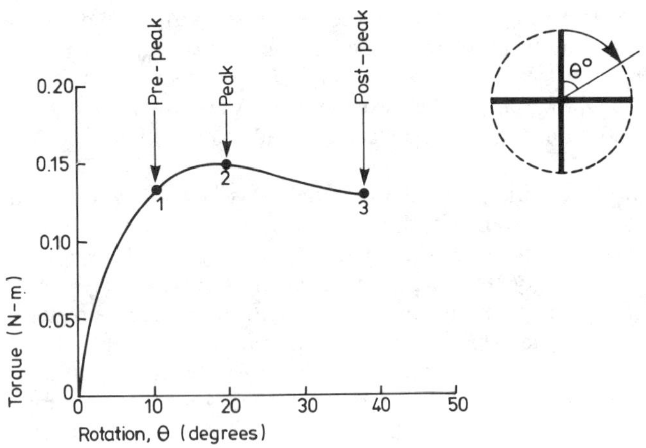

FIG. 13—*Development of shear structures in the vane test. Figures 14, 15, and 16 show the structures developed in reconstituted kaolin at Stages 1, 2, and 3.*

FIG. 14—*Pre-peak vane shear structures in kaolin, Stage 1, Fig. 13.*

FIG. 15—*Vane shear structures at peak shear stress in kaolin, Stage 2, Fig. 13.*

FIG. 16—*Post-peak vane shear structures in kaolin, Stage 3, Fig. 13.*

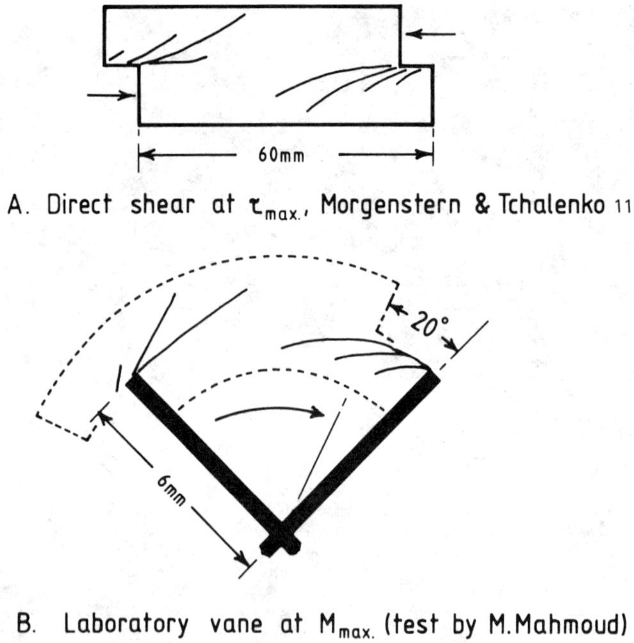

FIG. 17—*Comparison of shear structures in direct shear and vane test at maximum shear stress. Reconstituted kaolin in thin section, crossed polars.*

Undrained Strength Anisotropy

Anisotropy in soils is either inherent, resulting from depositional processes and thus reflecting grain characteristics, or is stress induced, a consequence of the strains engendered during deposition and any subsequent erosion. All soils can thus in general be expected to behave anisotropically and in particular will exhibit anisotropy with respect to undrained strength. This latter phenomenon results not only from anisotropy with respect to changes in the effective stress strength parameters with the direction of the applied shear stresses, but also from pore pressure effects. Useful reviews of the undrained strength anisotropy of soft clays are given by Bjerrum [*19*], Ladd et al. [*9*], and Jamiolkowski et al. [*23*].

Examples given by Jamiolkowski et al. [*23*] of the magnitude of undrained strength anisotropy are shown in Fig. 18 where the results of K_0-consolidated triaxial compression and extension tests and direct simple shear tests are plotted against the plasticity index. All the tests were carried out on normally consolidated vertical specimens. The triaxial compression tests are seen to yield the highest strengths, with the ratio of undrained strength to vertical effective stress, c_u/σ_v', averaging about 0.32, independent of I_P. The direct simple shear test yields c_u/σ_v' values that average between 0.22 and 0.28, increasing with I_P. The lowest strengths are obtained from the triaxial extension tests, where the average values of c_u/σ_v' range from 0.14 to 0.21, again increasing with the plasticity index. The undrained strength anisotropy of normally consolidated clays is thus greatest at low values of plasticity.

While the above figures illustrate the likely magnitude of undrained strength anisotropy in soft clays, they are not immediately relevant to the field vane test for which the dominant stress conditions at failure seem likely to be those of simple shear around the vertical periphery of the vane. This condition is shown schematically for the vane test in Fig. 19,

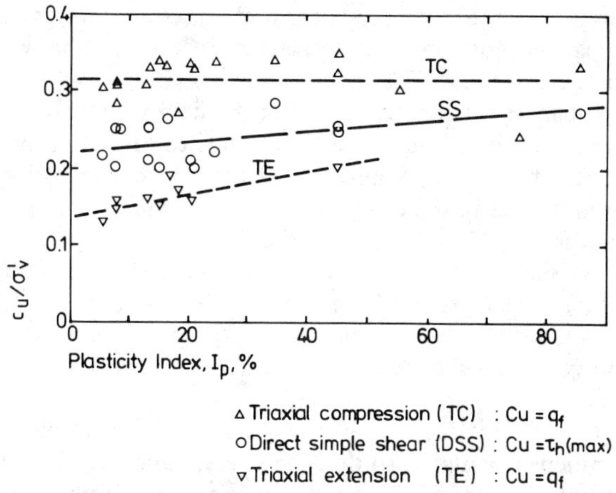

FIG. 18—*Undrained strength anisotropy from CK_0UC tests on normally consolidated clays;* [23]; *TC triaxial compression, SS simple shear, and TE triaxial extension.*

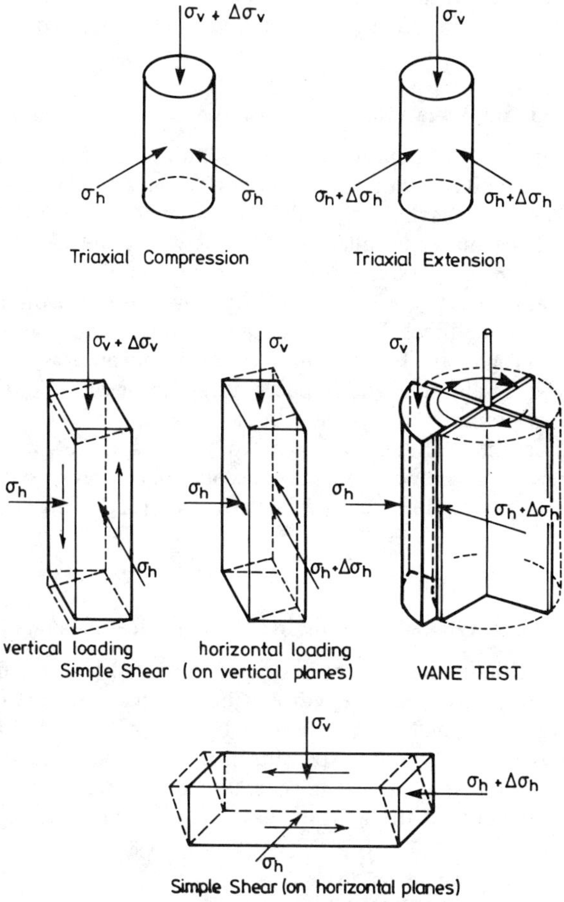

FIG. 19—*Undrained strength anisotropy, stress systems in different tests.*

where it is compared with other simple shear and triaxial loading systems. The stresses shown are the vertical and horizontal stresses acting in situ, together with the direction in which the stress increase to failure is applied in the laboratory.

The simple shear loading analogous to that of the field vane is one where plane-strain conditions are maintained vertically under the influence of σ_v and the shear stresses are increased in one of the two orthogonal directions in which σ_h acts, the constant normal stress of σ_h being maintained in the other direction. The more usual test to be carried out on specimens of this orientation is one where the shear stress is applied in the σ_v direction, and the normal stress and intermediate principal stress are both equal in value to σ_h. The results of such tests have been reported, for example, by Bjerrum [*19*]. It will be noted that the suggested field vane simple shear mechanism has some similarities with the more usual simple shear test on vertical specimens, since in this test, shearing also occurs in one of the σ_h directions.

The writer is aware of only one instance where a direct or simple shear test has been carried out on specimens orientated so that the stresses are comparable with those of the field vane simple shear mechanism. This test is reported by Karube et al.,[3] who found that they obtained the same undrained direct shear strength for a vertically oriented, horizontally sheared specimen as with similar tests on horizontally oriented, horizontally sheared specimens. This suggests that there may be similarities between the field vane strength and the simple shear strength on horizontal planes; this point is returned to later.

Progressive Failure in the Vane Test

There is little in the way of data to provide an indication of the extent of progressive failure in the vane test, and hence of its influence on the measured undrained strength. The work of Alencar et al.[4] suggests that only soils showing a particularly rapid post-peak reduction of strength will be markedly affected. Since few soft soils show such stress-strain behavior, it follows that progressive failure will be a relatively unimportant influence on the measured undrained strength. Potts et al.'s [*22*] analysis of the direct shear test led to the conclusions that in the direct shear test simple shear conditions applied at peak stress, and that little or no progressive failure occurred. If it is accepted, as previously discussed, that the rotation of the vane similarly imposes simple shear conditions, then presumably progressive failure is similarly unimportant in the vane test.

Whatever the situation eventually transpires to be, users of the vane test can always be reassured that the recorded strength must take account of any progressive failure that may have occurred. Consequently the measured strength will (neglecting strain-rate effects) be conservative.

Field Vane Strength and Triaxial Compressive Strength Relationships

The relationship of the undrained shear strength obtained with the field vane to other measurements of undrained shear strength is widely discussed in the literature. In early work it was generally assumed that agreement between vane and triaxial compressive undrained shear strengths was to be expected. In more recent years it has been realized that many factors are involved and that in general there will not be agreement between the two different measurements of strength. Indeed, it is not difficult to find in the literature

[3] Karube et al., in this publication, pp. 131–149.
[4] De Alencar et al., in this publication, pp. 150–165.

instances where, for example, field vane strengths exceed those of unconsolidated undrained compression tests, and other cases where the reverse is true.

In attempting to assess the magnitude of the strength measured by the field vane there is obvious merit in comparing the field vane undrained shear strength with some other "standard" measurement of the strength of the same clay. The most obvious choice for this comparison is the undrained shear strength obtained in triaxial compression tests. Unfortunately, sampling disturbance [*23,24*] is likely to result in an underestimate of the strength if UU triaxial tests are used for this comparison. Thus it is preferable to use CK_0UC tests on good quality specimens, where the measured undrained strength is obtained after reconsolidation to the in-situ effective stresses. It is accepted that, particularly with sensitive clays, this procedure may on occasion result in a lower voids ratio, and hence higher strength, than is relevant to the in-situ state.

Alternatively, the SHANSEP procedure [*25*] may be used. This requires that, following reconsolidation, a series of K_0-consolidated triaxial undrained compression tests is carried out over a range of over consolidation ratios (OCRs) that include the in-situ stress states. The in-situ strength is then determined indirectly, knowing the magnitude of the vertical effective stress σ'_v and OCR. This method is most successful with clays that do not have a structure that may be damaged by reconsolidation. With both techiques, K_0 in-situ must be known or estimated. These tests are usually carried out at comparatively low rates of compression, so that times to failure are of the order of 100 min.

A survey of the literature shows that there are a number of instances where both field vane and CK_0UC tests (or their equivalent) have been carried out at the same locality. Nineteen such case records are given in the Appendix, forming a data base. It is believed that c_u measured with the field vane has in each case been obtained using the "standard" procedures (Fig. 1) and the conventional interpretation (Eq 7) of the test. Note that the data base includes both normally consolidated and also lightly overconsolidated clays with OCRs up to 7.5.

This data base is not only of value for comparing field vane and triaxial compressive strengths, but also for examining other relationships. It is thus useful, in the first instance, to examine the basic trends of the data to check that they conform with previously established correlations. This is most conveniently done in terms of the ratio c_u/σ'_v, plotted against I_p, as shown in Figs. 20 and 21.

Such relationships have been plotted for the field vane test by many authors, following Skempton [*2,26*] who proposed the empirical expression

$$c_u/\sigma'_v = 0.11 + 0.0037\, I_p \qquad (11)$$

for normally consolidated clays. Another widely quoted relationship is that given by Bjerrum [*19*] for "Young" clays; that is, clays with OCR $\cong$ 1.0. A recent comprehensive compilation of such field vane data is that made by Leroueil et al. [*27*]. The approximate range of their data is indicated, together with Bjerrum's and Skempton's relationships, in Fig. 20, where the corresponding data from the Appendix are also given. Bjerrum's curve forms the lower bound to all the data, as would be expected since it relates to clays with OCR = 1. Where OCR < 1.5, the data from the Appendix agree well with all three relationships, but, of course, clays with higher values of OCR yield higher values of c_u/σ'_v.

The companion plot of the CK_0UC data from the Appendix (Fig. 21) shows a superficially similar pattern, except that the data for OCR < 1.5 are apparently independent of I_p. The nine lowest data points yield the average value of $c_u/\sigma'_v = 0.31$. It has already been seen (Fig. 18) that an identical value of c_u/σ'_v, similarly independent of I_p, was obtained by Jamiolkowski et al. [*23*] for a different set of CK_0UC data. Thus for both vane and triaxial

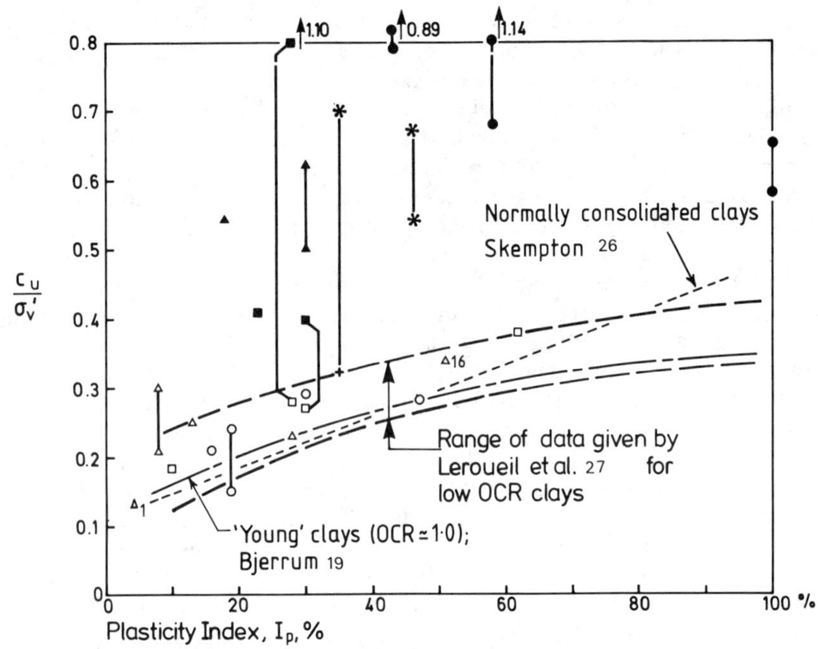

FIG. 20—c_u/σ_v' *versus* I_p *for field vane tests. The plotted points are obtained from the data given in the Appendix, though some cases where* $c_u/\sigma_v' > 0.8$ *are omitted. The key is given in Fig. 22.*

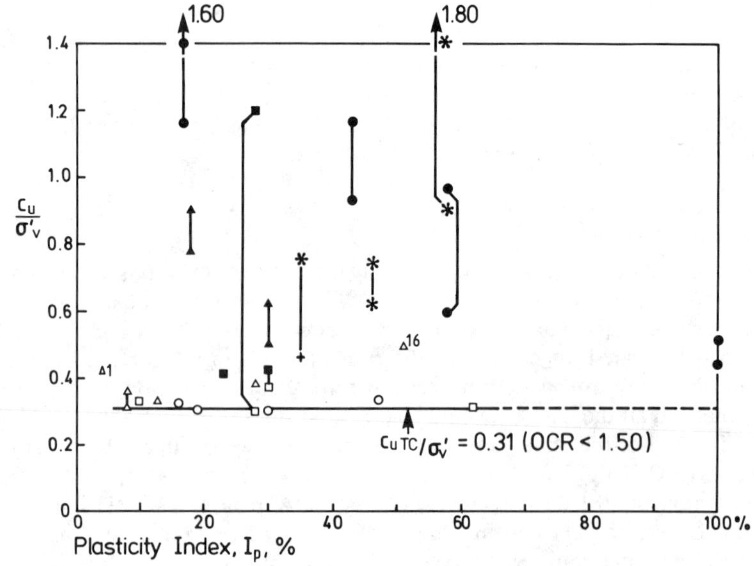

FIG. 21.—c_u/σ_v' *versus* I_p *for* CK_0UC *triaxial compression tests. The plotted points are obtained from data given in the Appendix; the key is given in Fig. 22.*

strengths the data base compares well with previous compilations and may be regarded as a reasonably representative data set.

Having checked the make-up of the data base, the relationship between undrained strength measured by the field vane and CK_0UC tests may now be tested. This is done in Fig. 22 where the ratio of the vane to the CK_0UC strength V_r is plotted against I_p. The best linear fit to the data is given by

$$V_r = 0.55 + 0.008\ I_p \tag{12}$$

The coefficient of correlation $r = 0.90$.

Various factors may be expected to be important in controlling the magnitude of the ratio V_r. These include sensitivity S_t, OCR, shear-strain rate, and anisotropy with respect to undrained strength. Approximate values of sensitivity and OCR in each case are indicated in Fig. 22. As would be expected, the more sensitive clays are those with low I_p. Sensitivity provides an indication of the likely effect of disturbance, either during vane penetration, sampling or with triaxial reconsolidation at the estimated in-situ stresses. There is evidence of this latter effect in Fig. 21, particularly for Cases 1 and 16. In both instances the triaxial compressive strengths seem atypically high, perhaps because of sampling disturbance giving rise to excessive compression on reconsolidation. These two cases thus apparently yield unrealistically low values of V_r, and these have been discarded in computing Eq 12.

Though it is not immediately evident in Fig. 22, there is some indication that V_r is related in some manner to OCR. If all the data points for OCR < 1.5 are compared with those for OCR > 1.5, then no significant difference is found between the two data sets.

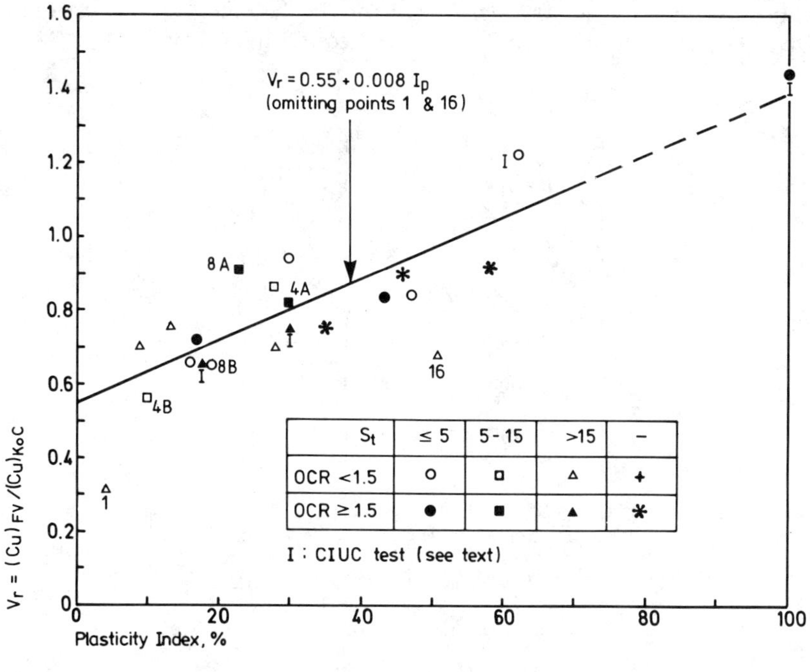

FIG. 22—*Relationship between* V_r *(the ratio of field vane and* CK_0UC *strengths) and* I_p.

However, if comparison is made between the two members of the pairs of data points for Cases 4A and 4B and 8A and 8B, where different OCRs apply at the same locality, then it is seen that in each instance the value of V_r is proportionately higher for the higher value of OCR. The possibility that V_r is influenced by OCR is returned to later.

Relationships Between c_u/σ'_v and OCR

Jamiolkowski et al. [23] present the results of the field vane undrained strength plotted as the ratio c_u/σ'_v against OCR. The latter parameter was in each case obtained from a companion series of oedometer tests. Data from nine different sites, for a range of different clays, some of which are given here in Fig. 23, show that the results yield reasonably linear relationships on a log-log plot. These can be expressed in the form

$$c_u/\sigma'_v = S_1(\text{OCR})^m \tag{13}$$

where S_1 is the undrained strength ratio for normally consolidated clay (that is, OCR $\cong$ 1). Similar relationships have been presented for various laboratory undrained strength tests by Ladd et al. [9], who concluded that m is typically 0.8 in simple shear tests.

The range of values of S_1 and m given by Jamiolkowski et al. for field vane tests are summarized in Table 3. With one or two exceptions the values obtained are quite similar. The exceptions are a high value of S_1 (0.74; typical values being 0.16 to 0.33) for an organic shelly clay, and a high value of m (1.51; typically 0.80 to 1.35) for a cemented, varved clay. The average of the values of S_1 and m are 0.22 and 0.97, respectively, if in each case the extreme values are discarded. Note that the value of m, as Jamiolkowski et al. observed, is greater for the field vane test than for direct simple shear tests. This is probably the consequence of the higher strain rate used in the field vane test. The different values of m for the two different tests are probably the reason for the ratio V_r of the vane to triaxial strength being partly dependent on OCR. This is because at higher values of OCR the vane

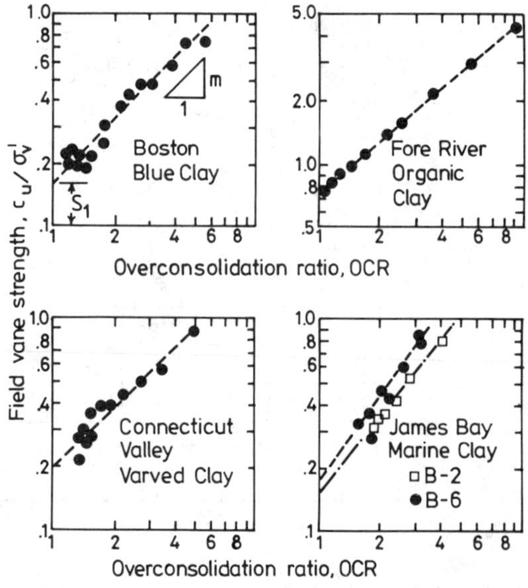

FIG. 23—*Undrained strength ratio versus OCR from field vane tests* [23].

TABLE 3—*Values of parameters* S_1 *and* m *at nine sites obtained by Jamiolkowski et al.* [23].

Parameters	S_1	m
Typical range of values (all sites)	0.16 to 0.33	0.80 to 1.35
Extreme value (one in each case)	0.74	1.51
Mean (all values)	0.28	1.03
Mean (discarding extreme value)	0.22	0.97

strength is proportionally greater, since from Eq 13, the undrained strength is a function of $(\text{OCR})^m$.

The field vane data from the Appendix may be plotted in a similar manner (Fig. 24*a*). In several of the case records a range of data applies; this is indicated in Fig. 24*a* by plotting the extreme points, joined by a line. Overall, it is seen in Fig. 24*a* that there is a trend for the data relating to the more plastic clays to yield higher values of c_u/σ'_v, as would be expected from the relationships shown in Fig. 20. In Fig. 24*b*, the field vane strength ratios have been normalized with respect to the values of S_1, at corresponding values of I_p, obtained from Bjerrum's relation shown in Fig. 20. This plot is independent of I_p; for simplicity the ranges of values shown in Fig. 24*a* are plotted as average points in Fig. 24*b*. Linear regression analysis yields the equation

$$(c_u/\sigma'_v) = S_1(\text{OCR})^{0.95} \tag{14}$$

for both Figs. 24*a* and *b*. In Fig. 24*a*, $S_1 = 0.25$, which may be regarded as the overall average value of S_1 for all the data in the Appendix.

Several points arise from these relationships. First, the values of S_1 and m (0.95) are very close to the average values obtained by Jamiolkowski et al. (0.22 and 0.97, respectively),

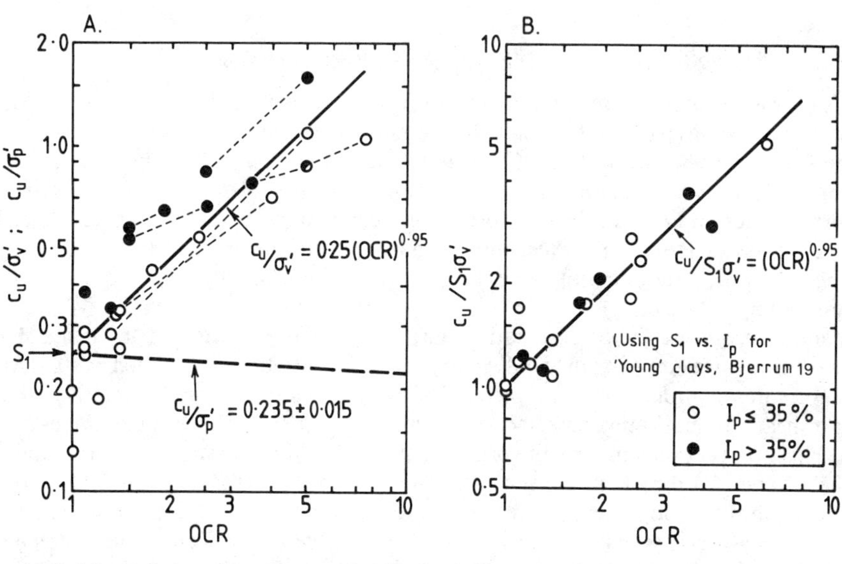

FIG. 24—*Relations between* c_u *and OCR for field vane and* CK_0UC *tests. The data are taken from the Appendix.*

in spite of the fact that only one case record is common to the two data sets. This consistency suggests that Eq 14, if combined with S_t from Bjerrum's "Young" clay relationship (Fig. 20), might be used in the absence of other data for the estimation of OCR. If this procedure is followed using the data in the Appendix, it transpires that in most cases an accuracy of about ±25% is obtained. Of course, cemented, strongly structured, organic, or otherwise unusual clays are unlikely to yield satisfactory values of OCR from this procedure.

Second, if Eq 14 is normalized by dividing by OCR, as is also shown in Fig. 24*a*, the resulting strength ratio (c_u/σ'_p, where σ'_p is the pre-consolidation stress) has a nearly constant value of 0.23 (±0.015). This analysis is similar to that of Mesri [*28*] who additionally multiplied the c_u/σ'_p relation by Bjerrum's factor μ (Fig. 9), at corresponding values of I_p. This resulted in the almost constant "field" strength ratio, independent of plasticity, of $\mu c_u/\sigma'_p = 0.22$ (±0.03). The data of Fig. 24*a* can be similarly treated, a value of $\mu = 0.90$ corresponding to the average plasticity index of the data base (33%). This yields an almost identical value of $\mu c_u/\sigma'_p$ to that obtained by Mesri, namely, 0.21 (±0.01). The data from the Appendix suggest that the overall range of scatter in the value of $\mu c_u/\sigma'_p$ about Mesri's often quoted field strength ratio is of the order of ±0.05.

The use of Bjerrum's factor μ to obtain a field strength ratio involves all the uncertainties of the back-analysis techniques that were used to establish the field strengths on which the factor μ is based. These uncertainties include such factors as the "three-dimensional" or "end" effects, which are not usually considered in stability analyses, the effect of tensile crack depths, which are rarely known with any certainty, and the consequences of progressive failure.

The important three-dimensional effects in stability analyses have been considered by Azzouz et al. [*29*], who recalculated Bjerrum's μ factor for a series of case records for which three-dimensional stability analyses were carried out. This resulted in the relationship shown as a dashed line in Fig. 9, typically reducing the value of μ by about 9%. The field strength relationship is thus more correctly given by the Azzouz et al. value of μ, which yields $\mu c_u/\sigma'_p = 0.20$, rather than Mesri's 0.22. Similarly, a value of 0.19 is obtained from the present data.

Field Vane Strength Versus Simple Shear Strength Relationships

It is now appropriate to return to the relationship between the field vane and the simple shear undrained strengths. The results of normalizing the field vane strength ratios given in the Appendix with respect to OCR are plotted in Fig. 25*a*. As can be seen, Bjerrum's Young clay c_u/σ'_p versus I_p relationship is a good fit to these data. This relationship must be corrected for strain-rate effects before it can be compared with simple shear data. This is done, assuming that simple shear time to failure was of the order of 100 min, by adjusting the field vane data to comparable failure times by multiplying by μ_R from Fig. 10*b*, at corresponding values of I_p.

In addition, since a factor 0.91, rather than 0.86, in Eq 7 is believed to yield a slightly more accurate measure of c_u; the field vane strength has also been multiplied by (0.91/0.86). This gives the relationship shown in Fig. 25*b*, which is seen to compare closely with the simple shear strength ratios reported by Jamiolkowski et al. [*23*], except at values of $I_p <$ 30%. This observation supports the previous suggestion that the stress systems applied by the field vane and direct simple shear test are perhaps similar. That the field vane data yield comparatively low strengths at low values of I_p is attributed, at least in part, to the effect of vane insertion disturbance in sensitive clays (which typically have low values of I_p), resulting in low strengths at low values of plasticity index. These low strengths are

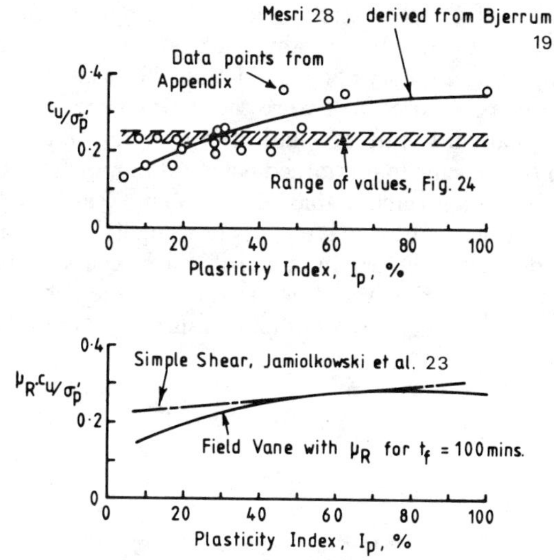

FIG. 25—*Relations between* c_u/σ'_v *and plasticity index. Note that* $\sigma'_p = \sigma'_v$ *(OCR). The field vane strength relation in* (b) *has been calculated using a factor (1.06 = 0.91/0.86) to obtain the best estimate of* c_{uv}.

consistent with the inference, previously drawn from Aas's [*15*] work, that the effect of disturbance may, in sensitive clays, result in the field vane measuring an undrained strength that is as much as 25% below the undisturbed value.

Conclusions

1. The essential features, dimensions, and procedures of the field vane test for the measurement of undrained strength appear to be more or less internationally agreed. These features of the "standard" test are summarized in Fig. 1.

2. The distribution of shear stress around the vane may be assumed to be uniform on the vertical edges of the vane blades but are probably highly nonuniform on the top and bottom surfaces. As a consequence the conventional interpretation of the test, given by

$$c_u = 0.86M/\pi D^3 \tag{7}$$

is probably conservative; it is likely that a more accurate interpretation of c_{uv} can be obtained by increasing the factor 0.86 to 0.91.

3. Vane insertion causes disturbance that results in underestimation of the in-situ undrained strength. This disturbance is most severe in sensitive clays and is probably the reason for the comparatively low strengths at low values of plasticity shown in correlations of c_u/σ'_v versus I_p. The only published estimates of the effects of insertion disturbance indicate a maximum strength loss of about 15%, but a loss of about 25% may be inferred for some of the highly sensitive, low plasticity Norwegian clays.

4. Allowing a "rest" period after vane insertion before shear will result in some consolidation, and hence in the subsequent measurement of an enhanced undrained strength. With many soils a "rest" period of more than an hour or so may result in an increase of strength greater than the reduction resulting from vane insertion disturbance. Conse-

quently it is recommended that the "rest" period should not exceed 5 min, thus ensuring a conservative estimate of the in-situ undrained strength.

5. Blight's [*20*] approximate theory of consolidation allows an estimate of the maximum time to failure that can be allowed to ensure the maintenance of undrained conditions during the test. The "standard" rate of rotation of 6 to 12°/min will result in failure in about 1 min. This in turn results in undrained behavior in most soils, providing they have a coefficient of consolidation no greater than about 100 m^2/year (3 × 10^{-2} cm^2/s. Thus the "standard" test should yield undrained strengths in virtually all reasonably uniform clays.

6. Since failure in the field vane test in most soft clays occurs in about 1 min, there are significant strain-rate effects in the "standard" test. Field vane undrained strengths may be adjusted to the longer times to failure t_f of other measurements of undrained strength using the multiplying factor

$$\mu_R = 1.05 - b\,(I_p)^{1/2} \tag{8}$$

where

$$b = 0.015 - 0.0075 \log t_f \tag{9}$$

These equations apply to clays where $I_p > 5\%$ and where t_f lies between 10 and 10 000 min; the values of b are probably conservative (leading to rather low estimates of strength) in the higher range of t_f.

7. The ratio of the field vane undrained strength to CK_0UC triaxial strength measurements is given approximately by

$$V_r = 0.55 + 0.008\,I_p \tag{12}$$

This relationship is only marginally dependent on the overconsolidation ratio.

8. It is suggested that the strength measured in the vane test is primarily that mobilized in simple shear around the vertical portion of the vane's periphery. This is comparable to (though not identical with) the stress system operating in horizontal simple shear tests, and at values of $I_p > 30\%$ the two tests record closely similar undrained strengths. At values of $I_p < 30\%$ the field vane test yields strengths believed to be reduced by vane insertion disturbance.

9. The field vane strength may be given in the form

$$c_u/\sigma'_v = S_1\,(\mathrm{OCR})^m \tag{13}$$

For "normal" clays $m \cong 0.95$; S_1, the undrained strength ratio at OCR = 1, may be estimated with adequate accuracy using Bjerrum's correlation between S_1 and I_p for "Young" clays (Fig. 20). For such clays Eq 13 provides a method for checking field vane data, or estimating the overconsolidation ratio. The accuracy of the method will be about ±25%. It is not possible to make suggestions as to the likely values of m or S_1 for strongly structured, organic, or other unusual clays, though for most of these clays Bjerrum's correlation will yield lower bound values of S_1.

Acknowledgments

The author gratefully acknowledges the assistance of Professors J. B. Burland, S. Leroueil, and C. P. Wroth, and of M. Mahmoud, all of whom commented on an earlier draft of this paper.

APPENDIX

TABLE A1—*Field vane and CK_0UC triaxial test data for various sites.*

Location	Depth, m	w_L, %	w_P, %	I_P, %	I_L	S_t	OCR	c_{ufv}/σ'_v	c_{utc}/σ'_v	c_{ufv}/c_{utc}	Reference
1. Ellingsrud, Norway	7 to 14	24	20	4±1	3 to 5	20 to 65	±1.0	0.13	0.42	0.31	30
2. Mastemyr, Norway	5 to 10	30	22	8±3	±3.0	>50	1.1	0.30 to 0.21	0.36 to 0.31	0.70±0.09	31
3. Fredrikstad, Norway	5 to 8	34	21	13±1	±1.5	±20	1.1	0.25	0.33	0.75±0.02	32
4A. Drammen,	5 to 10	. . .	. . .	30	. . .	5 to 7	1.4	0.40 to 0.27	0.44 to 0.37	0.82±0.10	33
4B. Norway	11 to 15	. . .	. . .	10	. . .	5 to 7	1.2	0.20 to 0.17	0.36 to 0.31	0.56±0.02	
5. Vaterland, Norway	12 to 16	40	24	16±3	0.6	3.0	. . .	0.23 to 0.19	0.33 to 0.30	0.66±0.03	34
6A. Tuckerton,	−14 to −16	34	17	17±8	0.9	5.0	7.5 to 5.0	1.06 to 0.89	1.60 to 1.16	0.72±0.05	35
6B. NJ[a]	−17 to −21	76	33	43±12	0.6	4.0	5.0 to 3.5	0.89 to 0.79	1.18 to 0.93	0.84±0.05	
7. Saint-Alban, P.Q., Canada	2 to 10	39	21	18±8	2.3	>15	2.4	0.56 to 0.53	0.90 to 0.78[b]	0.65±0.06	36
8A. Studenterlunden	6 to 8	47	24	23±5	0.63	7.5	>1.0	0.37	0.41	0.91	34, 37
8B. Oslo, Norway	10 to 20	44	25	19±3	0.65	4.0	±1.0	0.24 to 0.15	0.31 to 0.28	0.65±0.10	
9. Onsøy, Norway	1 to 9	. . .	. . .	28±8	. . .	5 to 7	5.0 to 1.3	1.10 to 0.28	1.20 to 0.30	0.86±0.06	33
10. S. Gloucester, Ontario, Canada	8 to 13	58	28	30±3	1.7	80	1.75	0.48 to 0.37	0.62 to 0.50[b]	0.75±0.01	38
11. Skå-Edeby, Sweden	10	51	23	28	1.43	20	1.4	0.26	0.38	0.70	39
12. Porto Tolle, Italy	16 to 28	52	22	30±2	0.47	2 to 3	1.1	0.29	0.30	0.94±0.01	40
13. Khor Al-Zubair, Iraq	+1 to 10	56	21	35±2	0.64	. . .	4.0 to 1.4	0.70 to 0.32	0.75 to 0.46	0.75±0.05	41
14. Natsushima, Japan	10 to 20	82	36	46	0.89	. . .	2.5 to 1.5	0.67 to 0.54	0.74 to 0.62	0.89±0.02	42
15. Panigaglia, Italy	2 to 17	72	25	47±3	0.75	5	. . .	0.28	0.33	0.84±0.02	40

TABLE A1—(*Continued*)

Location	Depth, m	w_L, %	w_P, %	I_P, %	I_L	S_t	OCR	c_{ufv}/σ'_v	c_{utc}/σ'_v	c_{ufv}/c_{utc}	Reference
16. Lilla Mellösa, Sweden	8 to 10	76	25	51±3	0.88	15 to 20[c]	1.3	0.34	0.49	0.68±0.02	43
17. unnamed, Japan	2 to 13	103	45	58±4	0.71	. . .	5.0 to 2.5	1.60 to 0.84	1.80 to 0.90	0.92	44
18. Queenborough, Kent, UK	−3 to 9	92	30	62±20	0.8	±5	1.1	0.38	0.31[b]	1.22	45
19. Launceston, Tasmania, Australia	3 to 9	143	43	100±10	0.95	2 to 3	1.9 to 1.5	0.65 to 0.58	0.51 to 0.44[b]	1.60±0.12	46

[a] In cases 6A & B, 14 and 18 elevations rather than depths are given.
[b] Triaxial test data for CIUC tests, multiplied by 0.9 where I_P<35% (see explanatory notes below).
[c] Sensitivity measured with fall-cone.

NOTE: In the data given in the Appendix four cases have been included where the triaxial tests were carried out on specimens isotropically consolidated to an all-round stress equal to the vertical effective stress. For normally and lightly over-consolidated soils, where $K_0 < 1.0$ (which occurs at values of OCR typically less than 3 to 5), this procedure will result in an overestimate of the K_0-consolidated strength. An arbitrary multiplying factor of 0.9 has been applied to the isotropic triaxial strength in the two cases where $I_p < 30\%$. The difference between isotropic and K_0-consolidated strengths will be less with more plastic clays, and no correction factor has been applied to the other two cases.

In cases 6A and B, 18 and 19 SHANSEP [25] procedures were used to obtain the CK_0UC or CIUC strengths.

Wherever possible the values of sensitivity that are quoted are those determined with the field vane.

References

[*1*] Carlson, L., "Determination In Situ of the Shear Strength of Undisturbed Clay by Means of a Rotating Auger," *Proceedings of the 2nd International Conference on Soil Mechanics and Foundation Engineering,* Vol. 1, 1948, pp. 265–270.

[*2*] Skempton, A. W., *Geotechnique,* Vol. 1, 1948, pp. 111–124.

[*3*] Cadling, L. and Odenstad, S. "The Vane Borer," Royal Swedish Geotechnical Institute, Proceedings No. 2, 1948.

[*4*] Flodin, N. and Broms, B., *Soft Clay Engineering,* E. W. Brand and R. P. Brenner, Eds., Elsevier, Amsterdam, The Netherlands, 1981, pp. 27–308.

[*5*] Aas, G., Lacasse, S., Lunne, T., and Høeg, K., "Use of In situ Tests for Foundation Design on Clay," *Proceedings of the Conference on Use of In Situ Tests in Geotechnical Engineering,* ASCE Special Publication 6, 1986, pp. 1–30.

[*6*] Wroth, C. P., *Geotechnique,* Vol. 34, 1984, pp. 449–489.

[*7*] Donald, I. B., Jordan, D. O., Parker, R. J., and Toh, C. T., "The Vane Test—A Critical Appraisal," *Proceedings of the 9th International Conference on Soil Mechanics and Foundation Engineering,* Vol. 1, 1977, pp. 81–88.

[*8*] Menzies, B. K. and Merrifield, C. M., Geotechnique, Vol. 30, 1980, pp. 314–318.

[*9*] Ladd, C. C., Foott, R., Ishihara, K., Schlosser, F., and Poulos, H. G., "Stress-Deformation and Strength Characteristics," *Proceedings of the 9th International Conference on Soil Mechanics and Foundation Engineering, Tokyo,* Vol. 2, 1977, pp. 421–494.

[*10*] La Rochelle, P., Roy, M., and Tavenas, F., "Field Measurements of Cohesion in Champlain Clays," *Proceedings of the 8th International Conference on Soil Mechanics and Foundation Engineering, Moscow,* Vol. 1, 1973, pp. 229–236.

[*11*] Morgenstern, N. R. and Tchalenko, J. S., *Geotechnique,* Vol. 17, 1967, pp. 309–328.

[*12*] Kimura, T. and Saitoh, K., *Soils and Foundations,* Vol. 23, No. 2, 1983, pp. 113–124.

[*13*] Matsui, T. and Abe, N., *Soils and Foundations,* Vol. 21, No. 4, 1981, pp. 69–80.

[*14*] Flaate, K., *Canadian Geotechnical Journal,* Vol. 3, 1966, pp. 18–31.

[*15*] Aas, G., "Study of the Effect of Vane Shape and Rate of Strain on Measured Values of In Situ Shear Strength of Clays," *Proceedings of the Conference on Shear Strength of Soils, Oslo,* Vol. 1, 1965, pp. 141–145.

[*16*] Torstensson, B. A., "Time-Dependent Effects in the Field Vane Test," *International Symposium on Soft Clay, Bangkok,* 1977, pp. 387–397.

[*17*] Perlow, M., Jr. and Richards, A. F., *Geotechnical Division, Proceedings of the ASCE,* Vol. 103, 1977, pp. 19–32.

[*18*] Weisel, C. E., "Some Factors Influencing In Situ Vane Test Results," *Proceedings of the 8th International Conference on Soil Mechanics and Foundation Engineering, Moscow,* Vol. 1.2, 1973, pp. 475–479.

[*19*] Bjerrum, L., "Problems of Soil Mechanics and Construction on Soft Clays," *Proceedings of the 8th International Conference on Soil Mechanics and Foundation Engineering,* Vol. 3, 1973, pp. 111–159.

[*20*] Blight, G. E., *Canadian Geotechnical Journal,* Vol. 5, 1968, pp. 142–149.

[*21*] Leroueil, S., Tavenas, F., Trak, B., La Rochelle, P., and Roy, M., *Canadian Geotechnical Journal,* Vol. 15, 1978, pp. 54–65.

[*22*] Potts, D. M., Dounias, D. T., and Vaughan, P. R., *Geotechnique,* Vol. 37, 1987, pp. 11–23.

[*23*] Jamiolkowski, M., Ladd, C. C., Germaine, J. T., and Lancellotta, R., "New Developments in Field and Laboratory Testing of Soils," *Proceedings of the 11th International Conference on Soil Mechanics and Foundation Engineering,* Vol. 1, 1985, pp. 57–153.

[*24*] Schjetne, K., "The Measurement of Pore Pressure During Sampling," Norwegian Geotechnical Institute Publication 94, 1972, pp. 1–5.

[*25*] Ladd, C. C. and Foott, R., *Geotechnical Division, Proceedings of the ASCE,* Vol. 100, 1974, pp. 763–786.

[*26*] Skempton, A. W., *Proceedings of the Institution of Civil Engineers,* Vol. 7, 1957, pp. 305–307.

[*27*] Leroueil, S., Magnan, J. P., and Tavenas, F., *Remblais sur Argiles Molles, Lavoisier, Paris,* 1985, pp. 1–342.

[*28*] Mesri, G., *Geotechnical Division, Proceedings of the ASCE,* Vol. 101, 1975, pp. 409–412.

[*29*] Azzouz, A. S., Baligh, M. M. and Ladd, C. C., *Journal of Geotechnical Engineering, Proceedings of the ASCE,* Vol. 109, 1983, pp. 730–734.

[*30*] Karlsrud, K., Aas, G., and Gregersen, O., "Can We Predict Landslide Hazards in Soft Sensitive Clays? Summary of Norwegian Practice and Experiences," Norwegian Geotechnical Institute Publication 158, 1985, pp. 1–24.

[*31*] Clauson, C-J., Graham, J., and Wood, D. M., *Geotechnique,* Vol. 34, 1984, pp. 581–600.
[*32*] Karlsrud, K., "Analysis of a Small Slide in Sensitive Clay in Fredrikstad, Norway," Swedish Geotechnical Institute Report 17, 1983, pp. 175–184.
[*33*] Lacasse, S., Jamiolkowski, M., Lancellotta, R., and Lunne, T., "In Situ Characteristics of Two Norwegian Clays," *Proceedings of the 10th International Conference on Soil Mechanics and Foundation Engineering,* Vol. 2, 1981, pp. 507–511.
[*34*] Aas, G., "Stability of Slurry Trench Excavations in Soft Clay," *Proceedings of the 6th European Conference on Soil Mechanics and Foundation Engineering,* Vienna, Vol. 1.1, 1976, pp. 103–110.
[*35*] Koutsoftas, D. and Fischer, J. A., *Journal of Geotechnical Engineering, Proceedings of the ASCE,* 1976, pp. 989–1005.
[*36*] Trak, B., La Rochelle, P., Tavenas, F., Leroueil, S., and Roy, M., *Canadian Geotechnical Journal,* Vol. 17, 1980, pp. 526–544.
[*37*] Eide, O., Aas, G., and Josang, T., "Special Application of Cast-in-place Walls for Tunnels in Soft Clay in Oslo," *Proceedings of the European Conference on Soil Mechanics and Foundation Engineering,* Vol. 1, 1972, pp. 485–498.
[*38*] Eden, W. J. and Law, K. T., *Canadian Geotechnical Journal,* Vol. 17, 1980, pp. 369–381.
[*39*] Holm, G. and Holtz, R. D., "A Study of Large Diameter Piston Samplers," *Proceedings of the International Conference on Soft Clay,* 1977, pp. 375–386.
[*40*] Ghionna, V., Jamiolkowski, M., Lacasse, S., Lancellotta, R., and Lunne, T., "Evaluation of Self-boring Pressuremeter," *Proceedings, International Symposium on In situ Testing, Paris,* 1983, pp. 294–301.
[*41*] Hanzawa, H., Matsuno, T., and Tsuji, K., *Soils and Foundations,* Vol. 19, No. 2, 1979, pp. 1–14.
[*42*] Hanzawa, H., *Soils and Foundations,* Vol. 19, No. 4, 1979, pp. 69–84.
[*43*] Larsson, R., "Basic Behaviour of Scandinavian Soft Clays," Swedish Geotechnical Institute Report 4, 1977, pp. 1–108.
[*44*] Kishida, T., Hanzawa, H., and Nakanowatari, M., *Soils and Foundations,* Vol. 23, No. 2, 1983, pp. 69–82.
[*45*] Nicholson, D. and Jardine, R. J., *Geotechnique,* Vol. 31, 1981, pp. 67–90.
[*46*] Parry, R. H. G., *Geotechnique,* Vol. 18, 1968, pp. 151–171.

DISCUSSION

Julio Augusto De Alencar, Jr.[1] (written discussion)—The ASTM International Symposium on Laboratory/Field Vane Shear Strength Testing presented studies on various factors that influence the execution and interpretation of the vane test as well as its advantages and limitations. It has shown there is still a lot of work to be done in order to properly understand the shearing mechanism generated during the test and adequately quantify, though on an empirical basis, the influence of different factors, like disturbance caused by insertion, rate of shearing, plasticity, sensitivity of the material being tested, and progressive failure.

Within this context, more emphasis should be given to influence of the structure of the material in results obtained from the test. Theoretical analysis presented by De Alencar, Chan, and Morgenstern as well as field data presented by other authors, in which it is observed that materials of higher sensitivity yield lower maximum measured torque, suggest, in our opinion, that progressive failure may have significant influence on the value of maximum torque measured, lowering its value. Therefore, some skepticism should be exercised with respect to considering the strength calculated for these materials as "peak strength."

Several correlations have been presented involving strength and index properties, especially the plasticity index. To obtain these indexes, it is necessary to completely destroy the structure of the material. It is well known today that the structure of the material exerts an extremely important influence on its strength (drained or undrained). Therefore it is not meaningful to compare S_u times index properties correlations involving materials of completely different structures. It is very important, then, whenever this kind of correlation is presented that some effort is spent on describing the structure of the material.

R. J. Chandler (author's closure)—There seems to be agreement that the field vane test is likely to underestimate the undrained shear strength of high sensitivity, low plasticity clays. The reasons for this remain uncertain. It will certainly be helpful to have good descriptions of the clay structure, where this can be done. However, for correlative purposes, at present the most satisfactory method is to relate natural water content and index properties, expressing the result as a liquidity index. Unfortunately, clay structure descriptions are not capable of quantification in any more convenient manner.

[1] 222 Michener Park, Edmonton, Alberta, Canada.

Alan G. Young,[1] Bramlette McClelland,[2] and Gerardo W. Quiros[2]

In-Situ Vane Shear Testing at Sea

REFERENCE: Young, A. G., McClelland, B., and Quiros, G. W., **"In-Situ Van Shear Testing at Sea,"** *Vane Shear Strength Testing in Soils: Field and Laboratory Studies, ASTM STP 1014,* A. F. Richards, Ed., American Society for Testing and Materials, Philadelphia, 1988, pp. 46–67.

ABSTRACT: In-situ vane shear testing to measure undrained shear strength in the offshore environment has experienced increased use and significant equipment improvements over the last decade. These changes arose from (1) the offshore petroleum industry's need to improve the quality of foundation site investigations for their structures, (2) technological advancements in mechanical and electrical systems, and (3) recognition by the geotechnical profession of the benefits of acquiring vane shear testing data for better interpretation of cohesive soil strength properties. This paper provides insight into the current international practice of in-situ vane shear testing by presenting results of a questionnaire to international experts. The paper also describes the historical developments that led to the current state of practice. A review of field vane deployment and important operational details is presented along with an appraisal of the applications of vane shear testing for foundation design purposes. The paper also presents various case studies demonstrating how vane shear testing data proved beneficial for offshore foundation design. Finally, the paper summarizes the author's opinions relative to the current state of the art of offshore vane shear testing and includes comments on possible future developments and applications.

KEY WORDS: investigations, soil properties, offshore drilling, field tests, soil tests, clay soils, shear strength, vane shear tests, shear rate, reviews, comparison, questionnaire, statistical distribution, design criteria

The use of in-situ vane shear testing (VST) for measurement of undrained shear strength s_u, has experienced a strong impetus in the offshore environment over the last decade for a number of reasons. First, major expansion in the installation of offshore petroleum exploration and production facilities has provided economic incentive to improve the quality of site investigations by expanding use of in-situ testing. Second, major technological developments in both mechanical and electrical systems have allowed VST to be performed in the offshore environment in a more practical, reliable, and efficient manner. Third, the advantages of VST for various fine-grained soil types and its benefits for engineering analyses and design have been demonstrated by a number of investigators [*1–3*]. This paper provides insight into the current international practice of offshore VST by presenting the following: (1) results of a questionnaire on the current practice of offshore VST submitted to an international group of geotechnical engineering experts, (2) an account of the historical developments that has led to the current state of practice, (3) a review of the field deployment and operational procedures of systems commonly used, (4) an appraisal of the offshore geotechnical applications of VST data for foundation design purposes, (5) various case studies where VST data were important for offshore foundation design, and

[1] Vice president, Marine Geosciences, Fugro-McClelland, Inc., 6100 Hillcroft, Houston, TX 77081.
[2] Fugro McClelland, Inc., P.O. Box 740010, Houston, TX 77274.

(6) the authors' conclusions concerning the state of the art, including comments concerning possible future development and application of VST.

International Survey of Practice

To develop a better understanding of the current practice of offshore VST, as well as opinions regarding application of the data, a questionnaire was sent to 47 international experts in geotechnical engineering. Seventy-four percent of these experts responded to the questionnaire. Of those that responded, 39% were from the United States, 19% from Norway, 9% from France, 6% each from the Netherlands, Italy, Canada, and United Kingdom, and 3% each from Australia, Denmark, and Belgium. The respondents were affiliated with four different groups as follows: (1) 35% worked with oil companies, (2) 11% were associated with research centers, (3) 34% worked with consulting firms, and (4) 20% taught in universities. Ten questions were presented in the questionnaire covering a wide range of subjects related to offshore VST. A summary of the responses to the ten questions will be presented in the following sections of this paper. Where distinct regional differences were recognized in the questionnaire responses, these will also be highlighted.

Historical Development

The use of VST originated onshore in 1928 when the Swedish engineer, John Olsson, devised an in-situ tool [*4*] for the purpose of avoiding disturbance associated with sampling the sensitive marine clays found in the Scandinavian countries. The practice of measuring undrained strength in situ with the vane shear tool spread as many groups recognized its benefits. Finally, the Swedish Geotechnical Institute began an extensive study in 1947 that culminated in 1950 with a comprehensive report by Cadling and Odenstad [*1*], which established a standard test procedure that was widely adopted throughout that region and elsewhere in the world.

The practice of offshore VST has been promoted primarily by the international petroleum industry and the U.S. Navy. Since the goals of the two groups had been substantially different, each group pursued different courses in their research and development activities. The international petroleum industry has promoted development of VST through international geotechnical engineering service companies while the U.S. Navy relied upon universities and government research centers. Considering that the petroleum industry's interest was directed towards geotechnical design information for axially and laterally loaded piles, the developmental efforts focused on equipment capable of obtaining in-situ data from the seafloor down to as deep as 200 m. On the contrary, the naval activities were directed towards sampling and VST equipment designed to provide information to only 3-m depth, which has been sufficient to provide information on sediment characteristics for mooring analysis and foundation design of small instrument packages and cable installations.

Since many of the equipment systems developed for U.S. Naval activities are not in extensive use today, this paper will give only a brief account of the historical developments in this area. The major focus will be to describe the evolution of equipment and operational practice of VST that has been applied to petroleum industry projects.

Offshore VST with the petroleum industry began in the Gulf of Mexico in 1952 when the California Company (Chevron) employed the services of Greer and McClelland, Inc., to fabricate and use a field vane device at a location in 8 m of water offshore Louisiana. The tests were performed in a cased boring made from a fixed platform 5 m above the water. An 80- by 145-mm vane was attached to the bottom of a string of NW rod that was rotated by a motorized torque device at the platform desk. Efforts to improve on the slow and cumbersome procedure were made in 1956 when McClelland Engineers developed a "vane borer" device as reported by Fenske [*5*]. The vane borer, which was supported by Humble Oil and Refining Company (Exxon), was designed to save time associated with removing the vane device after each test. The 76-mm-diameter × 152-mm-long vane could be recessed into the drill bit during drilling operations, and did not have to be retrieved after each test. The operation again required a fixed platform, and torque was applied at deck level. Following the advent of floating drilling and wireline sampling procedures in the early 1960's as described by McClelland [*6*], the vane borer method became obsolete.

A number of VST devices that could be operated from a submersible were developed for U.S. Navy work in the 1960s as described by Winget [*7*] and Inderbitzen et al. [*8*]. These devices were used on the submersibles, Deep Quest, and Alvin, in water depths up to about 1300 m; however, their testing depth was limited to 3 m. In 1965 the U.S. Navy built and operated a submerged tethered platform called the Deep Ocean Tests In Place and Observation System (DOTIPOS) capable of performing VST to 3-m depth in water depths up to 1800 m. Richards and Keller [*9*] at Lehigh University developed a tethered seafloor platform called the "Underwater Tower" capable of performing VST at 0.3-m intervals to a penetration of 3 m in 3500 m of water. Another tethered seafloor platform called the Multi-Purpose In Situ Testing System (MITS) [*10*], which was developed by Woodward-Clyde Consultants, allowed VST to be performed to a penetration of 6 m in water depths up to 450 m.

Additional use of VST was not initiated until the late 1960s, when the design of costly platforms to be installed in the soft soils of the Mississippi River delta front justified special in-situ testing measures to avoid the disturbance associated with commonly used sampling techniques. This required development of a method for performing such tests from a floating vessel. In 1970, McClelland Engineers collaborated with Shell Development Company to develop a field vane device called the "Remote Vane" [*2*] that could be deployed and operated downhole through the bore of ordinary oilfield drill pipe using a wireline operation. The resulting test system is shown in Fig. 1.

The outstanding success of the Remote Vane resulted in its use at over 60 sites in the Gulf of Mexico in the early 1970s. Fugro International, Inc., placed a similar tool in commercial service in the Gulf of Mexico in 1976 and gave further impetus to use of the in-situ vane for Gulf of Mexico geotechnical investigations.

Expansion of offshore petroleum activities worldwide throughout the late 1970s further

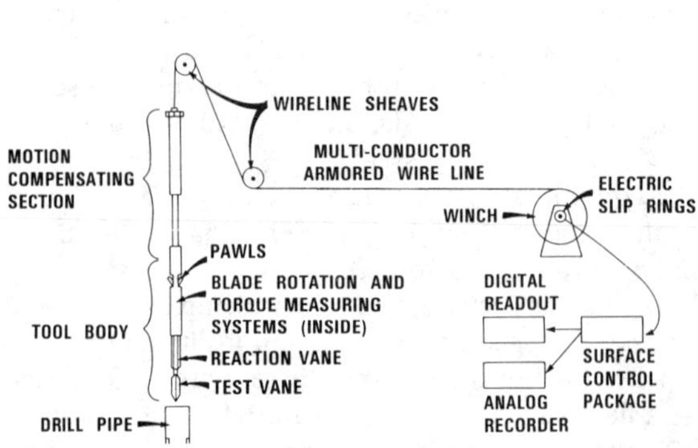

FIG. 1—*McClelland wire line field vane system* [3].

increased the use of VST. The sedimentary basins of the North Sea, the South China Sea, and the Indonesian continental shelf were found to hold clay soils where VST would be directly applicable. This worldwide increase in VST was confirmed by the respondents to the questionnaire. While only 13% of the 22 respondents had any experience with the test before 1970, another 60% gained experience with the test during the 1970s and another 27% began offshore use of the test in the 1980s.

The increased use of VST in the early 1970s was accompanied by further improvements in the electro-mechanical systems of the test devices. In 1978, McClelland deployed its fifth-generation Remote Vane, which achieved greatly improved dependability through modular construction and a reduction in the number of moving parts. Fugro in Holland redesigned their tool (Fig. 2) in the early 1980s to be fully compatible with their other seafloor testing systems, allowing it to be used in both downhole and seabed operations. In 1985, McClelland developed a free-falling downhole vane device as part of a wireless system called Dolphin [*11*] as shown in Fig. 3. Power for this fully self-contained test device is provided by a battery pack, while test commands and data acquisition are both provided by its electronic memory system. By 1987, this technological advancement had allowed the use of VST on projects where combined water depth and boring penetration exceeded 1200 m.

Field Deployment and Operation

In the offshore environment, operation with the VST is almost always performed from a floating vessel, such as a geotechnical drillship, an oilfield supply vessel with a portable drilling rig, or an offshore drilling rig such as those found on semisubmersibles or drill-

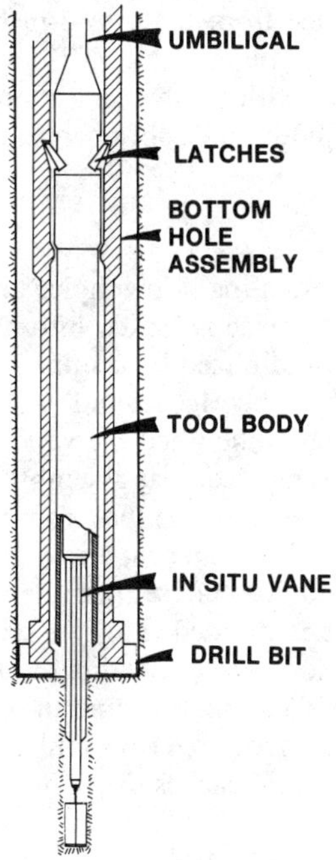

FIG. 2—*Fugro field vane (see footnote 3).*

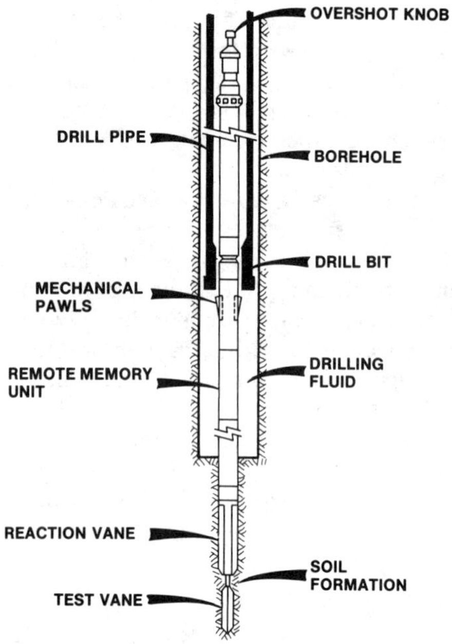

FIG. 3—*McClelland wireless field vane* [11].

ships. For this reason, the field vane must be part of a total system that includes some means for isolating the test device from the 1-to 3-m heave typically experienced by a drillship operation in moderately turbulent seas. VST devices are deployed both downhole, as part of a drilling operation, and with a tethered seafloor platform. Methods for isolating ship movement are required for both modes of operation.

Downhole Operation

There are two primary ways of isolating a downhole vane device from vessel heave. With the Fugro vane, the wireline test device is locked into the drill pipe at the bottom of the borehole, and the drill pipe is immobilized by a shipboard motion-compensation system as described by Geise et al.[3] Both of the McClelland vane types avoid the influence of ship motion by avoiding direct connection between the vane tool and the drill string. With the original wireline vane, a telescoping electrical assembly, identified as the "motion compensating section" in Fig. 1, isolates the vane tool from the ship's motion as described by Ehlers et al. [*3*]. The wireless Dolphin vane shown in Fig. 3 has no attachment either to the drill string or to a wireline and does not require the telescoping assembly.

The McClelland wireline vane is divided into two sections, the motion compensating section and the tool body. The tool body contains the test vane and reaction vane, the electrical motor that causes vane blade rotation, and the strain-gage-type torque transducer as shown in Fig. 1. The entire unit is lowered on a multi-conductor armored wireline cable by a mechanical winch. Once the tool reaches the bottom of the borehole with the drill bit

[3] Geise et al., in this publication, pp. 318–338.

suspended off the bottom, vane pawls are extended electrically or mechanically, making it possible for the drill pipe to push the tool about 1.0 to 1.5 m below the bottom of borehole. The bit is again raised and suspended off the bottom to isolate the tool from the moving drill string. At this point, the test is started using an instrument package located shipboard that also records data in both digital and graphic form.

With McClelland's wireless Dolphin vane, the entire tool free falls down the drill string until its pawls open mechanically below the drill bit, which is suspended off the bottom of the borehole. The vane is then pushed to the desired test depth, and the drill bit is raised off the bottom before testing, in the same manner as described for the McClelland wireline vane. The Dolphin System uses a remote data acquisition system to sample, digitize, and record data downhole as it is generated. Upon completion of the test, the tool is retrieved with a high speed wireline winch using an overshot assembly, and the test results are retrieved from the memory and printed by an on-board personal computer.

The Fugro vane is deployed by lowering the tool body on an umbilical cable down the bore of the drill pipe until it automatically latches into the grooves in the bottom hole assembly as shown in Fig. 2. Its hydraulic motor then pushes the test vane 1.0 to 1.5 m below the bottom of the borehole. At this point an electric motor within the tool body rotates the test vane while a torque transducer continuously monitors the soil resistance. Upon completion of the test, the entire vane is withdrawn to the surface without retraction of the vane back into the tool body.

Seabed Operation

The low soil strengths often encountered at shallow penetrations will not allow accurate strength data to be obtained with downhole VST devices because of lateral instability of the drill pipe and the tendency for the vane device to settle under its own weight. To overcome these problems, a small seabed template called Halibut [*3*] was designed in 1976 to allow a single vane test to be performed without a drill rig or drill pipe. The 1.3-m^2 template illustrated in Fig. 4 clamps the vane test body for testing at a selected penetration below its base. The system is lowered over the side of the vessel by an auxiliary crane, and the required ballast within the template provides the weight needed to push the vane to any selected test depth down to 7-m penetration. Initially, the Halibut used the McClelland wireline vane, but in 1985 the wireless Dolphin vane was introduced and eliminated the need for an electrical wireline.

Fugro[3] uses the seabed jacking device called Seacalf to push the vane tool below the seafloor to the desired test depth. Seacalf provides the seabed reaction required to push the vase as deep as 25 m in weak clay soils. After the vane test is performed at a selected depth, the vane can be pushed again to a deeper penetration and the testing cycle repeated.

Operational Details

Although the electro-mechanical systems described here are highly sophisticated and capable of producing high-quality test data, close attention to numerous operational details is required to realize the full potential of these tools. Some of these are as follows: (1) vane blade geometry, (2) vane rotation rate, (3) test penetration, (4) drilling fluid weight and pump pressure, and (5) torque calibration. Inadequate attention to or control of these operational details may adversely influence or invalidate strength data. The following discussion of these operational details is directed primarily to the downhole mode of operation, although many of the considerations are equally applicable to seabed applications.

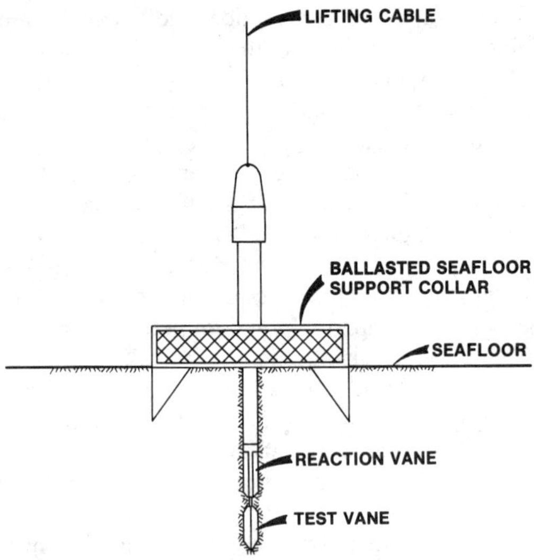

FIG. 4—*Halibut seafloor template* [11].

Vane Blade Geometry

The test vanes used offshore have consisted primarily of two configurations as shown in Fig. 5. The rectangular design was developed initially in Sweden and has been widely used onshore in Scandinavia. This geometry is also used with the Fugro vane. McClelland's vane generally uses blades with tapered ends. Since s_u is computed from the torsion required to rupture the soil inscribed by the vane blade edges, the equations used to convert observed torque to s_u must take into consideration blade geometry. Equations for computing s_u for various vane configurations have been presented by Perlow [*12*].

In the questionnaire, 74% of the respondents indicate that they specify geometry of the vane blades in their work. ASTM Method for Field Vane Shear Test in Cohesive Soils (D 2573) specifies a height-to-diameter ratio equal to 2.0 and area ratios ranging from 10 to 18% for large to small size vane blades, respectively. ASTM also states that the vane blade ends may be tapered. A close review of the questionnaires indicates that specifications used by European respondents generally follow Norwegian practice as described by Aas et al. [*13*] whereas the responses from the United States either omitted comments as to vane geometry or indicated acceptance of both configurations.

Several test vane sizes may be needed depending upon the strength range of the soil. A larger vane, typically about 65-mm maximum diameter caused by drillpipe limitations, is preferred in weaker soils. A smaller vane, seldom smaller than about 35 mm, is usually chosen for stronger soils to maintain a consistent level of resolution in the torque measurement. The authors believe that international standardization of the vane blade geometry will be beneficial. A height-to-diameter ratio of 2.0 and area ratios in the range of 12 to 15% are desirable from the authors' viewpoint.

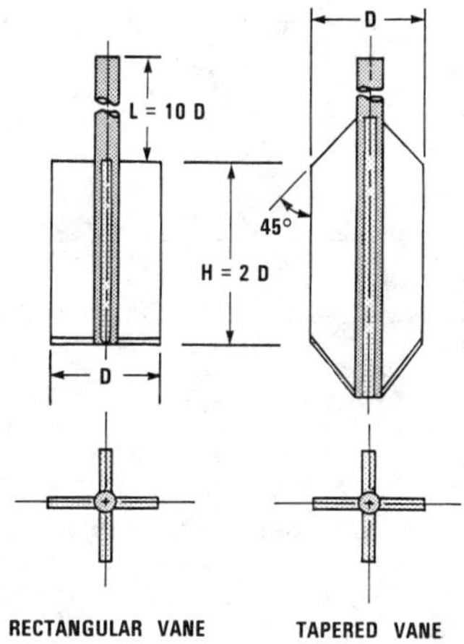

FIG. 5—*Geometry of field vane (D 2573).*

Vane Rotation Rate

Many investigators [*1,14–16*] have demonstrated that variations in the vane rotation rate, or strain rate, influence the observed shearing resistance of the soil. A change of one order of magnitude in the rotation rate, however, has been found to produce only a 5 to 15% change in the observed strength of normally consolidated clays [*15*]. Although this indicates only a moderate sensitivity to strain rate, there is general concurrence among engineers that either the vane rotation rate or the shear velocity at the vane edge should be standardized in the interest of consistency. European respondents indicate that they generally follow the test practice described by Aas [*17*] and specify a rotation rate at the tip of the vane of 12°/min, which results in a time to failure in the soil of about 2 to 5 min. Practice in the United States with the McClelland vane is to use a rotation rate of 18°/min, resulting in soil failure in about 1.5 to 3.5 min. The ASTM standard calls for a rotational rate of 6°/min, resulting in a time to failure of about 4 to 10 min, and corresponding to the rotation rate adopted for the Fugro vane.[3]

The sensitivity of a soil to loading rate is a function of the permeability of the soil, the length of the drainage path, and the gradient of induced pore pressure. In an undrained test, the primary objective is to perform the test fast enough so that drainage does not occur. Therefore, the VST requires that the shear velocity along the cylindrical failure surface be rapid enough that excess pore pressures do not dissipate. The shear velocity during a test varies not only with vane rotation rate but also with vane diameter. Figure 6 shows the relationship between shear velocity and rotation rate for vane diameters of 38.1 and 63.5 mm, and the shaded area encompasses the approximate range of rates existing in current marine geotechnical practice.

There is generally wide support for narrowing the range of strain rates used in current practice. Perlow and Richards [*16*] recommended that the VST be performed with a standard shear velocity of 0.15 mm/s, and Perlow [*12*] later recommended a rate of 0.10 mm/s. The authors, however, strongly recommend defining the standard as a reduced range of acceptable shear velocities, rather than one specific rate, and suggest a range of 0.10 to 0.15 mm/s, for example. This range would retain some flexibility in the mechanics of tool design and in the selection of vane size, and yet would significantly reduce the present wide range of shear velocities.

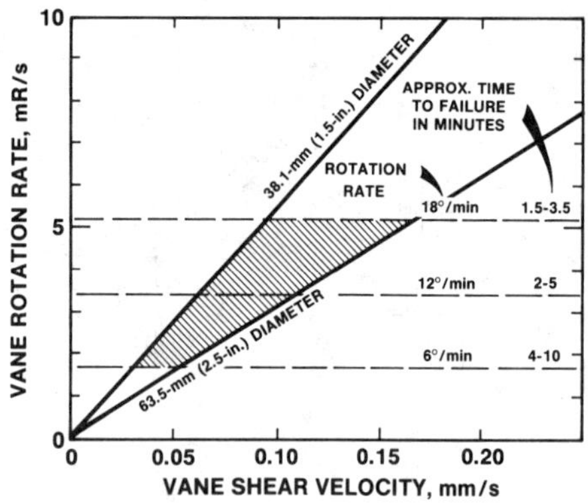

FIG. 6—*Relationship between vane rotation rate, shear velocity, and vane diameter.*

Test Penetration

Even though motion compensation of the vane tool is achieved, the sea state or vessel heave still plays an important operational consideration. The depth and degree of disturbance is influenced by the variable bit load on the bottom of the borehole during drilling, which is caused by vessel heave during drilling. Even with motion compensation systems, the cyclic load on the drill bit is not eliminated, and the depth of disturbance may extend to at least 1 m [*18,19*].[3] Experience has shown that both test and sample quality deteriorate with increasing severity of the sea state. For this reason, a minimum test vane penetration of 1 m should be used, and the field engineering supervisor should increase this to as much as 1.5 m in heavy seas or suspend operations, as judgment requires.

Fluid Weight and Pump Pressure

Excessive fluid pressure at the bottom of the borehole, whether from the mud column weight or from the pump pressure during drilling, can cause hydrofracture of the formation at and below that level. Similarly, insufficient fluid pressure can lead to formation heave and consequent disturbance of the soil before sampling. Bottom hole pressure can be controlled by careful monitoring of drilling rate, drilling fluid weight, and pump pressures. Thus, the drilling crew must exert proper care in advancing the borehole as well as in controlling the drilling fluid weight and pump pressure. The authors' experience indicates that operational problems can be avoided if proper care is exercised during the drilling operations to maintain a mud weight that will not induce hydraulic fracture, and to keep pump pressures high enough to produce an upward flow of cuttings but low enough to avoid jetting into the soil formation.

Torque Calibration

Measurement of the torque applied to the vane blade is another operational consideration that may strongly influence the measurement of s_u. Many investigators, including Perlow [*12*] and Kraft et al. [*15*], have recognized that s_u values measured using a calibrated torque spring will differ from those measured with an electrical torque transducer. Although earlier VST tools used the calibrated torque spring, practice since 1971 has been to use an electrical torque transducer. The advantage of this system is the constant rate of shear maintained on the failure surface. The electrical torque transducer also is generally located a short distance away from the test vane, thereby reducing test error caused by the vane rod friction. Offshore VST systems also generally include a torque calibration system that is used immediately before the tool is lowered downhole or inserted into the soil. The authors believe attention to details such as torque calibration and careful deployment results in reliable and consistent VST data.

Design Applications of VST

Since 1928 when the VST was first introduced, the primary purpose of the tool has been to provide a practical means for measuring s_u in situ, for direct use in geotechnical engineering design. Earlier uses of VST data onshore were aimed at solving slope stability and bearing capacity problems [*20–24*] using data free from the effects of disturbance associated with sampling and sample handling. Introduction of the tool offshore had the additional incentive of avoiding severe disturbance associated with removal of samples from great depths where total stresses were quite large in comparison to most geotechnical engineering problems. Results were desired for use in designing pile foundations and bottom supported systems and for solving marine slope stability problems. Many of the methods of analyses in geotechnical engineering rely upon empirical correlations based on a particular type of shear strength measurement. Use of VST data with such correlations requires application of an adjustment factor that interrelates the two test types. The purpose of the following section is to review prior experience in applying in-situ vane shear data to engineering design problems, drawing on both published references and the results of the international questionnaire.

Respondents to the questionnaire indicated, as shown in Fig. 7, a very favorable disposition towards obtaining VST data during a geotechnical investigation when underconsolidated or normally consolidated clays exist at a site. When the overconsolidation ratio (OCR) exceeds a value of four, indicating a highly overconsolidated clay, then the confidence level in obtaining and applying VST data is reduced so that only 6% of the respondents indicated VST would be useful.

Many geotechnical engineers believe that there is a limiting range of shear strength within which VST is an appropriate test procedure. Figure 8 shows the results to a question seeking to define that range. Although a small minority of the responses indicate VST could be performed in soils stronger than 225 kPa, 60% of the respondents believe the practical limit of the test is 150 to 200 kPa.

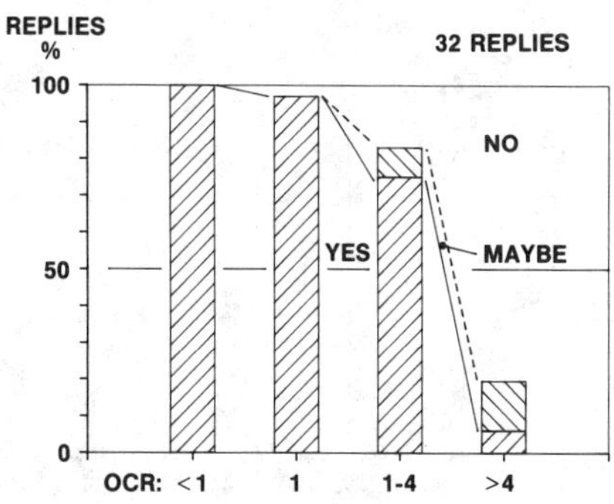

FIG. 7—*Opinions concerning vane test validity as a function of overconsolidation ratio.*

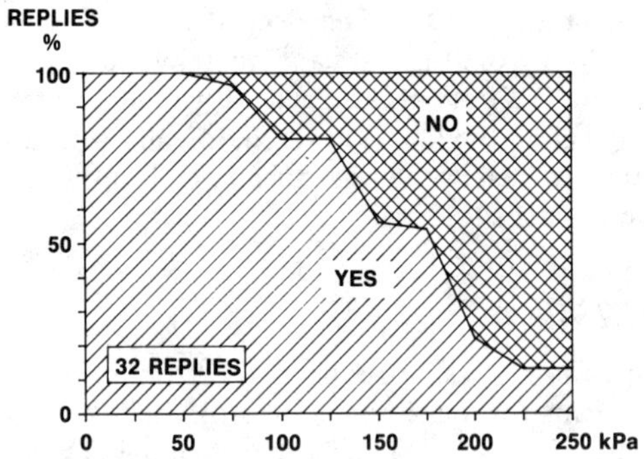

FIG. 8—*Opinions concerning vane test validity as a function of undrained shear strength.*

Since current design practice allows for a large number of different sampling procedures and in-situ and laboratory testing techniques, the authors surveyed the international group of experts as to their preferences. Assuming a site in moderate water depth where normally consolidated clays are known to exist, the questionnaire asked the experts to identify their preferred sampling and in-situ testing methods. Figure 9 shows the results of the survey based on a numerical ranking in which the most preferred method received a ranking of 6, the least a ranking of 1.0. Based on a numerical average of all responses, piston sampling received the highest ranking with an average of 5.4. This indicates that laboratory testing on the high-quality samples, as would be taken with a piston sampler, remains the most preferred method of determining the undrained shear strength profile. Laboratory testing on pushed samples, VST, and the cone penetrometer test (CPT) received nearly equal ranking with an average score of about 4.5. Laboratory testing on driven samples and use of the pressuremeter were the least preferred methods with average scores of 2.5.

The authors' assessment of the current state of international practice is that piston sampling with conventional laboratory or stress history and normalized soil engineering properties (SHANSEP) testing remains the most widely used method for selecting an s_u profile appropriate for geotechnical engineering design. Seventy percent of the respondents indicated that they currently use the vane on less than 50% of their offshore projects as shown in Fig. 10. However, many experts realize the difficulties associated with obtaining high-quality samples even with the piston sampler and are making greater use of in-situ testing

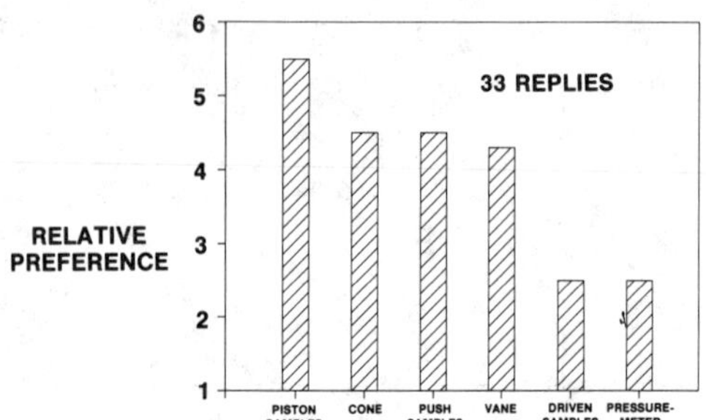

FIG. 9—*Relative preference for in-situ testing and sampling.*

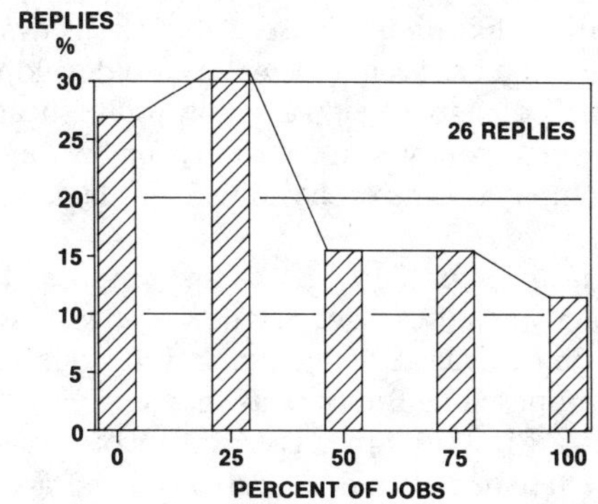

FIG. 10—*Frequency of field vane use on offshore projects.*

using either the CPT or VST. Although laboratory testing on piston samples remains the preferred method for acquiring geotechnical strength data, 97% of the respondents expressed the opinion that VST is an important investigative tool for normally to slightly overconsolidated clays.

Another survey question examined the level of confidence associated with design analyses using only VST data. As shown in Fig. 11, slightly over 50% of the responses indicated willingness to use VST data alone for mud mat bearing capacity design. Thirty to forty percent of the respondents would rely solely on vane data for analyses of axial or lateral pile capacity, jack-up rig footing penetration, or pipeline design. Most respondents believe VST strength data serve a useful purpose in assessing questionable or unusual strength data on samples, especially when the results are influenced by unusual geologic conditions. A majority therefore prefer that VST strength data be acquired, but they prefer to use the data in conjunction with other types of shear strength measurement.

Adjustment Factors

Since the earliest use of the VST in Scandinavia, it has been recognized by a number of investigators that VST strength data are usually higher than laboratory strengths on recovered samples. Extensive experience [*2,3,20,21*] demonstrates that VST strength data

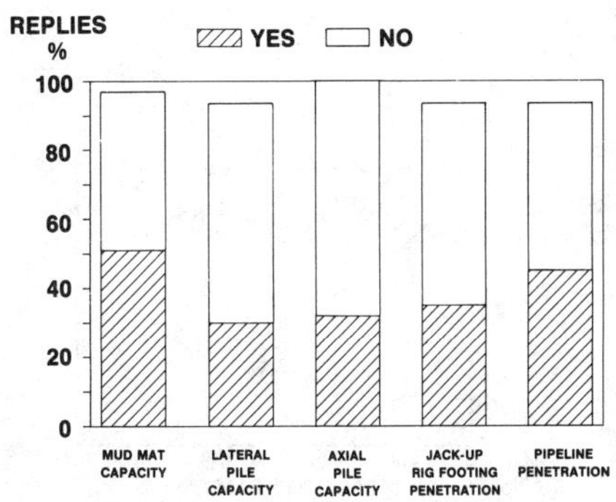

FIG. 11—*Opinions regarding engineering design with field vane data alone.*

are inappropriate without modification for use with various design procedures. For this reason, much research attention has been directed toward developing an adjustment factor that could be applied to VST data to develop an appropriate shear strength for a particular design problem, such as pile capacity, slope stability, or bearing capacity. The following discussion reviews and compares some of the adjustment factors that have been proposed for various design analyses.

A strength adjustment factor μ for the VST was first proposed by Bjerrum [*20*] for slope stability and bearing capacity analyses. A correlation of μ with plasticity index (PI) of the clay was proposed based on experience with a number of slope stability and embankment failures. The values of μ proposed by Bjerrum, as shown in Fig. 12, vary from 1.0 for a PI of 20 to as low as 0.7 for a PI of 70. Since Bjerrum's work was published, a number of other investigators have evaluated VST data and correlated μ with the following: liquid limit [*22*]; s_u/p_c' ratio where p_c' is the preconsolidation pressure from the consolidation test [*23–25*]; and strain rate, anisotropy [*26*], and liquidity index.[4] Within all these various correlations, the adjustment factor varies from about 0.5 to 1.0. The most recent work by Aas et al. [*13*] proposed values of μ as shown in Fig. 13, which are a function of the strength ratio $s_u/\bar{\sigma}_{vo}$, where $\bar{\sigma}_{vo}$, is the effective overburden pressure.

Most of the efforts to develop values μ for use in pile design have taken place in the United States [*15*]. Emrich [*27*] found in a study on Mississippi Delta clays that an average μ of 0.70 was required to adjust VST data to equivalence with s_u as determined by unconfined compression tests on high-quality piston samples. In another study conducted at the same site in Venice, LA, Doyle et al. [*2*] reported that a μ of 0.66 was required with VST data to correctly predict the capacity of a 30.5-cm-diameter open-end pile driven to 46-m depth.

In a later study of 14 sites in the Gulf of Mexico around the Mississippi River Delta, Ehlers et al. [*3*] concluded that an average μ of 0.75 as shown in Fig. 14 was appropriate for VST data used in axial pile design. This value of μ was considered applicable to highly plastic, normally consolidated clays, for which unconfined compression tests on high-quality piston samples had been found to give s_u values closely comparable to unit skin friction in pile load tests. In an extensive correlation study of pile load tests and analytical procedures, Olson [*21*] reported that values of μ of 0.75 and 0.70 were required with VST data to correct to UU-triaxial and unconfined compression test data on high-quality samples, respectively.

The authors' questionnaire also sought opinions as to the appropriate μ to be applied when using VST data in various design problems (Fig. 15). Depending upon the particular design problem identified, 35 to 48% indicated that they use some adjustment factor,

[4] Johnson et al., in this publication, pp. 293–305.

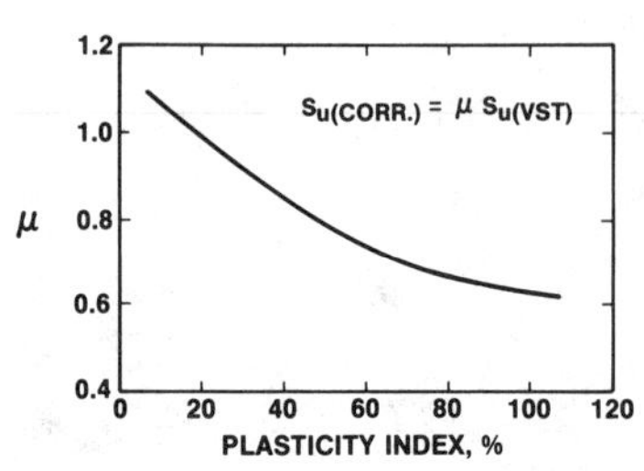

FIG. 12—*Field vane correction factor as a function of plasticity* [20].

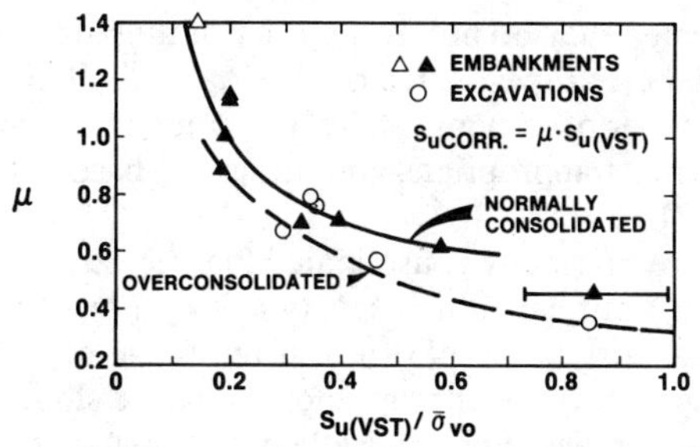

FIG. 13—*Field vane correction factor as a function of stress history* [13].

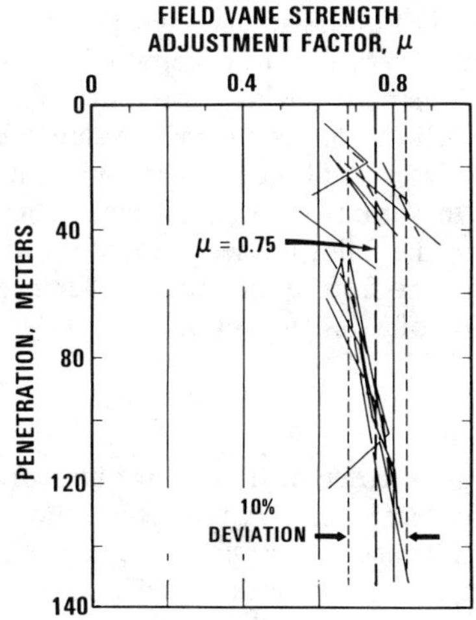

FIG. 14—*Field vane strength adjustment factor based on ratio of unconfined compression test shear strength to field vane shear strength* [3].

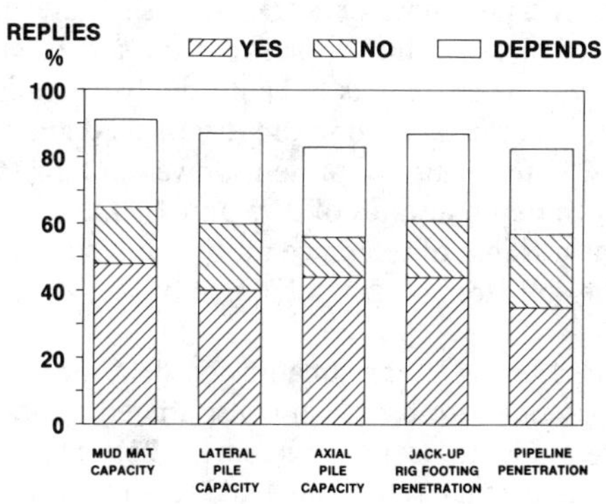

FIG. 15—*Opinions regarding advisability of using adjustment factors for various design applications.*

whereas 13 to 22% indicated they do not. A majority of those respondents who use adjustment factors, especially the Europeans, referred to Aas et al. [*13*] as the preferred source for selecting them. A number of responses from the United States indicated μ in the range of 0.70 to 0.90 would be most appropriate, and this group based their comments primarily on the work of Olson [*21*].

The $s_u/\bar{\sigma}_{vo}$ ratio for many normally consolidated marine clays generally fall in the range of 0.20 to 0.30 depending upon laboratory test type [*28*]. For strength ratios in this range the μ obtained from Aas' correlation [*13*] lies in the range of 0.80 to 0.90. The authors agree with Aas that these factors are appropriate for marine slope stability studies in soils with strength ratios in this range. For axial pile capacity analyses, however, the authors believe that lower values of μ in the range of 0.70 to 0.80 may be more appropriate for normally to slightly overconsolidated clays as described above. For laterally loaded pile analyses, Matlock [*29*] indicates that his *p-y* (load-deflection) criteria for soft clay were developed with emphasis placed on the s_u from the VST; thus, adjustment factors are probably not appropriate for this type of design analyses. For design analyses associated with mud mat bearing capacity, jack-up rig footing penetration, or pipeline bearing capacity, the authors suggest using values of μ based on Aas' correlation [*13*].

In summary, the authors believe that confidence is slowly building in the appropriate selection of μ for design analyses in offshore areas. When using the data, the engineer should give close attention to various geologic characteristics of the sediments. Several examples will be presented in the following section that illustrate how differing sediment characteristics, probably deriving from unidentified differences in depositional environment, complicate the selection of a design s_u profile.

Review of Case Studies

The authors selected four sites to demonstrate how different depositional characteristics and sediment properties can affect laboratory test results and how field VST data can contribute to a more complete understanding of the strength profile. One of these case studies was used in the questionnaire, and the results of the survey related to that site will be presented.

Case Study I

The first site, in 200 m of water, offshore Texas, is situated at the continental shelf edge. The VST data, index tests, and laboratory strength results are presented in Fig. 16. The soils at this site consist of massive, highly plastic, very soft to very stiff clays. Results of consolidation tests indicate the sediments to be normally consolidated. The degree of saturation was close to 100% near the seafloor and diminished gradually but irregularly with depth to the low 90's at 135 m. Figure 16 presents lower and upper bound strength profiles for which s_u increases with depth at rates of 0.95 and 1.90 kPa/m, respectively. These two lines are shown solely as a frame of reference on this and subsequent figures, to facilitate visual comparison of the soil strength data obtained by the various laboratory and in-situ techniques.

The VST data presented in Fig. 16 are unadjusted field measurements. The SHANSEP [*30*] strength profiles presented are based on consolidated undrained direct simple shear (CK_0UDSS) tests. There evidently is good agreement between the miniature vane (MV), unconsolidated-undrained triaxial (UU), field VST, and SHANSEP strength measurements. Throughout most of the boring depth, the VST values for s_u are about 10 to 15% greater than the average of the SHANSEP measurements. There appears to be very close

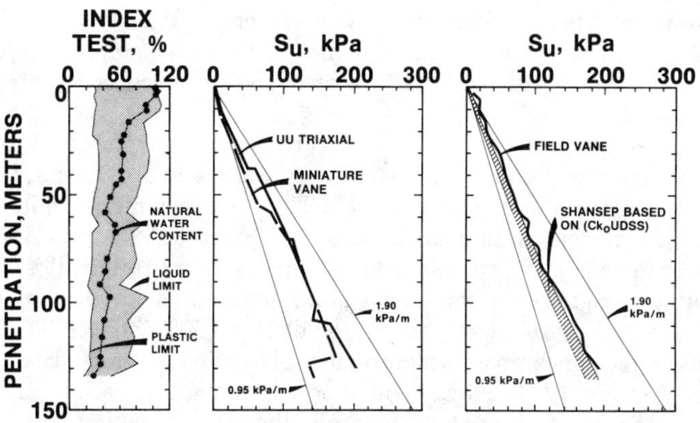

FIG. 16—*Case I: Normally consolidated clay profile depicting consistency in strength data. Site is in the Gulf of Mexico, offshore Texas, in about 200 m of water.*

agreement between the VST results and both the MV and UU strengths [*28,31*]. It is interesting to note that if the VST data had been adjusted by a factor of 0.75 to 0.8, then the adjusted values would have fallen well below the MV and UU strength profiles and below a significant portion of the SHANSEP strength profile.

Case Study II

The second site lies in about 400 m of water offshore from Louisiana. The soils are also normally consolidated clays with plasticity indices ranging from 35 to 40. Strength profiles for the various tests are presented in Fig. 17 along with the same reference lines previously introduced. The MV data obtained on high-quality pushed samples agree fairly closely with the VST data to about 80- to 90-m depth. The UU data are consistently lower than any of the other strength data and only slightly exceed the lower bound reference line throughout most of the boring. The average of the SHANSEP profile based on CK_0UDSS tests is about 15 to 25% less than the profile measured by the VST. The measured degrees of saturation

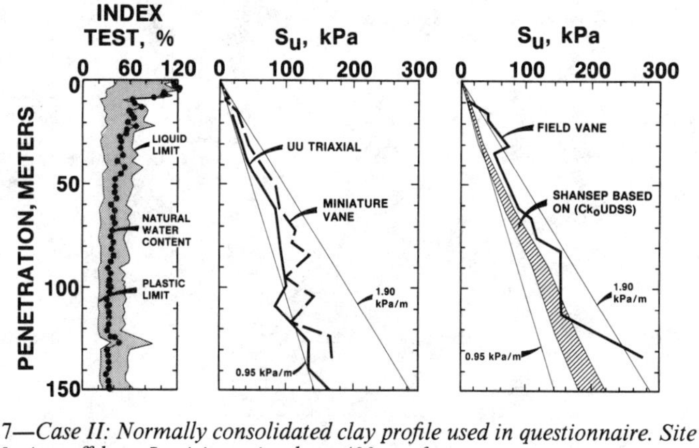

FIG. 17—*Case II: Normally consolidated clay profile used in questionnaire. Site is in the Gulf of Mexico, offshore Louisiana, in about 400 m of water.*

on samples from this site decreased from about 95% near the seafloor to about 85% at the terminal boring depth of 150 m. The gassy nature of these soils probably explains why low s_u strength measurements were obtained on recovered samples, especially with the UU tests.

Data obtained by laboratory and in-situ tests at this site were presented in the questionnaire to the international group of experts, who were asked to provide a s_u profile for axial pile design by the API method of analysis [*32*]. Their average interpretation of the data is a strength profile increasing with depth at a rate of about 1.39 kPa/m as shown in Fig. 18. Normally consolidated clays with similar plasticity characteristics in this offshore region usually exhibit a strength profile increasing at a rate of about 1.50 kPa/m. One standard deviation from the average of all responses extends about 30% above and below the survey average, a range which encompasses almost all of the various strength data obtained at the site, as indicated in Fig. 19. The lower bound of the standard deviation band is essentially equal to the UU profile on pushed samples, and the VST strength profile closely follows the upper bound.

The large spread in the standard deviation demonstrates the uncertainty associated with selecting an s_u profile for pile design when alternative methods for measuring s_u produce widely divergent results. Many respondents placed greater emphasis on the lower bound UU profile whereas other experts selected a s_u profile that represented the upper bound of the VST data. In the authors' opinion, a design profile based on the upper bound of the SHANSEP data would be best for this site. This profile would average about 10 to 20% less than the VST profile and would correspond to a μ equal to 0.9 to 0.8.

The wide divergence in opinion regarding a design s_u profile for this case clearly indicates the need for more pile load tests correlated with s_u data from a large suite of laboratory and in-situ tests. For a 183-cm-diameter pile, a standard deviation from the average s_u profile determined from the survey as shown in Fig. 18 would result in pile penetrations varying from 110- to 140-m penetration to achieve an ultimate pile capacity of 60 MN. The average s_u profile would result in the 60-MN capacity being predicted at 120-m penetration.

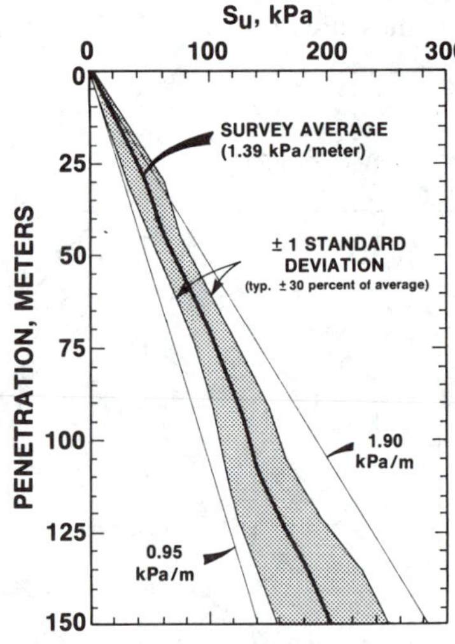

FIG. 18—*Shear strength interpretations from questionnaire for Case II site.*

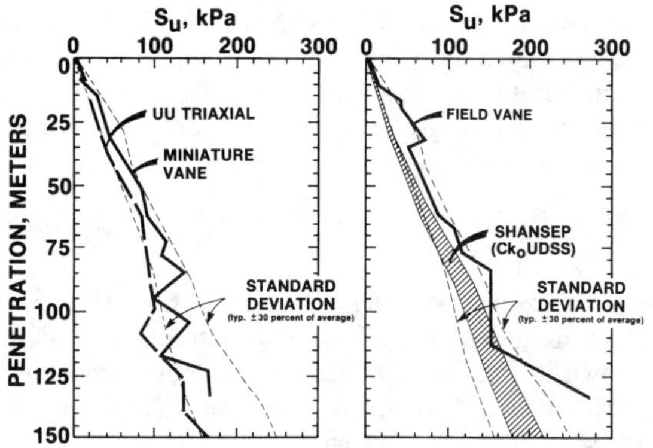

FIG. 19—*Comparison of questionnaire shear strength interpretation with shear strength measurements for Case II site.*

Case Study III

The third site is located offshore Louisiana in about 600 m of water. The soils consist of moderately plastic, slightly overconsolidated clays with plasticity indices ranging from about 40 to 50. The degree of saturation decreases from about 96% near the seafloor to as low as 86% at 150-m depth.

As shown in Fig. 20, the S_u trend lines for the UU and the SHANSEP data are generally parallel to the VST profile down to about 90 m depth, below which the soils begin to exhibit a very gassy and expansive nature. At this depth, the UU trend line decreases sharply in comparison to the other s_u profiles, showing effects of disturbance that results when high quality but poorly saturated samples are removed from a substantial confining pressure.

The SHANSEP profile based on CK_0UDSS tests closely parallels the VST profile even

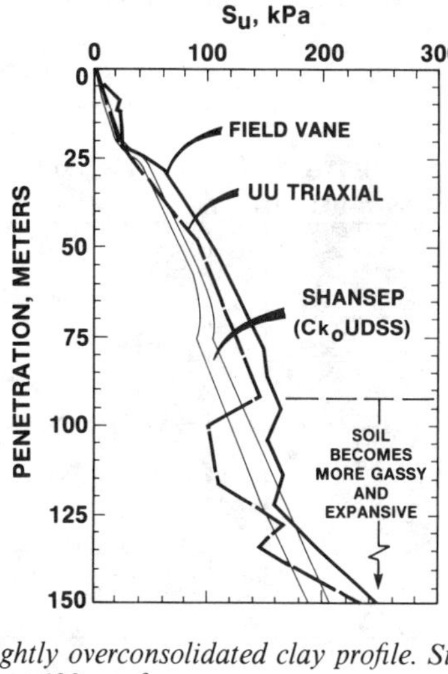

FIG. 20—*Case III: Slightly overconsolidated clay profile. Site is in the Gulf of Mexico, offshore Louisiana, in about 600 m of water.*

though the strengths are about 20 to 30% lower than the VST strengths. This case study confirms the difficulty associated with developing a s_u profile from conventional laboratory data on recovered samples of gassy and expansive soils. The VST data are unaffected by sample disturbance and therefore give a more representative indication of the strength trend with depth.

Case Study IV

The final case study, of a site located very close to the Case III site, concerns a normally consolidated clay with a unique and highly sensitive structure. Data from the upper 12 m of this deposit are shown in Fig. 21, and within that depth range the LI ranges from 130 to 100%. From the seafloor to 6-m depth, s_u from the VST is from 5 to 10 times greater than strengths measured in the laboratory with the MV test on fully remolded samples. In contrast to these high ratios, the MV strengths measured on piston samples are only about twice the remolded strengths, emphasizing the difficulty in avoiding disturbance of highly sensitive soils, even when careful piston sampling is used.

X-ray radiographs performed on the recovered samples indicated a high-density network of worm tracks formed by micro-organisms. The worm tracks were observed to be filled with pyrite crystals by using a scanning electron microscope. The cementation and bonding of the overall soil mass by the network of pyrite crystals contributes to the strength of this sensitive soil. Extremely low strengths result when additional stress is applied to this microstructure and breaks down the network of pyrite crystals.

Commentary

These four case studies illustrate the uncertainties associated with relying solely on undisturbed samples in the marine environment. From experiences such as these, the authors believe that VST data provide a valuable "frame of reference" to help evaluate the extent of sample disturbance that may have influenced conventional laboratory s_u data. The case studies also show that the VST often provides a better indication of the overall trend of strength with depth because of reduced data scatter. With certain types of sensitive or gassy soils, the VST may prove to be the most reliable, and possibly the only reliable, method for obtaining s_u data.

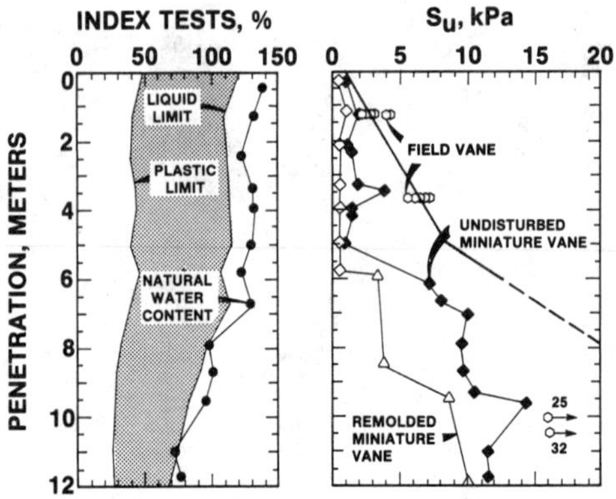

FIG. 21—*Case IV: Highly sensitive normally consolidated clay profile.*

Future Applications

Most field use and experimental testing with the VST has been to determine the s_u of cohesive soils. More recently, a number of investigators [*13,33,34*] have proposed expressions for computing the at-rest coefficient of lateral earth pressure K_0 from VST data. Aas [*13*] has presented results of field measurements of K_0 made at five sites with the VST and found close agreement with other in-situ measurements.

Other studies [*35–37*][5] have attempted to develop stress-deformation characteristics, such as the shear modulus G and the soil deformation modulus E, from the torque versus rotation measurements obtained from the VST. Use of the VST data for this application requires a very sensitive torque system to obtain accurate angle-torque time determinations that are not influenced by stiffness of the torque measuring system. The authors are not aware of the VST being used for these applications in the offshore environment; however, they are confident that existing mechanical and electrical systems can be used or modified for this purpose when research confirms applicability of these techniques for computing K_0, G, and E.

Conclusions

Based on the case studies presented here, the authors' experience, and the 34 responses to the authors' questionnaire, the following conclusions are offered as indicative of current practice in using VST for offshore engineering studies:

1. The state of the art of offshore vane shear testing has improved greatly in the last decade. Advanced electrical mechanical systems have operated reliably and efficiently in water more than 1000 m deep, independently of vessel motion, providing digital recording or data from tests performed at the seafloor and at subsea penetrations as great as 440 m.
2. While conventional sampling and laboratory testing continue to hold the dominant place in engineering practice for determining s_u strength profiles, investigators increasingly favor use of VST as an important additional investigative tool in normally consolidated clays.
3. Case studies demonstrate that low saturation and high sensitivity of some marine deposits may cause laboratory values of s_u to be significantly understated, even when high-quality sampling and testing procedures are employed. VST is a valuable tool for recognizing such discrepancies and for providing realistic strength trends with depth.
4. There are a number of field operational procedures and details that must be monitored or controlled to secure VST data of high quality, including vane blade geometry, vane rotation rate, bottomhole test penetration, drilling fluid weight and pressure, and torque calibration. Some of these operational details warrant further attention in developing practice standards.
5. Evidence continues to accumulate, and most experts agree, that an adjustment factor μ should be applied to VST data when selecting s_u values for design purposes. For normally consolidated clay deposits in the Gulf of Mexico and many other regions of the world where $s_u/\bar{\sigma}_{vo}$ ranges from 0.2 to 0.3, some approximate values for μ and their application in current offshore practice are as follows: 0.7 to 0.8 for designing axially loaded piles; 1.0 for development of *p-y* curves for laterally loaded piles; and 0.8 to 0.9 for bearing capacity and slope stability problems.
6. Future applications of VST, drawing on recent research, which offers a more fundamental understanding of the state of stress and pore-pressure generation in the failure zone,

[5] Pamukcu, S. and Sukayda, J. N., in this publication, pp. 193–208.

may include determinations of the at-rest coefficient of lateral earth pressure K_0, the shear modulus G, and the deformation modulus E. Recent progress in equipment development provides encouragement that any test equipment refinements required for these applications can be incorporated as part of offshore VST systems.

The authors would like to challenge the geotechnical engineering community to fully recognize the potential use of the VST in the offshore environment. As with all technologies, full benefit will only be realized when a serious effort is made to go beyond the precedent of past practice and to embrace those testing techniques best adapted to new and unusual environments. From a strong commitment to strive for excellence and to continually improve methodologies, VST in the future should become a standard test for offshore geotechnical investigations.

References

[*1*] Cadling, L. and Odenstad, S., "The Vane Borer: An Apparatus for Determining the Shear Strength of Clay Soils Directly in the Ground," *Proceedings, Royal Swedish Geotechnical Institute,* No. 2, 1950.

[*2*] Doyle, E. H., McClelland, B., and Ferguson, G. H., *Proceedings, 3rd Offshore Technology Conference,* Vol. 1, Houston, TX, 1971, pp. 21–32.

[*3*] Ehlers, C. J., Young, A. G., and Focht, J. A., Jr., "Advantages of Using In Situ Vane Tests for Marine Soil Investigations," *Proceedings, International Symposium on Marine Soil Mechanics,* Mexico City, Mexico, 1980.

[*4*] Osterberg, J. O., "Introduction" *Symposium on Vane Shear Testing of Soils, STP 193,* American Society for Testing and Materials, Philadelphia, 1956, pp. 1–7.

[*5*] Fenske, C. W., "Deep Vane Tests in Gulf of Mexico," *Symposium on Vane Shear Testing of Soils, STP 193,* American Society for Testing and Materials, Philadelphia, 1956, pp. 16–25.

[*6*] McClelland, B., "Trends in Marine Site Investigation: A Perspective," *Proceedings, Offshore Europe '75,* University of Aberdeen, Aberdeen, Scotland, 1975, pp. 220.1–220.9.

[*7*] Winget, C. L., "Hand Tools and Mechanical Accessories for a Deep Submersible," Woods Hole Oceanographic Institution, Reference 69-32, 1961.

[*8*] Inderbitzen, A. L., Simpson, F., and Gross, G., "A Comparison of In Situ and Laboratory Vane Shear Measurements," Lockheed Ocean Laboratory Report 681703, Lockheed Missile and Space Company, 1970.

[*9*] Richards, A. F. and Keller, G. H., "In Place Measurement of Shear Strength and Bulk Density in Gulf of Maine Clays," American Geophysical Union Transactions, Vol. 49, 1968, p. 221.

[*10*] Lowell, V. B. and Ehlers, C. J., "Development of Deepwater Soil Sampling and In Situ Testing Techniques Phase I: Survey of Current Practice and Plan for Future Development," McClelland Engineers Report 0181-0211 to Mobil Research and Development Corporation, 1982.

[*11*] Peterson, L. M, Johnson, G. W., and Babb, L. V., "High Quality Sampling and In Situ Testing for Deep Water Geotechnical Site Investigations," *Proceedings, ASCE Specialty Conference on Use of In Situ Tests in Geotechnical Engineering,* Blacksburg, VA, 1986, pp. 913–925.

[*12*] Perlow, M., Jr., "A Technical Report on the Vane Shear Test," submitted to ASTM Subcommittee D18.13 on Marine Geotechnics, Task Group on In Situ Testing, American Society for Testing and Materials, Philadelphia, 1984.

[*13*] Aas, G., Lacasse, S., Lunne, T., and Hoeg, K., "Use of In Situ Tests for Foundation Design on Clay," *Proceedings, ASCE Specialty Conference on Use of In Situ Tests in Geotechnical Engineering,* Blacksburg, Virginia, 1986, pp. 1–30.

[*14*] Whitman, R. V., Richardson, A. M., Jr., and Nasim, N. M., *The Response of Soils to Dynamic Loadings, Report No. 10,* Massachusetts Institute of Technology, Cambridge, Contract No. DA-22-079-eng-224 with U.S. Army Engineers Waterways Experimental Station, 1962.

[*15*] Kraft, L. M., Jr., Ahmad, N. and Focht, J. A., Jr., *Proceedings, 8th Offshore Technology Conference,* Houston, Texas, Vol. 3, 1976, pp. 75–96.

[*16*] Perlow, M., Jr. and Richards, A. F., *Journal, Geotechnical Engineering Division,* ASCE, Vol. 103, No. GT1, 1977, pp. 19–32.

[*17*] Aas, G., "A Study of the Effect of the Shape of Vane and Rate of Strain on In Situ Shear Strength of Clays," *Proceedings, Sixth International Conference on Soil Mechanics and Foundation Engineering,* Montreal, Vol. 1, 1965, pp. 141–145.

[18] Young, A. G., Quiros, G. W., and Ehlers, C. J., *Proceedings, 15th Offshore Technology Conference,* Vol. 1, Houston, TX, 1983, pp. 193–204.

[19] Richards, A. F. and Zuidberg, H. M., "In Situ Determination of the Strength of Marine Soils," *Strength Testing of Marine Sediments: Laboratory and In-Situ Measurements, STP 883,* American Society for Testing and Materials, Philadelphia, pp. 11–40.

[20] Bjerrum, L., "Embankments on Soft Ground," *Proceedings, ASCE Specialty Conference on Performance of Earth and Earth-Supported Structures,* Vol. 2, Purdue University, Lafayette, IN, 1972, pp. 1–54.

[21] Olson, R. E., "Analysis of Pile Response Under Axial Loads," Geotechnical Engineering Report GR84-18, University of Texas, Austin, TX, 1984.

[22] Pilot, G., "Study of Five Embankments on Soft Soils," *Proceedings, ASCE Speciality Conference on Performance of Earth and Earth-Supported Structures,* Vol. 1, Purdue University, Lafayette, IN, 1972, pp. 81–99.

[23] Mesri, G., *Journal of the Geotechnical Engineering Division, Proceedings of the ASCE,* Vol. 101, No. GT4, April 1975, pp. 409–412.

[24] Trak, B., LaRochelle, P., Tavenas, F., Leroueil, S., and Roy, M., "A New Approach to the Stability Analysis of Embankments on Sensitive Clays," *Proceedings, 32nd Canadian Geotechnical Conference,* Quebec, Canada, 1976, pp. 3.1–3.25.

[25] Larsson, R., *Canadian Geotechnical Journal,* Vol. 17, No. 4, 1980, pp. 591–602.

[26] Dascal, O. and Tournier, J. P., *Journal of the Geotechnical Engineering Division, Proceedings of the ASCE,* Vol. 101, No. GT3, March 1975, pp. 297–314.

[27] Emrich, W. J., "Performance Study of Soil Sampler Deep-Penetration Marine Borings," *Sampling of Soil and Rock, STP 483,* American Society for Testing and Materials, Philadelphia, 1971, pp. 30–50.

[28] Quiros, G. W., Young, A. G., Pelletier, J. H., and Chan, J. C., "Shear Strength Interpretation for Gulf of Mexico Clays," *Proceedings, Speciality Conference on Geotechnical Practice in Offshore Engineering,* ASCE, Austin, TX, 1983, pp. 144–165.

[29] Matlock, H., *Proceedings, 2nd Offshore Technology Conference,* Vol. 1, Houston, TX, 1970, pp. 577–594.

[30] Ladd, C. C. and Foott, R., "New Design Procedure for Stability of Soft Clays," *Journal of the Geotechnical Engineering Division, Proceedings of the ASCE,* Vol. 100, No. GT7, July 1974, pp. 763–786.

[31] Koutsoftas, D. and Fischer, J. A., *Journal of the Geotechnical Engineering Division, Proceedings of the ASCE,* Vol. 102, No. GT9, Sept. 1976, pp. 989–1005.

[32] American Petroleum Institute, *Recommended Practice for Planning, Designing, and Constructing Fixed Offshore Platforms,* API RP 2A, 16th ed., 1986.

[33] Law, K. T., *Canadian Geotechnical Journal,* Vol. 16, 1979, pp. 11–18.

[34] Wroth, C. P., *Geotechnique,* Vol. 34, No. 4, 1984, pp. 449–489.

[35] Stevenson, H. S., *Vane Shear Determination of the Viscoelastic Shear Modulus of Submarine Sediments,* M. A. thesis, Graduate College, Texas A&M University, Austin, TX, 1973.

[36] Selvadurai, A. P. S., "On the Estimation of the Deformability Characteristics of an Isotropic Elastic Oil Medium by Means of a Vane Test," *International Journal for Numerical and Analytical Methods in Geomechanics,* Vol. 3, 1979, pp. 231–243.

[37] Madhav, M. R. and Rama Krishna, K. S., *Journal of the Geotechnical Engineering Division, Proceedings of the ASCE,* Vol. 103, No. GT11, Nov. 1977, pp. 1337–1340.

Part II: Field Vane Theory and Interpretation

Dennis E. Becker,[1] *J. H. A. Crooks,*[2] *and Ken Been*[2]

Interpretation of the Field Vane Test in Terms of In-Situ and Yield Stresses

REFERENCE: Becker, D. E., Crooks, J. H. A., and Been, K., **"Interpretation of the Field Vane Test in Terms of In-Situ and Yield Stresses,"** *Vane Shear Testing in Soils: Field and Laboratory Studies, ASTM STP 1014,* A. F. Richards, Ed., American Society for Testing and Materials, Philadelphia, 1988, pp. 71–87.

ABSTRACT: Interpretation of undrained shear strength data requires knowledge of the effective stress regime that is controlling the strength on the failure surface. Approximately 90% of the total resistance measured by the standard vane is provided by the vertical circumscribed failure surface. Consequently, vane test results are dominated by the strength mobilized on the vertical plane. Therefore, it is the horizontal in-situ effective and yield stresses that predominantly control the vane shear strength. Yet, to date, vane strength correlations have been considered in terms of vertical stresses only.

This paper discusses an interpretation for the field vane test within generalized state concepts. State can be characterized by an overconsolidation ratio that accounts for both vertical and horizontal stresses. The controlling influence of the in-situ effective and yield stresses on measured field vane strength is demonstrated using data for 14 clay deposits. Vane strength normalized with respect to in-situ effective stresses are correlated with overconsolidation ratio to confirm the generalized state concepts and importance of accounting for horizontal stresses. Normalized vane strength correlations are compared to normalized strength ratios back-analyzed from field failures. It is shown that strength ratios normalized using vertical preconsolidation pressures are not sufficiently refined to provide a good basis for comparison. Strength ratios based on horizontal yield stresses provide a more rational basis for comparison.

KEY WORDS: vane, normalized shear strength, state, overconsolidation ratio, in-situ horizontal stresses, yield, correlations, correction factors

- C Ratio of σ'_{hy}/σ'_p
- K_0 Ratio of in-situ horizontal to vertical effective stress, $\sigma'_{ho}/\sigma'_{vo}$
- I_p Plasticity index
- I'_0 In-situ mean stress $= (\sigma'_{vo} + 2\sigma'_{ho})/3 = \sigma'_{vo}(1 + 2K_0)/3$
- I' $(\sigma'_v + 2\sigma'_h)/3$
- I'_y $(\sigma'_v + 2\sigma'_h)y/3 = (\sigma'_p + 2\sigma'_{hy})/3$
- OCR Overconsolidation ratio in terms of vertical stresses (that is, σ'_p/σ'_{vo})
- OCR_I Overconsolidation ratio in terms of mean stresses, that is $(\sigma'_p + 2\sigma'_{hy})/(\sigma'_{vo} + 2\sigma'_{ho})$
- S_t Sensitivity

[1] Associate, Golder Associates, Mississauga, Ontario L4X 2B6, Canada.
[2] Principal and associate, respectively, Golder Associates, Calgary, Alberta T2H 0T3, Canada.

VCL	Virgin consolidation line
e	Void ratio
m	Exponent
p	$(\sigma_1' + \sigma_2' + \sigma_3')/3$
q	$\sigma_1 - \sigma_3$
s_u	Undrained shear strength
$s_{u(\text{vane})}$	Vane shear strength
σ_{vo}'	In-situ vertical effective stress
σ_{ho}'	In-situ horizontal effective stress
σ_p'	Pre-consolidation pressure
σ_{hy}'	In-situ horizontal yield stress
μ	Empirical vane correction factor

For more than 60 years the field vane test has been used as a convenient method of measuring the undrained shear strength of clays. In the last 10 to 15 years the test has been studied critically, and a better appreciation of some of the factors that influence the test results has been developed. However, incorporation of these factors into the analyses of the field vane test is still incomplete. Until this is possible, empirical correlations will play an important role in the "calibration" of field vane test results. Many empirical correlations involve the use of index properties and vane strengths normalized with respect to vertical stresses only. Few of these correlations take into account the factors that predominantly control, from a physical viewpoint, the undrained strength measured in the vane test.

Numerous reports indicate that the strength measured in the field vane test is not always directly applicable for the assessment of the stability of fills/cuts and foundation bearing capacity. Instances are also reported in which the undrained shear strength measured by the field vane and other in-situ and laboratory tests do not agree. In some cases strength gain as a result of consolidation has been indicated by the results of triaxial tests but not by the results of the field vane test.

These discrepancies are to be expected because of the differences that exist between the boundary conditions associated with the field vane test and those prevailing in other types of tests and during field loading. The strength response of a soil during any particular test will depend on the effective stresses imposed on the soil. A rational understanding of the undrained shear strength measured in a particular test and comparison with strengths obtained from other tests requires that the effective stress regime that controls the strength on the failure surface be known. Because of the geometry of the failure surface imposed by the standard vane with a height to diameter ratio of two, conventional interpretation (that is, uniform distribution of shear stress) indicates that 86% of the total resistance is generated on the vertical failure surface. If the shear stress distribution on the imposed failure surface is not uniform, as is likely the case, the vertical plane could contribute up to approximately 94% of the total measured resistance [*1*]. Therefore the vane strength is dominated by the strength mobilized on the vertical failure surface.

In many cases, strength is related to the preconsolidation pressure in the vertical direction. However, Law [*2,3*] demonstrated in laboratory triaxial tests that the strength measured by the vane test was predominantly controlled by the horizontal consolidation pressure. Only slight increases in vane resistance were observed for the case of increasing vertical consolidation pressure while the horizontal consolidation pressure was maintained constant.

Given the above observations and the mode of failure associated with the field vane test, it is evident that the interpretation of the field vane should account for the effective horizontal stresses. Therefore the in-situ effective current stresses (that is, σ'_{ho} and σ'_{vo}) and the effective yield (consolidation) stresses, in both the vertical and horizontal directions, must be included in the analysis of the field vane test. Yield is defined, in this paper, as the stress at which a change from small strain to large strain response is observed, under a given direction of loading. In the vertical sense, the conventional preconsolidation pressure corresponds to the vertical yield stress. Similarly, the horizontal yield stress reflects the stress at which a marked change in deformation/strain is observed under loading in the horizontal direction.

This paper discusses an interpretation of the field vane test within the context of the "state" of the clay. State incorporates the influence of void ratio and in-situ stress conditions; these can be characterized by K_0 and an over-consolidation ratio, which accounts for both vertical and horizontal stresses. To demonstrate the controlling influence of the in-situ effective stress regime on the vane strength, data for 14 clays, which exhibit both strain-softening and nonstrain-softening stress-strain characteristics are examined. Information on the clay type, plasticity index, sensitivity, vane strength, preconsolidation pressure, K_0, and horizontal yield stress is presented. Based on these data, correlations of vane shear strength normalized with respect to in-situ effective stresses and overconsolidation ratio are presented. These confirm the generalized state concepts used in interpretation and the importance of accounting for the in-situ horizontal effective stress regime. Comparisons between normalized vane strength ratios and those operational in field failures are also made.

The Behavior of Clay in Terms of State

In general terms, the state of a soil is represented by a point in three-dimensional space reflecting void ratio, normal effective stress, and deviator stress (that is, *e-p-q* space). However, in practical terms, this is cumbersome, and for convenience the current state of a soil can be represented by a point in void ratio-log effective stress space. To quantify current state, it is also necessary to relate it to a reference condition, which can also be represented in void ratio-log effective stress space. The virgin consolidation line (VCL), under one-dimensional conditions, has traditionally been used as a reference condition for clays, engineering behavior often being assessed on the basis of the degree of over-consolidation.

Figure 1 illustrates how the current state of a clay can be quantified in terms of degree of overconsolidation. On Fig. 1*a*, a general three-dimensional stress condition involving the three principal normal stresses is examined using the VCL defined by isotropic consolidation (that is, $e - \log I'$, where I' is the mean normal stress). Figure 1*b* presents the conventional void ratio versus log effective vertical stress associated with one-dimensional consolidation. In both cases, the clay has been subjected to a maximum consolidation stress condition represented by point P and then unloaded to its current condition represented by point O. Traditionally, only the in-situ vertical stresses are considered, and the state of the clay can be represented as log OCR = (σ'_p/σ'_{vo}). For the three-dimensional stress condition, state can be defined by log OCR_I, which accounts for both vertical and horizontal effective in-situ and yield stresses (Fig. 1), and as such, is considered to be a more general description.

For the clays examined in this study, $e - \log I'$ data are not available and the required I'_y values cannot be determined. Thus, it is assumed that $I'_y = (\sigma'_p + 2\sigma'_{hy})/3$, which is a simplification of the general case but is probably reasonable.

Extensive data and experience relating clay behavior to "state" exist for log OCR defined in terms of vertical stresses alone. It has been well established, both theoretically [*1*] and experimentally [*4*] that the undrained strength ratio for over-consolidated clays can be rep-

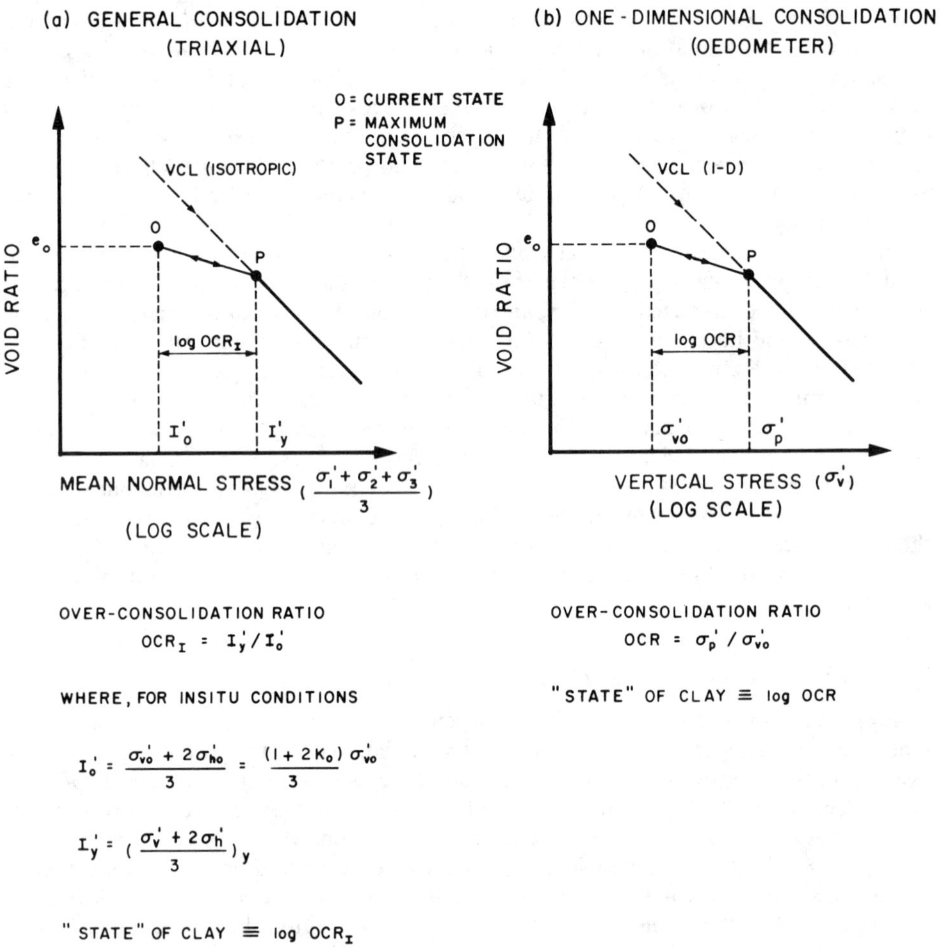

FIG. 1—*Generalized state concepts for clays.*

resented by the following expression

$$(s_u/\sigma'_{vo})_{oc} = (s_u/\sigma'_{vo})_{nc}\mathrm{OCR}^m \quad (1)$$

The value of m can vary from one type of test to another. For direct simple shear tests (DSS) the magnitude of m, based on the available evidence, does not vary significantly for a wide range of clays and is approximately 0.8 [*4*].

The important influence of stress history in terms of K_0 and OCR on the engineering properties of clays is well appreciated [*1,4,5*]. The statement that log OCR represents state and that state controls behavior is therefore not new. However, it would appear to be more appropriate to redefine Eq 1 in terms of OCR_I for vane test interpretation.

It is recognized that other factors, such as fabric, mineralogy, and physio-chemical conditions also influence the engineering behavior of soils. However, in the development of meaningful vane strength correlations, first order physical effects, such as in-situ stresses, must first be properly taken into account to provide a rational basis for understanding both soil behavior and the effects of the above factors.

Data Base Description

A review of available information was carried out to identify those clays for which sufficient data were available to assess the significance of K_0 and horizontal yield stress on vane shear strength within the above general concepts. The descriptions and properties of the various clays, which constitute the data base for this study, can be found in the appropriate references. Additional field and laboratory investigations were not carried out.

A summary of the geotechnical data for the 14 clays examined in this study is presented on Table 1. The data base includes sites from five different countries and incorporates a wide range of material types, this diversity being indicated by OCR values varying from 1.0 to 12.3 and K_0 values ranging from 0.5 to 2.8. It is noted that in the cases of the Genesee and Beaufort Sea clays, the measured values of K_0 differ significantly from the traditional expectation of the magnitude of K_0 based on existing well-known correlations [*4,6*]. The measured values of K_0 in the Genesee and Beaufort Sea clays are significantly higher than would be anticipated based on these empirical correlations. However, the data base also includes cases in which the measured K_0 values agree with those expected based on the conventional correlations.

The clays included in the data base also exhibit a considerable range in plasticity index and sensitivity values. In general, the clays from Sweden are sensitive to moderately sensitive and exhibit moderate strain-softening behavior. The Canadian clays range from lean, insensitive clays with nonstrain-softening behavior to highly sensitive strain-softening Leda clay. The other clays from Norway, the United States, and Hong Kong are generally insensitive to mildly sensitive clays that exhibit nonstrain-softening to slightly strain-softening behavior.

Determination of K_0

A variety of techniques were employed to determine K_0 values for the clays. Basically, the tests were carried out in situ and included hydraulic fracturing (HF), self-boring pressuremeter (SBP), total stress cells (TSC), and to a lesser extent the Menard pressuremeter (MPM) and the flat plate dilatometer (DM). At several sites, a combination of these methods were employed and compared by the original reporters. In this study, the value of K_0 selected at any particular depth is as reported in the original references. No separate interpretation for K_0 has been made. In the case of reported discrepancies between K_0 determined by different methods at the same depth, a judgment was made as to which value appeared to be more consistent. If the discrepancy could not be resolved, the data set at that particular depth was not included in the data base. In general, only the reported data (OCR, K_0, and $S_{u(\text{vane})}$) available from specific measurements at a particular depth were used.

Determination of In-Situ Horizontal Yield Stress

The current data base does not include all the necessary information to permit interpretation of horizontal yield stress from the results of SBP tests using refined techniques such as those described by Jefferies et al. [*7*]. Thus, for the purposes of this study, simple graph-

TABLE 1—*Geotechnical data for clays.*

Location	I_p	S_t	$S_{u(\text{vane})}$ (kPa)	σ'_p, kPa	OCR (σ'_p/σ'_{vo})	$K_{0\ (\text{METHOD})}$[a]	σ'_{hy}, kPa	Reference
Ska-Edeby (Sweden)	29 to 55	10 to 23[b]	7 to 12	25 to 55	1.0 to 1.5	0.5 to 0.6 (TSC/HF)	. . .	Massarsch et al. [22]
Backebol (Sweden)	48 to 63	13 to 28[b]	16 to 18	40 to 55	1.3 to 1.8	0.6 to 0.7 (TSC/HF)	. . .	Massarsch et al. [22]
Kalix (Sweden)	95 to 112	10 to 15[b]	14 to 17	35 to 45	1.4 to 2.1	0.7 to 0.9 (TSC/HF)	. . .	Massarsch et al. [22]
Jarva Krog (Sweden)	26 to 50	18 to 26[b]	19 to 24	70 to 80	1.1 to 1.5	0.9 to 1.0 (TSC/HF)	. . .	Massarsch et al. [22]
Ursvik (Sweden)	25 to 28	17 to 26[b]	9 to 16	50 to 75	1.3 to 1.9	0.7 to 0.8 (TSC/HF)	. . .	Massarsch et al. [22]
Onsoy (Norway)	23 to 26	5 to 7[c]	11 to 26	50 to 110	1.0 to 4.0	0.6 to 1.0 (SBP)	55 to 135 (SBP)	Lacasse et al. [23] (1981); Lacasse and Lunne [24]; Lunne et al. [25]
Drammen (Norway)	10 to 37	7[c]	18 to 30	70 to 130	1.1 to 1.6	0.6 to 0.7 (SBP)	100 to 170 (SBP)	
Gloucester (Canada)	20 to 33	30 to 100[c]	15 to 45	55 to 155	1.5 to 1.9	0.7 to 0.8 (SBP/HF)	145 (SBP)	Bozozuk and Leonards [26]; Bozozuk [27]; unpublished data

St. Albans (Canada)	20 to 30	16 to 20[c]	11 to 20	40 to 85	2.0 to 2.7	0.9 to 1.2 (MPM/TSC/HF)	. . .	Tavenas et al. [28], Roy et al. [29]; unpublished data
Wallaceburg (Canada)	20 to 27	6 to 8[c]	25 to 75	125 to 400	1.0 to 10.0	0.7 to 1.5 (HF/Lab)	120 to 400 (Lab)	Becker [30]
Beaufort Sea (Canada)	20 to 30	<6[c]	40 to 80	220 to 300	3.0 to 6.0	1.0 to 2.0 (SBP/Lab)	200 to 340 (SBP)	Jefferies et al. [21]
Genesee (Canada)	32 to 68	2 to 6[c]	45 to 155	140 to 600	1.0 to 12.3	0.9 to 2.8 (TSC/HF/DM)	. . .	Chan [20]
Boston Blue Clay	20 to 30	3 to 6[c]	*e*	*e*	1.0 to 4.0	0.6 to 1.2 (Lab)	. . .	Ladd [31]; Ladd et al. [4]
Hong Kong	25 to 65	<10[c]	7 to 100	30 to 480	1.8 to 3.0	0.7 to 1.0 (Lab)	. . .	Foott and Koutsoftas [32]

[a] HF = hydraulic fracturing. MPM = Menard pressuremeter. SBP = self-bored pressuremeter. TSC = total stress cell. DM = dilatometer. Lab = laboratory.
[b] Based on fall-cone.
[c] Based on field vane.
[d] Based on results of oedometer tests on "horizontal" specimens.
[e] Data presented normalized as S_u/σ'_{vo} against OCR.

ical construction was used; the results of SBP tests reported in terms of applied radial pressure and induced radial strain interpreted as indicated on Fig. 2. Horizontal effective yield stresses were determined in this manner for the Onsoy, Drammen, Gloucester, and the Beaufort Sea clays.

For the Wallaceburg case record, the horizontal effective yield stress was taken as the "horizontal pre-consolidation pressure" determined by the work interpretation [8] of the results of oedometer tests on "vertically trimmed" specimens (that is, trimmed such that the long axis of the specimen is parallel to the depth [vertical] axis in situ).

Vane Strength Normalized with Respect to In-Situ Current Effective Stresses

The data for the 14 clays are summarized on Fig. 3*a* in terms of vane strength normalized with respect to the in-situ mean normal current effective stress I_0', plotted against conventional overconsolidation ratio (that is, in terms of vertical stresses). The strength ratio is observed to increase with increasing OCR, consistent with previous discussions. There is limited scatter in the data, and two distinct trends are evident. Sensitivity or stress-strain behavior, or both, of the clay appears to be the factor in causing the different relationships shown on Fig. 3*a*. In general, the data from the insensitive clays constitute the lower relationship while the upper relationship tends to be defined by data from the more sensitive clays.

This observation suggests that an appreciation of the stress-strain behavior of the clay is an important factor in the interpretation of the field vane. A similar conclusion applies to any other in-situ test and is consistent with the findings presented by Wroth [1].

For comparison, Fig. 3*b* presents the relationship between field vane strength normalized with respect to in-situ vertical effective stress and OCR. The data indicate a similar increase in normalized strength ratio with increasing OCR as discussed above, but exhibit more scatter. Further, the distinction based on sensitivity differences is not as evident as

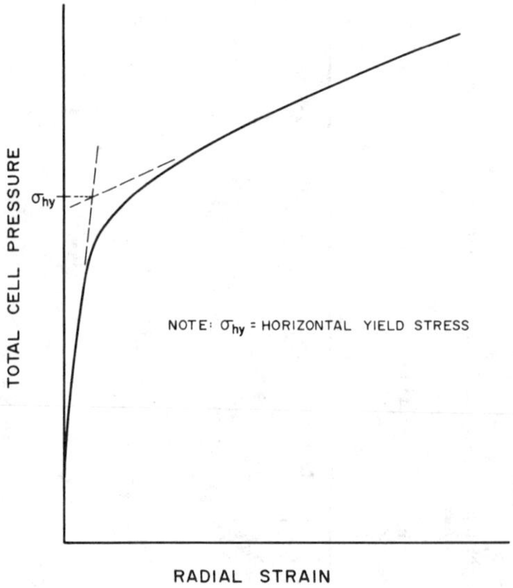

FIG. 2—*Graphical construction for the determination of horizontal yield stress from the results of self-boring pressuremeter tests.*

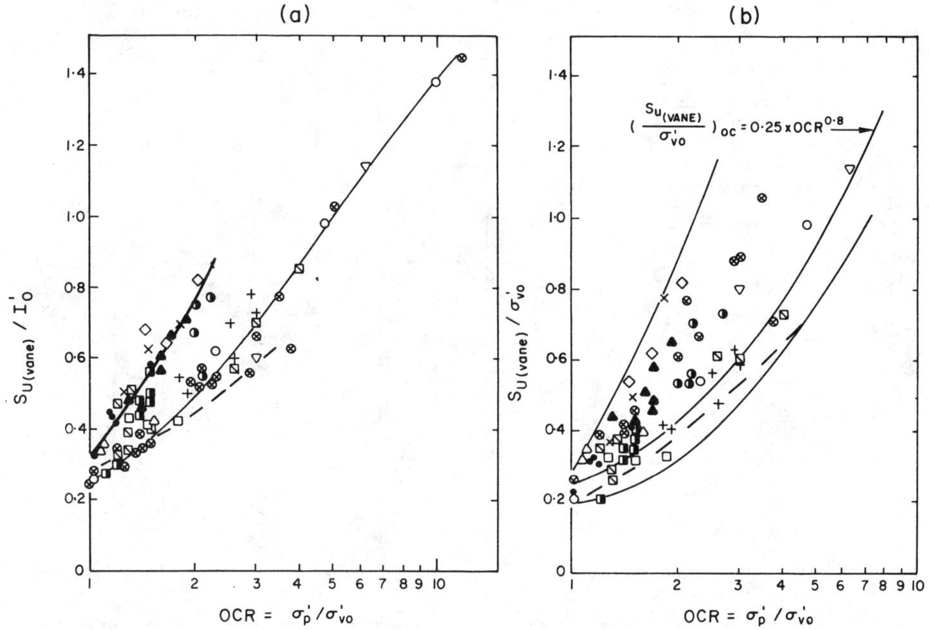

SITE	K_o	SENSITIVITY S_t	I_p (%)
• SKA-EDEBY	0·5-0·6	10-23	29-55
× BACKEBOL	0·6-0·7	13-28	48-63
◇ KALIX	0·7-0·9	10-15	95-112
△ JARVA KROG	0·9-1·0	18-26	26-50
□ URSVIK	0·7-0·8	17-26	25-28
◑ ST. ALBANS	0·8-1·2	16-20	20-30
▲ GLOUCESTER	0·7-0·8	30-100	20-33

SITE	K_o	SENSITIVITY S_t	I_p (%)
○ WALLACEBURG	0·7-1·5	6-8	20-27
◨ DRAMMEN	0·6-0·7	7	10-37
⧄ ONSOY	0·6-1·0	5-7	23-36
+ HONG KONG	0·7-1·0	<10	25-65
▽ BEAUFORT SEA	1·0-2·0	<6	20-30
⊗ GENESEE	0·9-2·8	2-6	32-68
— — BOSTON BLUE CLAY	0·6-1·2	3-6	20-30

(AVERAGE OF RANGE INDICATED BY LADD, 1975)

FIG. 3—*Correlations between normalized vane shear strength and OCR.*

in the case where the mean normal effective stress was used to normalize undrained strength. The relationship given by Eq 1 is also indicated on Fig. 3*b* and lies close to the lower bound of the observed correlation for the clays studied. In general, the shape of the theoretical curve is similar to the observed trend, which suggests that a value of $m = 0.8$ is also reasonable for the field vane test.

The scatter in the data summarized on Fig. 3*b* compared to the lesser scatter in Fig. 3*a* indicate the importance of horizontal stresses. The important influence of K_0 is clearly demonstrated on Fig. 4, which presents the normalized strength values for the field vane test for only the Genesee, Wallaceburg, Beaufort Sea, and Onsoy sites. At these sites the ranges in OCR and K_0 are significant, but the values for plasticity index, in particular, sensitivity (reflection of stress-strain behavior), are quite similar (Table 1). The correlation of $S_{u(\text{vane})}/I_0'$ and OCR (Fig. 4*a*) is a well-defined narrow band compared with the considerable scatter, in particular at high values of OCR, evident in the $S_{u(\text{vane})}/\sigma_{vo}'$ correlation (Fig. 4*b*). Further, the observed correlation between $S_{u(\text{vane})}/I_0'$ and OCR agrees reasonably well

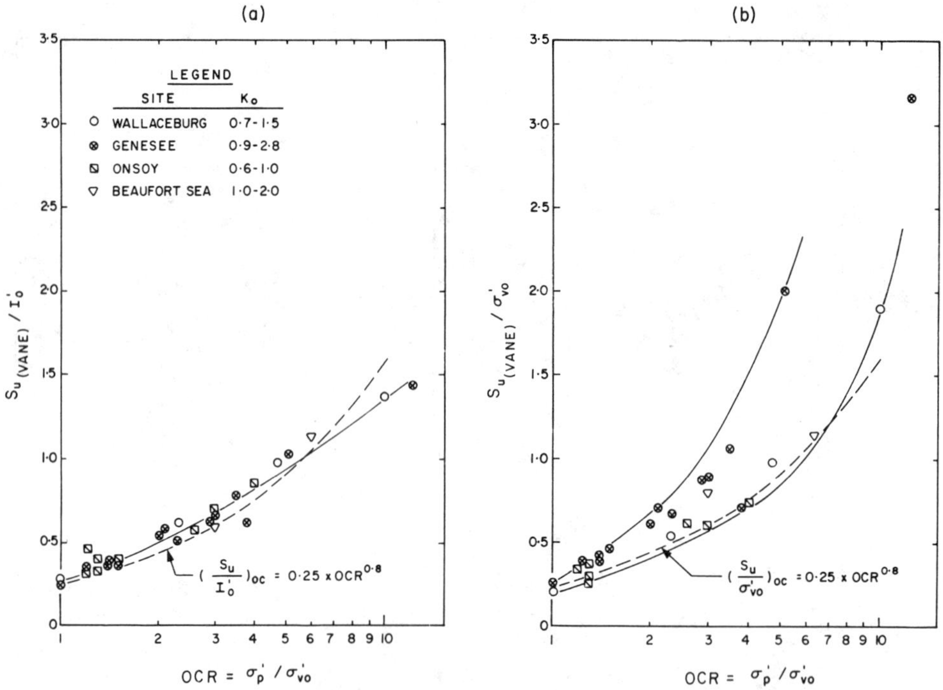

FIG. 4—*The effect of K_0 on normalized vane shear strength and OCR correlation.*

with the theoretical and experimental relationship between the undrained strength and OCR [*1,4*]. A normalized strength of approximately 0.25 is indicated for normally consolidated clays.

The correlations presented thus far have used the conventional definition of OCR in terms of vertical stress alone. However, as discussed previously, it is more appropriate to define state expressed as overconsolidation ratio in terms of the in-situ effective mean normal stress (that is, OCR_I).

Sufficient information was available to determine the horizontal yield stress σ_{hy}' for five clays (Table 1). Figure 5 presents the correlation between $S_{u(\text{vane})}/I_0'$ and OCR_I. Actual measurements of σ_{hy}' were carried out at Onsoy, Drammen, Gloucester Wallaceburg, and the Beaufort Sea sites; consequently the data for these clays are represented by the large symbols. The correlation is identified as a well-defined relationship, which shows an increase in normalized vane shear strength with increasing OCR_I as was observed for the other correlations. A normalized strength ratio of approximately 0.22 is indicated for $OCR_I = 1$.

To further increase the data base, values for σ_{hy}' were assumed for the other sites as indicated on Fig. 5. It was noted that for Gloucester, Onsoy and Drammen clays, the ratio of σ_{hy}'/σ_p' is relatively constant at about 1.3. This ratio in yield stresses was assumed to apply to the other sites where measured σ_{hy}' values were not available. Values of normalized strength and OCR_I based on this assumption are represented by the small symbols on Fig. 5. These data are consistent with the relationship defined using measured values of σ_{hy}'; the scatter in the data is not significantly increased.

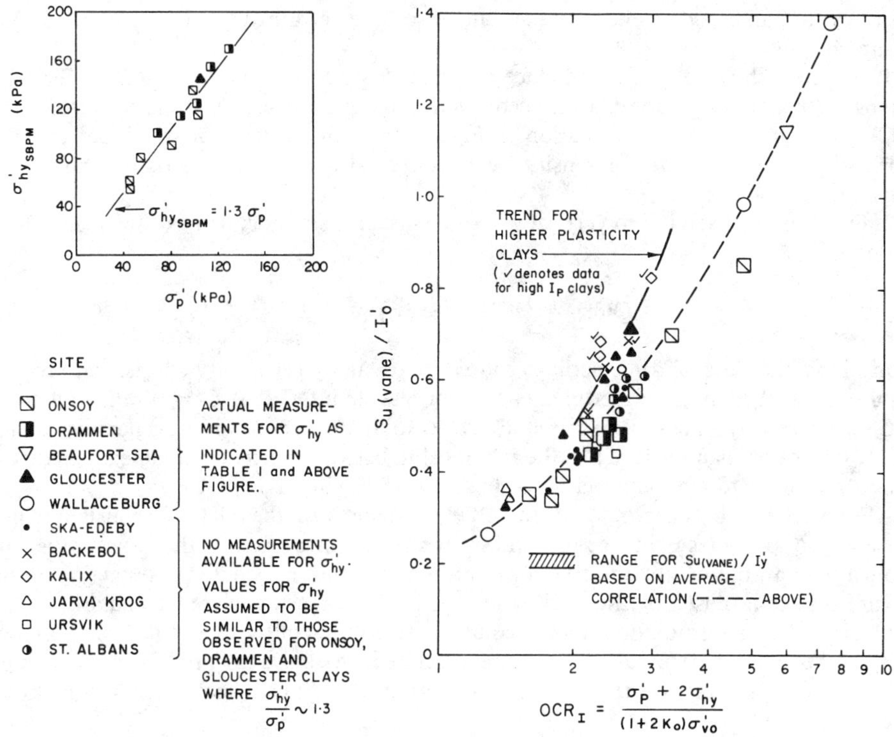

FIG. 5—*Correlation between normalized vane shear strength and* OCR_I.

The normalized vane strength with respect to the effective mean yield stress I'_y can also be determined from the data. Both the normalized strength ratio and OCR_I include I'_0 in the denominator. Therefore, by dividing the normalized vane strength ratio by OCR_I, values of $S_{u(\text{vane})}/I_y$ are obtained, which decouples the correlation from OCR_I. As indicated on Fig. 5, the resulting ratio is independent of OCR_I and lies between 0.19 and 0.22. Further, given the range in plasticity index for the data base, it appears that the ratio is also independent of plasticity index.

It is important to appreciate that the relationship $\sigma'_{hy} = 1.3\ \sigma'_p$, applied to the data base, is an assumption based on data for three clays. There is no evidence to verify that this relationship will apply to all material types outside of the current data base, and in fact this is unlikely. The magnitudes of σ'_{hy} and σ'_p will depend on the K_0 value and the shape of the yield envelope for the clay. Assuming that K_0 for a plastic clay will be higher than for a lean clay and that the yield envelope is approximately symmetrical about the K_0 line, it would be expected that, for more plastic clays, σ'_{hy} would be greater than 1.3 σ'_p. Indirect evidence supporting this expectation may be discernible from Fig. 5. As indicated, the higher plasticity clays lie above the normalized strength-OCR_I relationship. If the actual σ'_{hy} values are higher than that assumed (that is, greater than 1.3 σ'_p) these data points would move further to the right and possibly correspond with the trend indicated by the remainder of the data.

A second possibility is that the sensitivity of clays also influences the shape of the yield envelope and may influence the relationship between σ'_{hy} and σ'_p. Clarification of this point requires that actual σ'_{hy} values for the data base be determined (that is, measured), not assumed.

It has been shown that the influence of plasticity and sensitivity, on normalized vane strength-stress history correlations, becomes less distinct when strength is correlated with OCR_I instead of OCR. Examination of Figs. 3 and 5 indicates that the upper trend in normalized vane strength for sensitive clays essentially moves to the right. This can be explained as follows.

If the ratio of horizontal to vertical yield stress is a constant defined as $C = \sigma'_{hy}/\sigma'_p$, OCR_I is defined as

$$OCR_I = OCR\,(1 + 2C)/(1 + 2K_0) \tag{2}$$

Variations in both C and K_0, perhaps caused by plasticity/sensitivity effects, can produce values of OCR_I that are either greater than or less than OCR. If $C = K_0$ then OCR_I = OCR. For the clays included in this study, the combinations of C and K_0 generally resulted in OCR_I greater than OCR. Thus the relationship between normalized vane strength and OCR_I will move to the right relative to the relationship involving OCR.

The correlation between $S_{u(\text{vane})}/I'_0$ and OCR_I, together with the relatively constant values for $S_{u(\text{vane})}/I'_y$, provides evidence that the shear strength measured by the vane test is controlled by both the in-situ effective mean normal current and yield stresses acting on the failure surface imposed by the field vane. The horizontal stresses are particularly important, since in this formulation, they account for two-thirds of the I'_0 and I'_y values. Knowledge of the in-situ horizontal effective stress regime is therefore essential for rational interpretation of field vane test results. Interpretation in terms of vertical stresses alone is not meaningful.

Comparison of Vane Strength with Operational Strength in the Field

Because of differences in the conditions associated with any in-situ test type and those prevailing in field loading situations, it is necessary to compare the in-situ test strength with the strengths that are mobilized in clays stressed in the field. The latter can only be defined at failure and a considerable data base related to embankment failure on clays, for example, has been developed [*9,10*]. In making such comparisons it is convenient to express strength in a normalized form. It is considered that for vane strengths the comparison should be based on strength values plotted in terms of S_u/I'_0 versus OCR_I. Unfortunately, reported back-analyses of field failures generally do not include the data (that is, K_0 and σ'_{hy}) necessary to prepare this form of relationship. Therefore a comparison between the S_u/I'_0 versus OCR_I relationship based on field failures with that for the field vane (Fig. 5) is not possible at this time.

The traditional approach to comparing vane strengths and those operational in field failures is based on (S_u/σ'_p) ratios normally plotted against plasticity index. The $S_{u(\text{vane})}/\sigma'_p$ versus plasticity index data for the 14 clays included in the present study as shown on Fig. 6*a*. These data generally fall within the range identified by Larsson [*10*] for field vane tests in a wide variety of clay types. They also exhibit the same wide range in scatter. Based on the arguments presented in this paper, this large scatter is to be expected since the ratio $S_{u(\text{vane})}/\sigma'_p$ does not take into account the important horizontal in-situ and yield stresses.

Also shown on Fig. 6*a* is the range in average undrained strength operational at failure in the field as back-calculated from case records. The $S_{u(\text{operational})}$ values are also normalized with respect to σ'_p and plotted against plasticity index. Comparison of the $S_{u(\text{vane})}$ and $S_{u(\text{operational})}$ ratios indicate that except at low I_p values, the trends do not agree, with the

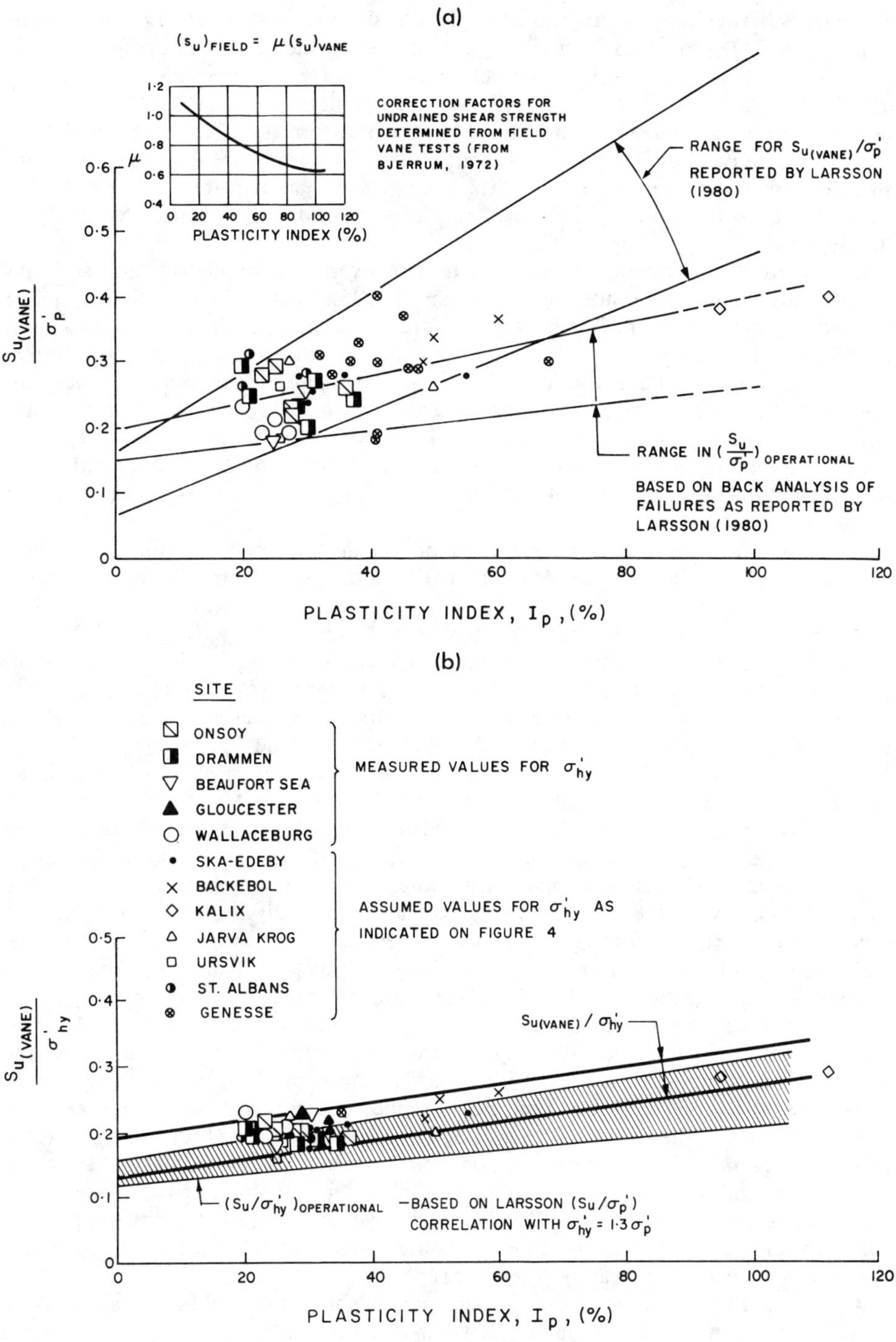

FIG. 6—*Comparison of normalized vane strength with strength operational at failure during field loading.*

discrepancy between the two increasing with increasing plasticity. Evaluation of the same general data base prompted Bjerrum [*9*] to propose that the field vane strength be "corrected" to provide the operational strength in the field. He proposed the use of $S_{u(\text{field})} = \mu S_{u(\text{vane})}$ where the magnitude of the correction factor μ decreases with plasticity as shown on the insert on Fig. 6*a*. Bjerrum justified the use of a correction factor from a physical viewpoint because of differences in strain rates between the field vane test and field loading situation, anisotropy, and to a lesser extent, progressive failure. Strain rate effects were considered to be the most significant factor. These correction concepts have recently been thoroughly discussed and updated by Aas et al. [*11*].

The validity of the use of a correction factor to convert vane strengths to field strengths has been the subject of considerable controversy [*12,13*]. Graham et al. [*14*] present data that indicate that the effect of strain rate on undrained shear strength, measured in a number of laboratory tests on many clays, did not increase with plasticity and presumed that the same finding would apply to field vane tests. They therefore questioned the basis for Bjerrum's correction factors. Mesri [*15*] pointed out that if Bjerrum's correction factors, typical OCR and typical $S_{u(\text{vane})}$ were combined, the (S_u/σ'_p) operational value was approximately constant at about 0.22 and was independent of plasticity. This interpretation was in turn questioned by Larsson [*10*] based on his correlation shown on Fig. 6*a*; however an overall average value of 0.22 seemed reasonable.

Notwithstanding this controversy, the use of Bjerrum's correction factors has become prevalent in practice and a number of alternative correction factors have been proposed [*16–18*].

It is considered that this situation, referred to by Schmertmann [*13*] as the "correction crisis," is the result of comparisons not being made on a completely rational basis; important factors are not taken into account. For example, Leonards [*19*] describes several cases of failures involving soft clays where failure was controlled by a discrete weak layer, which was not included in the measured strength profile but which, he argued, controlled stability. Back-analyses that do not recognize the presence of a weak layer would result in an incorrect assessment of the operational field strength.

In this context, it should also be appreciated that strength values back-calculated from field failures are not necessarily an ultimately reliable yardstick against which to judge the accuracy of various types of strength measurement. The undrained shear strength of the foundation clay is not the only sensitive parameter in the analysis of a failure. Other factors, such as the strength of the crust and fill materials and the assumed mode of failure, can have a significant effect on the results of the analysis [*12*]. Thus the strength value, quoted as being operational at the time of failure, must always be viewed critically.

An alternative view of the comparison presented on Fig. 6*a* is that, based on the arguments presented in this paper, S_u/σ'_p ratios are not a proper basis for comparison; the important horizontal stresses are not included. Therefore, the S_u/σ'_p versus plasticity relationships shown on Fig. 6*a* are not sufficiently refined to provide a basis for accurate comparison. Given that the vane strength is essentially controlled by the horizontal yield (consolidation) stress [*2,3*], it follows that, for the cases where the horizontal yield stress is greater than the vertical yield stress (preconsolidation pressure), the measured vane strength when normalized by only σ'_p will numerically produce a ratio that is too high. This may explain the observed trend of the normalized strength ratio, S_u/σ'_p for the field vane test, which increases significantly with increasing plasticity index.

As discussed previously, insufficient data are available for the field failures to make a more rational comparison of strengths plotted in the form shown on Fig. 5. However, some data are available to explore this possibility, albeit in a simpler form, by comparing (S_u/

σ'_{hy}) ratios. The rationale behind this approach is that σ'_{hy} represents a more appropriate normalizing parameter for field vane strengths than σ'_p. $S_{u(\text{vane})}/\sigma'_{hy}$ ratios for the clays included in this study are shown plotted against plasticity index on Fig. 6*b*. As was the case for data presentation on Fig. 5, clays with measured σ'_{hy} values are identified using the same large symbols. The small symbols indicate assumed values of σ'_{hy} based on the assumption that $\sigma'_{hy} = 1.3\ \sigma'_p$. The (S_u/σ'_{hy}) operational range also shown on Fig. 6*b* is based on Larsson's (S_u/σ'_p) operational range assuming that $\sigma'_{hy} = 1.3\ \sigma'_p$.

There are a number of observations that can be made based on Fig. 6*b*. First, the scatter in the $S_{u(\text{vane})}/\sigma'_{hy}$ data is substantially reduced from that evident on Fig. 6*a* and is even slightly less than exhibited by the operational strength data. Second, there is a much smaller difference between the vane and operational strengths normalized with respect to σ'_{hy} than is the case when the strengths are normalized with respect to σ'_p, although the vane strength ratios are still slightly higher than the operational strength ratios. Finally, the (S_u/σ'_{hy}) trends for the current data base do not diverge with increasing plasticity.

It is noted that the trends in both the vane and operational (S_u/σ'_{hy}) ratios increase with plasticity and by about the same degree. As discussed previously in the context of the $S_{u(\text{vane})}/I'_0$ versus OCR_I relationship (Fig. 5), this may be due to the $\sigma'_{hy} = 1.3\ \sigma'_p$ assumption being incorrect. Thus if σ'_{hy}/σ'_p increases beyond 1.3 with increasing plasticity, then the σ'_{hy} values for the more plastic clays would be greater than those assumed and the (S_u/σ'_{hy}) ratios would decrease. In this event, it is possible that the trend in (S_u/σ'_{hy}) ratios for both the vane and operational strengths would be relatively constant with increasing plasticity. Again, the validity of this hypothesis can only be examined if actual σ'_{hy} values are known.

There is a further important consequence of this hypothesis that should be appreciated: it is not the actual plasticity of the clay that is the controlling factor. Rather it is the changes in the in-situ conditions and physical characteristics of different materials (that is, in-situ effective current and yield stresses) that are important and which may not be wholly captured by plasticity index. While there may be a correlation between plasticity and horizontal current in-situ and yield stresses for many clays, this should not be assumed to be always the case. For example, Chan [*20*] and Jefferies et al. [*21*] demonstrate that for the lightly overconsolidated Genesee and Beaufort Sea clays, K_0 values do not vary with OCR and plasticity as would be traditionally expected based on the work of Brooker and Ireland [*6*].

Despite the apparently persuasive agreement between the vane and operation strengths shown on Fig. 6*b*, it is clearly evident from the above discussion that this should not be used as a justification (or otherwise) that the field vane produces the "correct" strength for use in analyses of field loading situations. Nor should it be used to develop a new generation of correction factors to be applied to the vane to obtain the "correct" strength. What the "agreement" shown on Fig. 6*b* means is that if strength data are compared in a form that accounts for the more important factors affecting the strength values, a more coherent understanding will emerge. While the form of data presentation including σ'_{hy} (Fig. 6*b*) is an improvement over that based on σ'_p (Fig. 6*a*), it remains to be seen if this improvement is further enhanced when the data are plotted in terms of S_u/I'_0 versus OCR_I.

Conclusions

Based on the data presented in this study, the following main conclusions can be drawn:

1. The vane strength normalized with respect to in-situ current effective stress (both in terms of σ'_{vo} and I'_o) increases with increasing OCR. The scatter in data for $S_{u(\text{vane})}/\sigma'_{vo}$ is, however, considerable.

2. K_0 has a significant influence on normalized vane strength as reflected in the relatively narrow scatter in the data correlations between $S_{u(\text{vane})}/I'_0$ and OCR for clays with similar sensitivity and stress-strain behavior.

3. Stress-strain characteristics and sensitivity influence the correlation between $S_{u(\text{vane})}/I'_0$ and OCR, which suggests that proper interpretation of the field vane or any in-situ test should address the stress-strain characteristics of the soil.

4. The relatively narrow band correlation obtained between $S_{u(\text{vane})}/I'_0$ and OCR_I and the relatively constant value of $S_{u(\text{vane})}/I'_y$ of about 0.20 clearly indicate that field vane strength is controlled by both the in-situ effective and yield stresses in both the vertical and horizontal directions.

5. The conventional S_u/σ'_p versus plasticity index approach for comparing vane strength and strength operational in field failures is not sufficiently refined to permit rational comparison because the horizontal stress components are not taken into account.

6. Comparison of vane and operational strengths should be based on S_u/I'_0 versus OCR_I plots, but these data are currently not available for field failures. A simplified approach based on S_u/σ'_{hy} indicate reasonable agreement between vane and operational strength with less data scatter. However, because of the assumptions required to develop σ'_{hy}, this agreement should only be viewed as an indication that inclusion of horizontal stresses in this type of analysis is essential. It should not be used as the basis for developing alternative vane correction factors.

References

[1] Wroth, C. P., *Geotechnique,* Vol. 34, No. 4, 1984, pp. 449–489.

[2] Law, K. T., *Canadian Geotechnical Journal,* Vol. 16, No. 1, 1979, pp. 11–18.

[3] Law, K. T., *Proceedings of Eleventh International Conference on Soil Mechanics and Foundation Engineering,* Vol. 2, 1985, pp. 893–898.

[4] Ladd, C. C., Foot, R., Ishihara, K., Schlosser, F., and Poulos, H. G., *Proceedings of Ninth International Conference on Soil Mechanics and Foundation Engineering,* Vol. 2, 1977, pp. 421–494.

[5] Jamiolkowski, M., Ladd, C. C., Germaine, J. J., and Lancellotta, R., *Proceedings of Eleventh International Conference on Soil Mechanics and Foundation Engineering,* Vol. 1, 1985, pp. 57–154.

[6] Brooker, E. W. and Ireland, H. O., *Canadian Geotechnical Journal,* Vol. 2, No. 1, 1965, pp. 1–15.

[7] Jefferies, M. G., Ruffell, J. P., Crooks, J. H. A., and Hughes, J. M. O., *Strength Testing of Marine Sediments: Laboratory and In Situ Measurements, STP 883,* American Society for Testing and Materials, Philadelphia, 1984, pp. 487–514.

[8] Becker, D. E., Crooks, J. H. A., Been, K., and Jefferies, M. G., "Work as a Criterion for Determining In Situ and Yield Stresses in Clays," *Canadian Geotechnical Journal,* Vol. 24, No. 4, 1987, pp. 549–564.

[9] Bjerrum, L., *Proceedings of Eighth International Conference on Soil Mechanics and Foundation Engineering,* Vol. 3, 1973, pp. 111–159.

[10] Larsson, R., *Canadian Geotechnical Journal,* Vol. 17, 1980, pp. 591–602.

[11] Aas, G., Lacasse, S., Lunne, T., and Hoeg, K., *Proceedings of ASCE Specialty Conference (In Situ '86) on Use of In Situ Tests in Geotechnical Engineering,* 1986, pp. 1–30.

[12] Milligan, V., *Proceedings of ASCE Conference on Performance of Earth and Earth-Supported Structures,* Vol. 2, 1972, pp. 31–48.

[13] Schmertmann, J. H., *Proceedings ASCE Conference on the In Situ Measurement of Soil Properties,* Vol. 2, 1975, pp. 57–138.

[14] Graham, J., Crooks, J. H. A., and Bell, A. L., *Geotechnique,* Vol. 34, No. 3, 1984, pp. 327–340.

[15] Mesri, G., *ASCE Journal of Geotechnical Engineering Division,* Vol. 101, No. 3, 1975, pp. 409–412.

[16] Helenelund, K. V., "Methods for Reducing Undrained Shear Strength of Soft Clay," Swedish Geotechnical Report, No. 3.

[17] Azzouz, A. S., Baligh, M. M., and Ladd, C. C., *ASCE Journal of the Geotechnical Engineering Division,* Vol. 109, No. 4, pp. 730–734.

[18] Pilot, G., *Proceedings of ASCE Conference on Performance of Earth and Earth-Supported Structures,* 1972, Vol. 1, pp. 81–99.
[19] Leonards, G. A., *ASCE Journal of the Geotechnical Engineering Division,* Vol. 2, pp. 185–246.
[20] Chan, A. C. Y., "Geotechnical Characteristics of Genesee Clay," Ph.D. Thesis, Faculty of Civil Engineering, University of Alberta, Edmonton, Canada, 1986.
[21] Jefferies, M. G., Crooks, J. H. A., Becker, D. E. and Hill, P. R., "On the Independence of Lateral Stress from Over-Consolidation in Some Beaufort Sea Clays," *Canadian Geotechnical Journal,* Vol. 24, No. 3, 1987, pp. 342–356.
[22] Massarsch, K. R., Holtz, R. D., Holm, B. G. and Fredriksson, A., *Proceedings of ASCE Conference on In Situ Measurement of Soil Properties,* Vol. 1, 1975, pp. 266–286.
[23] Lacasse, S., Jamiolkowski, M., Lancellotta, R., and Lunne, T., *Proceedings of Tenth International Conference on Soil Mechanics and Foundation Engineering,* Vol. 2, 1981, pp. 507–511.
[24] Lacasse, S. and Lunne, T., "In Situ Horizontal Stress from Pressuremeter Tests, 1983," *Norwegian Geotechnical Institute,* No. 1246, pp. 1–12.
[25] Lunne, T., Eide, O., and DeRuiter, J., *Canadian Geotechnical Journal,* Vol. 13, No. 4, 1976, pp. 420–441.
[26] Bozozuk, M. and Leonards, G. A., *Proceedings of ASCE Conference on Performance of Earth and Earth-Supported Structures,* Vol. 1, 1972, pp. 299–318.
[27] Bozozuk, M., *Proceedings of ASCE Conference on Subsurface Exploration for Underground Excavation and Heavy Construction,* 1974, pp. 333–349.
[28] Tavenas, F. A., Blanchette, G., Leroueil, S., Roy, M., and LaRochelle, P., *Proceedings of ASCE Conference on In Situ Measurement of Soil Properties,* Vol. 1, 1975, pp. 450–476.
[29] Roy, M., Juneau, R., LaRochelle, P., and Tavenas, F. A., *Proceedings of ASCE Conference on In Situ Measurement of Soil Properties,* Vol. 1, 1975, pp. 350–372.
[30] Becker, D. E., "Settlement Analysis of Intermittently-Loaded Structures Founded on a Clay Subsoil," Ph.D. thesis, Faculty of Engineering Science, University of Western Ontario, London, Canada, 1981.
[31] Ladd, C. C., *Proceedings of ASCE Conference on In Situ Measurement of Soil Properties,* Vol. 2, 1975, pp. 153–160.
[32] Foott, R. and Koutsoftas, D., "Geotechnical Engineering for the Replacement of Airport at Chek Lap Kok," *Symposium on Soil and Rock Improvement,* Arim Institute of Technology, 1982.

V. Silvestri[1] *and M. Aubertin*[2]

Anisotropy and In-Situ Vane Tests

REFERENCE: Silvestri, V. and Aubertin, M., **"Anisotropy and In-Situ Vane Tests,"** *Vane Shear Strength Testing in Soils: Field and Laboratory Studies, ASTM STP 1014,* A. F. Richards, Ed., American Society for Testing and Materials, Philadelphia, 1988, pp. 88–103.

ABSTRACT: This paper presents the results of a field investigation carried out to determine the undrained shear strength of a sensitive clay deposit. The tests were performed by means of a Nilcon vane borer in Louiseville, Quebec, Canada. In order to study the possible anisotropic nature of the undrained shear strength, the apparatus was equipped with both rectangular-shaped and diamond-shaped vanes. The rectangular vanes had a height/diameter ratio that varied between 0.5 and 2.0. The diamond-shaped vanes had their blades cut at an angle that varied between 30° and 60° with respect to the horizontal. The results of the tests carried out with the rectangular vanes indicate that the undrained shear strength in the vertical plane S_{uv} is lower than the undrained shear strength in the horizontal plane S_{uh}. Thus, the value of the degree of strength anisotropy, defined as S_{uh}/S_{uv}, varies between 1.14 and 1.41. For the diamond-shaped vanes, the inferred strengths are intermediate between S_{uh} and S_{uv}. In addition, an elliptical failure criterion is found to adequately describe the observed response.

KEY WORDS: undrained shear strength, soft clays, anisotropy, field tests, rectangular vanes, diamond-shaped vanes, elliptical failure criterion

Nomenclature

a, b, c, d, e	Parameters in Davis and Christian's equation
i	Inclination angle with respect to the horizontal
m_v	Torque on unit height of cylinder generated by rotation of vane
n	Coefficient related to the shape of stress distribution
p_n	Parameter related to the type of stress distribution
r	Radius
A, B	Parameters in Bishop's equation
D	Diameter of vane
H	Height of vane
M	Total Torque
M_h	Torque mobilized on each horizontal end surface of vane
M_v	Torque mobilized on vertical cylindrical surface
S_{uh}	Undrained shear strength on horizontal plane
S_{ui}	Undrained shear strength on plane inclined at an angle i with respect to the horizontal
S_{uv}	Undrained shear strength on vertical plane
$S_u(r)$	Horizontal undrained shear stress acting at a distance r from the central axis of the vane

[1] Professor, Department of Civil Engineering, Ecole Polytechnique, Montreal, Quebec, Canada.
[2] Professor, Department of Applied Sciences, Université du Québec en Abitibi-Témiscamingue, Rouyn, Quebec, Canada.

Introduction

The vane test is widely used for the in-situ determination of the undrained shear strength of clays [*1–5*]. The results obtained are directly used in short-term stability analyses [*6–8*] despite numerous questions that still remain unanswered about the interpretation of the test. Among the various factors that affect the interpretation of the test, it is recognized that remolding, pore-pressure buildup, strain rate effects, progressive failure, and shear strength anisotropy have a major influence [*9–12*]. This latter factor, namely, strength anisotropy, which affects both bearing capacity and slope stability calculations [*13–15*], is the subject of this study.

Keeping in mind the results obtained by other investigators [*16–22*], the field investigation was carried out by means of a Nilcon vane borer equipped with vane blades of various lengths and shapes. The tests were performed in a lightly overconsolidated sensitive clay of Eastern Canada. The results show that the degree of anisotropy, defined as the ratio of the undrained shear strength mobilized on the horizontal plane to that mobilized on the vertical plane, or S_{uh}/S_{uv}, varies between 1.14 and 1.41. However, it is also indicated that the precise value of S_{uh}/S_{uv} strongly depends on the shear stress distribution assumed to act along the edges of the blades. In addition, it is found that the anisotropic strength criterion that best describes the observed behavior is that suggested by Hill [*23*].

Undrained Shear Strength Anisotropy

It is useful to briefly consider the source of anisotropy in clays. First, there is inherent, or intrinsic, anisotropy, which is related to soil structure. It results from particle orientation during sedimentation [*15,24–26*]. A second, equally important component of anisotropy can only be determined if undrained shear is initiated from an anisotropic state of stress. This component termed, "stress system induced anisotropy," is induced by rotating the principal stresses during loading from their orientations at the end of consolidation [*15,27,28*]. It is evident that if induced anisotropy is included in the test procedure, then the combined effect of both induced and inherent anisotropy will actually be measured. Because such is often the case in the field testing of soils, their combined effect is generally not dissociated in the analysis of practical problems [*29*].

There have been numerous studies on clay strength anisotropy. However, most of these studies have been carried out in the laboratory [*11,14,15,25,30–34*]. It seems that Aas [*16*] has been the first investigator to propose a method by which the vane test could be used in the field to determine the degree of anisotropy of a clay deposit. Tests with rectangular vanes of different height/diameter ratios were performed in order to evaluate the anisotropy of the clay with respect to the undrained shear strength along vertical and horizontal shear planes. Denoting the total torque by M, the torque from the vertical surface by M_v, and that from each of the two horizontal end surfaces by M_h, the following relation is valid

$$M = M_v + 2M_h \tag{1}$$

Assuming that the maximum possible shear stress in the vertical plane S_{uv} and that on the horizontal plane S_{uh} occur at the same rotation of the vane, then the registered torque is equal to

$$M = \frac{\pi D^2 H}{2} S_{uv} + \frac{\pi D^3}{p_n} S_{uh} \tag{2}$$

where

D = diameter of vane,
H = height of vane,
p_n = factor that depends on the shear stress distribution assumed at top and bottom of vane-generated failure cylinder (Table 1).

Denoting $S_u(r)$ the value of the undrained shear strength mobilized at a distance r from the central axis on the horizontal end surface, one has

$$S_u(r) = \frac{S_{uh}}{\left(\frac{D}{2}\right)^n} r^n \tag{3}$$

TABLE 1—*End shear stress distribution in rectangular vanes.*

Type	Pattern	n	Pn
Uniform	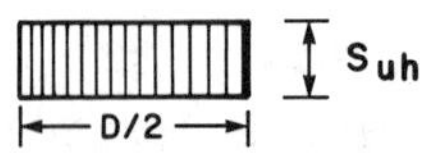	0	6
Parabolic		1/2	7
Triangular	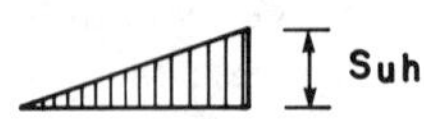	1	8
Square power	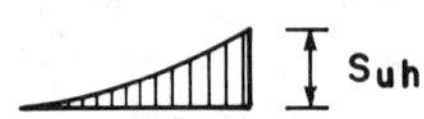	2	10
Cubic power	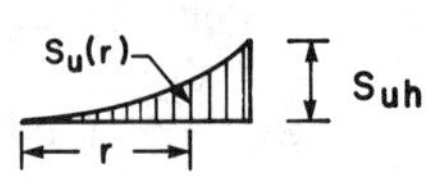	3	12
Fourth power		4	14
Bessel function (Cassan, 45)		–	6.77
Trapezoidal	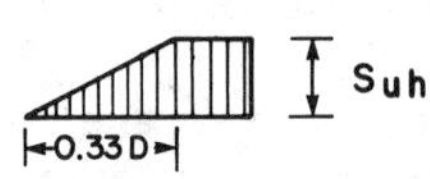	–	6.48
Strain-softening		–	7.27

where S_{uh} represents the maximum value of $S_u(r)$ and n is a coefficient that depends on the shape of the shear stress distribution, as shown in Table 1. In addition, because

$$2M_h = \int_0^{D/2} S_u(r)\, 2\pi r\, r\, dr \tag{4a}$$

or

$$2M_h = \int_0^{D/2} \frac{S_{uh}}{(D/2)^n}\, r^{n+2}\, dr \tag{4b}$$

and

$$2M_h = \frac{\pi D^3}{2} \frac{S_{uh}}{(n+3)} \tag{4c}$$

Then, from Eqs 1 and 4c, one gets

$$p_n = \frac{1}{2(n+3)} \tag{4d}$$

For a uniform shear stress distribution, $n = 0$ and $p_n = 6$, Eq 2 becomes

$$M = \frac{\pi D^2 H}{2} S_{uv} + \frac{\pi D^3}{6} S_{uh} \tag{5}$$

Hence

$$M \frac{2}{\pi D^2 H} = S_{uv} + S_{uh} \frac{D}{3H} \tag{6}$$

This means that in a graphical plot in which the vertical and horizontal axes represent $M(2/\pi D^2 H)$ and $(D/3H)$, respectively, Eq 6 describes a straight line intersecting the vertical axis at a value equal to S_{uv} and having a slope equal to S_{uh}. In addition, the intersection of the straight line with the negative $(D/3H)$ axis yields directly the value of the anisotropic strength ratio S_{uh}/S_{uv} [*16*].

Wiesel [*17*] discussed theoretically the problems arising from an Aas analysis [*16*] if the shear stress peaks on the vertical side and horizontal ends of the vane should occur at different rotations. He gave a simpler graphical method for the Aas analysis where, by plotting M against H, for vanes with the same D, he obtained

$$M = m_v H + 2M_h \tag{7}$$

where m_v = torques on unit height of cylinder side. These torques can be transformed into S_{uh} and S_{uv} once a proper shear stress distribution is retained. If the results are plotted at peak torque M, the Wiesel and Aas analyses are identical.

Another method for the evaluation of the degree of anisotropy has been proposed by Donald et al. [*18,19*] and is shown in Fig. 1. Shear stresses may be calculated from any assumed distribution and the inferred torque. The calculations can be carried out at any

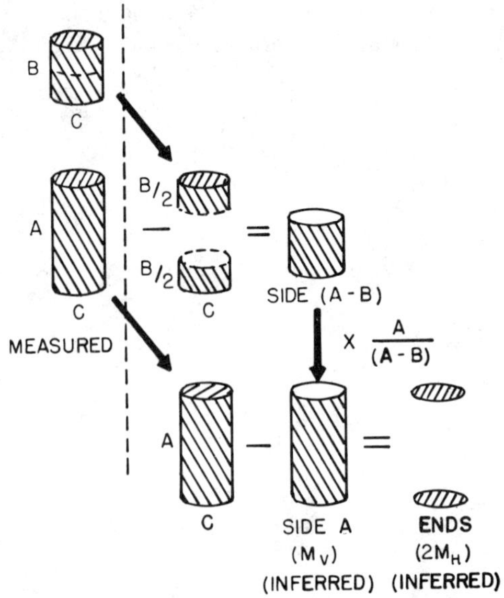

FIG. 1—*Anisotropic analysis* [18,19].

rotation, assuming that the mobilized shear stress patterns are similar for vanes of different lengths at the same rotation.

In order to study the undrained shear strength S_{ui} mobilized on planes inclined at an angle i with respect to the horizontal, it has been proposed to use diamond-shaped or rhomboidal vanes. Assuming a uniform shear stress distribution on the edges of the blades, one obtains [*16,20*]

$$S_{ui} = \frac{6M \cos i}{\pi D^3} \tag{8}$$

where D = diameter of rhomboidal vane.

For the interpretation of the variation of S_{ui} as a function of the inclination angle i, several strength criteria have been put forward in the past. Only some of them will be discussed herein. On the basis of anisotropic elasticity [*35*], Casagrande and Carrillo [*30*] proposed the following simple elliptical distribution

$$S_{ui} = S_{uh} \cos^2 i + S_{uv} \sin^2 i \tag{9}$$

For example, Reddy and Rao [*13*] have applied Eq 9 for the analysis of bearing capacity of anisotropic soils. It should be also noted that Murf [*36*] obtained a solution for the vane stress distribution in anisotropic soils.

Using the results of laboratory tests on overconsolidated London clay, Bishop [*27*] suggested the following criterion

$$S_{ui} = S_{uv} (1 - A \sin^2 i) (1 - B \cos^2 i) \tag{10}$$

where A and B are parameters for fitting the equation to experimental data.

The results obtained by means of vane tests in soft Bangkok clay led Richardson et al. [*20*] to propose the following relationship

$$S_{ui} = S_{uh}S_{uv}(S_{uh}^2 \sin^2 i + S_{uv}^2 \cos^2 i)^{-1/2} \tag{11}$$

And finally, using the criterion proposed by Hill [*23*], Davis and Christian [*24*] obtained the following relationship [*37,38*]

$$S_{ui} = \frac{e \cos 2i + a(1 + c \sin^2 2i)^{1/2}}{1 + d \sin^2 2i} \tag{12}$$

where

$$a = \frac{S_{uv} + S_{uh}}{2}, \qquad b = a\frac{S_u45}{(S_{uv}S_{uh})^{1/2}}$$

$$c = \frac{a^2 - e^2}{b^2} - 1, \qquad d = \frac{a^2}{b^2} - 1,$$

$$e = \frac{S_{uh} - S_{uv}}{2},$$

and S_{u45} represents the value of S_{ui} for $i = 45°$. Equation 12 will be used later to describe some of the results obtained on Louiseville clay.

Experimental Procedure

The Louiseville site is located at about 110 km northeast of Montreal, on the north shore of the St. Lawrence River, in the heart of the Champlain clay deposits. The general properties of these clays are well known and well documented in the literature [*39,40*]. The physical properties of the Louiseville clay deposit are presented in Fig. 2.

In order to obtain further information on the geotechnical properties of the Louiseville clay, undisturbed blocks were recovered below the weathered crust, at a depth varying between 3 and 5 m. The general properties of the Louiseville clay obtained from the block samples are shown in Table 2.

The field testing program has been carried out with both rectangular and diamond-shaped vanes as shown in Fig. 3. A Nilcon Vane Borer has been used for all the tests [*41,42*]. Three borings were made with each vane for a total of 21 borings. The values of the undrained shear strengths reported in this paper are the mean values of the three individual measurements. Individual values and experimental scattering may be found in Ref *37*. Measurements were made at depths of 2, 3, 4, 5, 6, 7, 8, 9, and 10 m. The borings were spaced at 1.5 m center to center. The vanes used were made of high ultimate strength tempered, chrome-nickel steel. A special sealed slip-coupling permitted an approximate 15° slip, or "play" between the rods and the vane. The slip-coupling enabled determination of the friction caused by the turning of the rods alone.

The vane tests were started 10 mn after the blades had been inserted at the desired depth. In order to achieve the standard vane rotation recommended with the Nilcon apparatus, the crank was rotated at an angular speed of 2 rps.

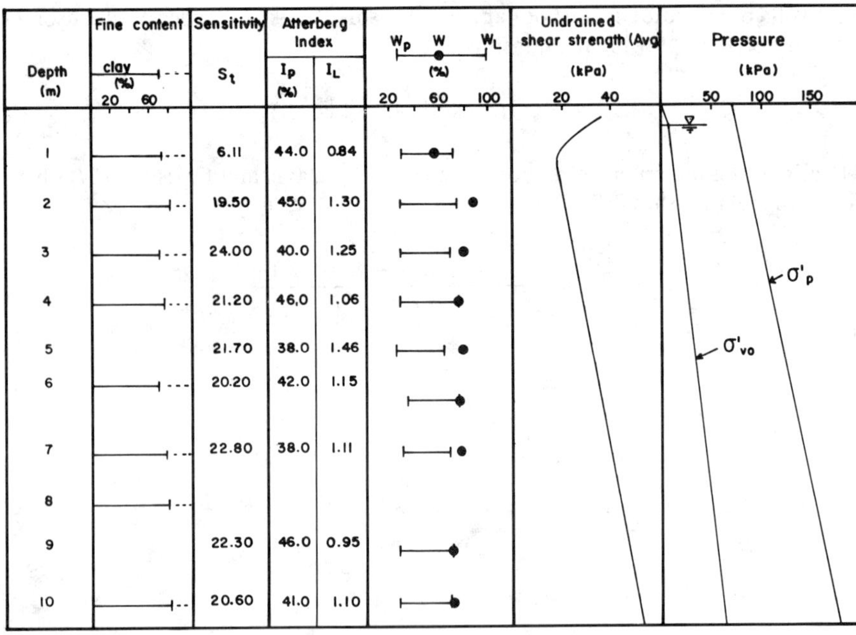

FIG. 2—*Geotechnical properties of Louiseville clay* [40].

Analysis of Test Results

Field test results obtained by means of the rectangular vanes are presented in Fig. 4, according to the graphical procedure of Wiesel [*17,19*]. The straight line relationships shown in this figure indicate that the distribution of S_{uv} is a linear function of H.

For the determination of the undrained shear strength from the measurement of the peak torque, a uniform shear stress distribution is usually assumed along the edges of the vane, resulting in Eq 3 [*5,8,9,43,44*]. However, it has been suggested that the shear stress distributions might not be quite uniform, especially regarding that for S_{uh} [*10*]. Then, by consid-

TABLE 2—*Geotechnical properties of the Louiseville clay.*

Property	Value
Water content, %	82.0
Liquid limit, %	59.0
Plastic limit, %	26.0
Liquidity index	1.7
Silt content, %	22.0
Clay content, %	78.0
Activity	0.4
Preconsolidation pressure, kPa	
vertical direction	86.0
horizontal direction	78.0
Undrained shear strength by direct shear, kPa	
vertical direction	32.0
horizontal standard sample	36.0

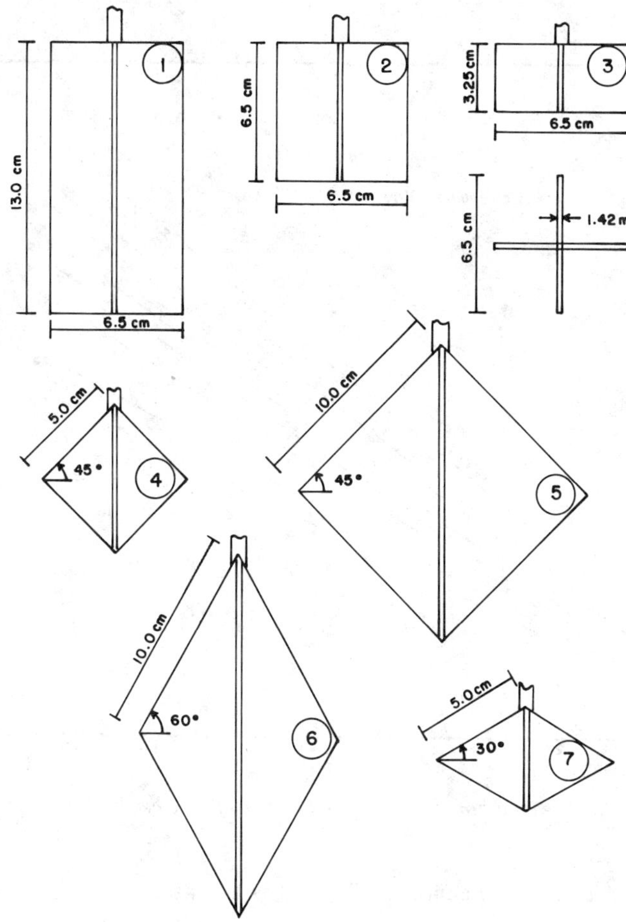

FIG. 3—*Vane configurations.*

ering other probable shapes, it is possible to determine S_{uh}, as indicated in Table 1 and Eq 2. For example, Table 1 shows that a triangular shear stress distribution for S_{uh} having $p_n = 8$ yields a value of S_{uh} equal to 8/6 of that obtained by assuming a uniform distribution in which $p_n = 6$. It is understood that such a conclusion is valid for the same shear stress distribution assumed to act along the vertical edges of the vane.

It has been suggested by Donald et al. [*8,19,22*] that for an elastic material, the shear stress distributions along the vane edges would be as shown in Fig. 5*a*. When considering such a shear stress distribution, the factor p_n in Eq 3 would be much greater than 10 or 12 and would, as such, yield a much higher value for S_{uh}. By considering the vane as applying pure torsion in an elastic medium, Cassan [*45*] has shown that the distribution of S_{uh} is somewhat like a parabola (Table 1). Using this author's approach yields a value of p_n equal to 6.77, whereas a parabola would give $p_n = 7$, as shown also in Table 1. In addition, it has been shown that for brittle clays, the peak strength would be first reached at the vane corners, and by the time the peak torque would be attained, the shear stress distribution might be as shown in Fig. 5*b* [*46*]. The end distribution shown in this figure agrees with

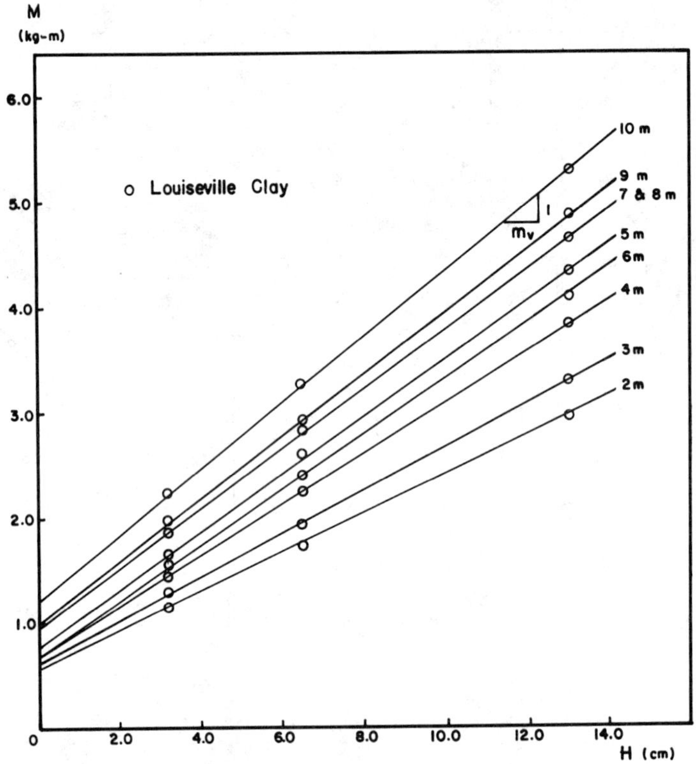

FIG. 4—*Mobilized maximum torque as a function of vane height (Wiesel's method).*

the experimental results of Lemasson [*21*] and is a consequence of progressive failure [*10,18*].

In a previous paper [*38*] the authors have presented an anisotropic analysis, assuming a uniform distribution for S_{uv} and a parabolic distribution for S_{uh}. The degrees of anisotropy thus obtained were much greater than the values one would get by using Bjerrum [*11*] results. The S_{uh}/S_{uv} ratios were also found to be much greater than the values obtained from direct shear tests, the results of which are also shown in Table 2. The discrepancy between the values obtained in the field and those measured in the laboratory may be partially due to soil disturbance and pore-pressure increase caused by the insertion of the vane [*9,12,47–52*].

From the values of m_v and M_h obtained in Fig. 4, the undrained shear strengths S_{uv} and S_{uh} were calculated by assuming a shear stress distribution similar to that shown in Fig. 5*b*. The equations used to calculate both S_{uv} and S_{uh} are the following:

$$S_{uv} = \frac{m_v}{0.4\pi D^2} \tag{13}$$

and

$$S_{uh} = \frac{M_h}{0.075\pi D^3} \tag{14}$$

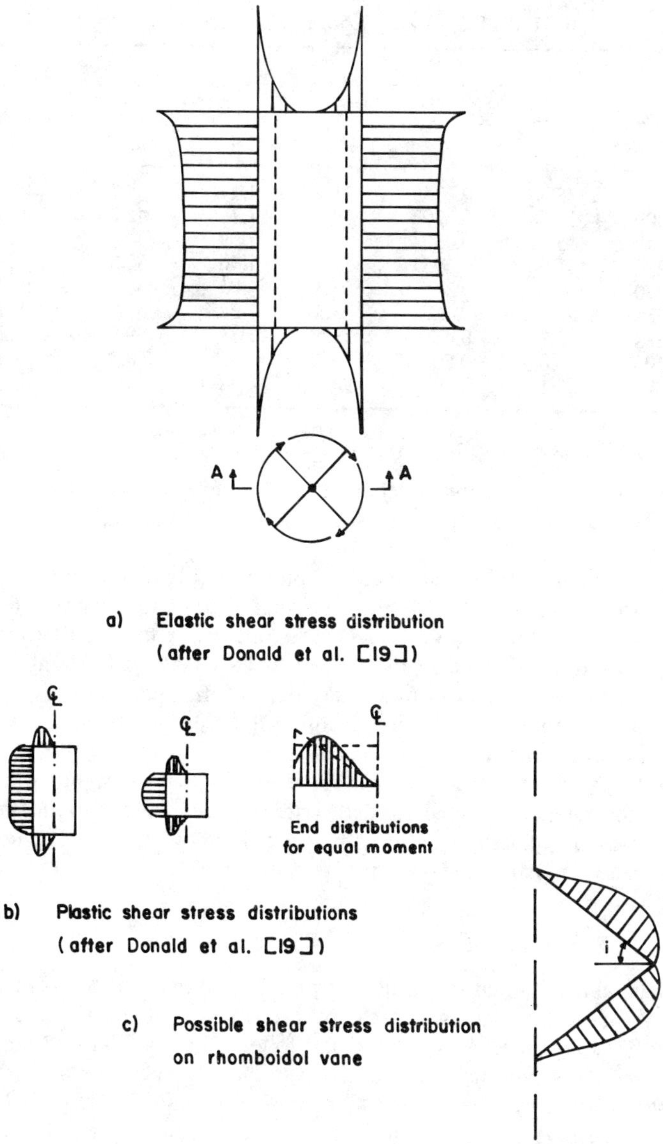

FIG. 5—*Shear stress distributions.*

By comparing Eq 13 with Eq 5, it is seen that Eq 13 gives a value for S_{uv} which is 25% greater than that inferred using Eq 5. Concerning S_{uh}, Eq 14 gives a value for S_{uh} which is about 11% greater than that given by Eq 5.

The values of S_{ui} that are also shown in Table 3 are calculated by assuming a shear stress distribution of the type shown in Fig. 5*c* and using the following approximate equation

$$S_{ui} \simeq \frac{8M \cos i}{\pi D^3} \tag{15}$$

TABLE 3—*Undrained shear strength values.*

Depth, m	m_v, kg·m/cm	S_{uv}, kPa	M_h, kg·m	S_{uh}, kPa	S_{uh}/S_{uv}	S_{ui}, kPa 45° Vane 5	60° Vane 6	30° Vane 7
2	0.192	35.4	0.275	41.6	1.17	38.2	34.7	37.7
3	0.209	38.6	0.320	48.5	1.26	42.2	40.0	44.6
4	0.248	45.9	0.330	50.0	1.09	44.8	47.3	39.7
5	0.279	51.5	0.395	59.8	1.16	52.8	48.3	54.0
6	0.270	49.9	0.345	52.3	1.05	48.2	47.2	48.9
7	0.289	53.3	0.485	73.5	1.38	57.9	58.8	66.9
8	0.289	53.3	0.485	73.5	1.38	59.0	54.2	73.2
9	0.303	56.0	0.500	75.7	1.35	60.4	54.2	68.2
10	0.320	59.1	0.595	90.1	1.52	64.6	60.5	77.5

It should be noted that the constant coefficient equal to 8 in this equation represents an average value between those given by the square power and the strain-softening models in Table 1.

The results presented in Table 3 indicate the presence of two distinct layers of clay having different degrees of anisotropy: (1) from the surface down to a depth of 6 m, S_{uh}/S_{uv} = 1.14 and (2) from 6 to 10 m, S_{uh}/S_{uv} = 1.41. It is interesting to note that the presence of these two layers is also indicated in the report by Tavenas and Leblond [*40*] about the geotechnical properties of the Louiseville clay deposit. Indeed, these investigators show that at a depth of 6 m there exists a discontinuity in the profile showing the coefficient of permeability as a function of depth.

Combining the values of S_{uv} and S_{uh} with those of S_{ui} yields the results shown in Figs. 6 and 7. Among the four anisotropic strength criteria presented earlier in the paper, it is found that the most adequate one is that due to Davis and Christian [*24*]. For the sake of clarity, only this last criterion is shown in Figs. 6 and 7.

Discussion

There are several factors that affect the interpretation of the vane test. The first concerns the shape of the failure surface generated by the rotation of the vane. Generally, it is assumed that the failure surface is circular and follows the boundaries established by the radius of the vane. This hypothesis is retained even though experimental evidence suggests that a thin, partially sheared zone exists surrounding the failure surface [*43,47,52*]. Thus some error is introduced if it is assumed that the total torque applied to the vane causes shear only on the vertical and horizontal surfaces bounded by the vane blades. In other words, correcting for this effect would produce greater values for S_{uv}.

A second factor is the progressive failure that takes place along the vane edges and the stress redistribution that occurs. Consider the development of shear stress along the edges of the vane blades as the vane is rotated. Assuming that the shaft diameter is infinitesimally small, then on the central axis of rotation no displacement occurs and no shear stress is mobilized. The shear strength in the horizontal plane is therefore progressively mobilized with increasing radius along the horizontal edge as the vane begins to rotate. The maximum possible shear stress in the horizontal plane S_{uh} will probably not occur at the same rotation, and so the maximum torque could well be registered when either S_{uv} or more probably S_{uh} has exceeded its peak value at the edges, the peak value being mobilized

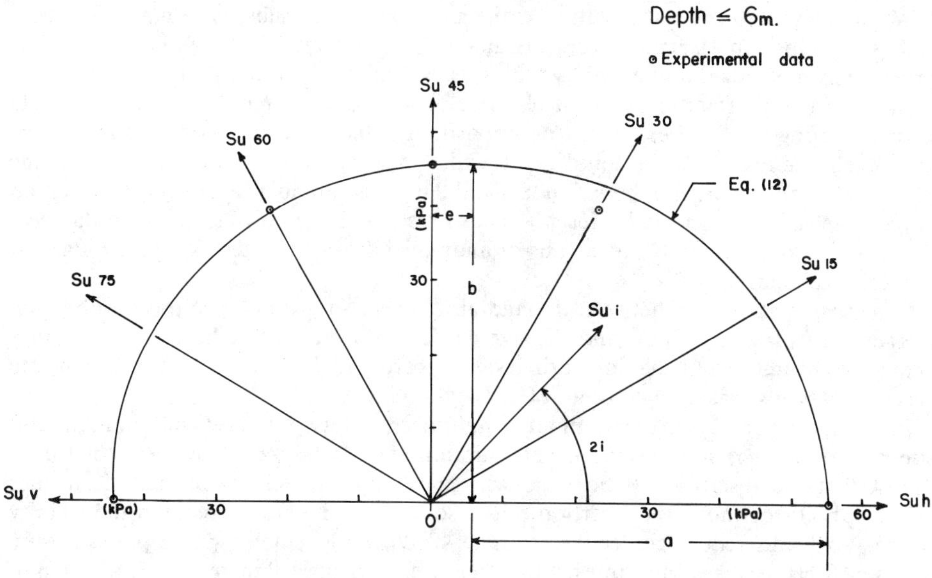

FIG. 6—*Variation of shear strength with orientation (2 to 6 m).*

nearer the axis of rotation, as shown in Fig. 5*b* [*46*]. The most surprising aspect of the test results is the straight line relationships of Fig. 4. As discussed by Donald et al. [*19*] the adoption of the shear stress distribution of Fig. 5*b* should result in a scale effect. However, the results shown in Fig. 4 indicate that the shear stress distributions for S_{uv} are of similar patterns and do not depend upon the value of the height/diameter ratio [*51*]. It is therefore concluded that the tests indicate an uniform distribuiton of S_{uv}.

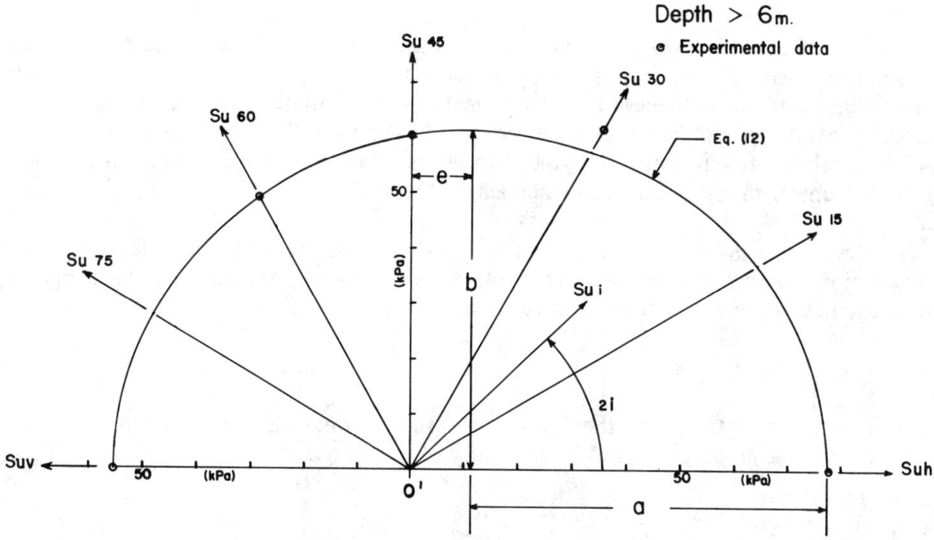

FIG. 7—*Variation of shear strength with orientation (7 to 10 m).*

When analyzing vane test results, a major uncertainty still subsists: What is the stress path, within the limit state surface, generated by the vane test? It is well known that the shear strength of clay is influenced by the stress path followed during shearing. This aspect is particularly important for the vane test, because of the rotation of the principal stresses during shearing [*8*], and because the exact position of the limit state surface is influenced by the geologic history of the soil deposit, which in turn, affects the strain history of the soil [*1,11,53–55*]. In order to better understand the stress path followed during the rotation of the vane, an investigation is presently taking place at Ecole Polytechnique on the measurement of both S_{uh} and S_{uv} by means of undrained direct shear tests on horizontal and vertical specimens.

Empirical relationships between the undrained shear strength S_u, the preconsolidation pressure, the plasticity index, and the overconsolidation ratio may be used to obtain a better insight into clay behavior during vane shearing [*1,11,25,39,53,56,57*]. Strain rate effects may be also taken into account [*8,11,19,43,53,58–61*].

In-situ testing of clays is of paramount importance because the effects of remolding and stress relief inherent of clay sampling are minimized [*3,8,11,39,62*]. However, it has been shown that the insertion of the vane will cause some remolding around the blades [*12,46,63*]. Furthermore, the insertion causes some pore-pressure increase in the vicinity of the vane, thus reducing effective stresses [*47,49*]. Both remolding and pore pressure buildup effects decrease with time, so that the shear strength will increase as a function of stand-by time before carrying out the test [*9,10,47,48*].

Considering the numerous factors presented here and the inherent variability of the vane test, it is surprising that S_{uv} has been compared with so much success to different laboratory tests, particularly with the direct shear stress, triaxial extension, and uniaxial compression [*1,8,19,42,47*].

It is obvious that the anisotropic analysis presented in this paper is restrained by many uncertainties, and that the results must be regarded with caution, because of their partly empirical nature.

Conclusion

The anisotropic analyses of the vane test is complicated by numerous factors, among which progressive failure, pore-pressure buildup, remolding, strain rate, and stress path effects have a major influence. The exact shear stress distributions along the vane blades are still not known, although they have been inferred empirically with success in this study.

The method presented in this paper, which uses both rectangular vanes of different height/diameter ratios and diamond-shaped vanes, can be used to determine S_{uv}, S_{uh}, and S_{ui}.

The test results show that the degree of strength anisotropy varies between 1.14 and 1.41. In addition, the elliptical failure criterion put forward by Davis and Christian [*24*] appears to adequately represent the observed variation of S_{ui}.

Acknowledgments

The authors wish to express their gratitude to the National Research Council of Canada for the financial support received in the course of this study.

References

[*1*] Hanzawa, H. and Kishida, T., "Determination of In-Situ Undrained Strength of Soft Clay Deposits," *Soils and Foundations,* Vol. 22, No. 2, 1982, pp. 1–14.

[2] Mlynarek, Z. B. and Sanglerat, G., "Relationship Between Shear Parameters and Cone Resistance for Some Cohesive Soils," *Proceedings, International Symposium on In Situ Testing,* Paris, Vol. 2, 1983, pp. 347–352.
[3] Frank, R. A., "Essais de chargement, de déchargement et de cisaillement en forage ou en place à l'aide d'appareils, spécifiques tels le pressiomètre, pénétromètre, scissomètre, etc.," Rapparts Généraux, Thème 5, *Proceedings, International Symposium on In-Situ Testing,* Paris, Vol. 3, 1983, pp. 61–75.
[4] Bowles, J. E., *Physical and Geotechnical Properties of Soils,* 2nd ed., McGraw Hill Book Co., New York, 1984, pp. 459–462.
[5] Canadian Geotechnical Society, *Canadian Foundation Engineering Manual,* 2nd ed, Montréal, 1985, pp. 53–55.
[6] Hanzawa, H. and Kishida, T., "Fundamental Considerations on Undrained Strength Characteristics of Alluvial Marine Clays," *Soils and Foundations,* Vol. 21, No. 1, 1981, pp. 39–50.
[7] Aas, G., "Stability Problems in a Deep Excavation in Clay," *Proceedings, International Conference on Case Histories in Geotechnical Engineering,* University of Missouri-Rolla, Vol. 1, 1984, pp. 315–323.
[8] Wroth, C. P., "The Interpretation of In-Situ Soil Tests," *Géotechnique,* Vol. 34, No. 4, 1984, pp. 449–489.
[9] Aas, G., "A Study of the Effect of Vane Shape and Rate of Strain on the Measured Values of In-Situ Shear Strength of Clays," *Proceedings, International Conference on Soil Mechanics and Foundation Engineering,* Montréal, Quebec, Canada, Vol. 1, 1965, pp. 141–145.
[10] Flaate, K., "Factors Influencing the Results of Vane Tests," *Canadian Geotechnical Journal,* Vol. 3, No. 1, 1966, pp. 18–31.
[11] Bjerrum, L., "Problems of Soil Mechanics and Construction in Soft Clays," State-of-the-Art Report, *Proceedings, 8th International Conference on Soil Mechanics and Foundation Engineering,* Moscow, Vol. 3, 1973, pp. 111–159.
[12] LaRochelle, P., Roy, M., and Tavenas, F., "Field Measurements of Cohesion in Champlain Clays," *Proceedings, 8th International Conference on Soil Mechanics and Foundation Engineering,* Moscow, U.S.S.R., Vol. 1, 1973, pp. 229–236.
[13] Reddy, A. S. and Rao, K. N. V., "Bearing Capacity of Strip Footing on Anisotropic and Nonhomogeneous Clays," *Soils and Foundations,* Vol. 21, No. 1, 1981, pp. 1–6.
[14] Nakase, A. and Kamei, T., "Undrained Shear Strength Anisotropy of Normally Consolidated Cohesive Soils," *Soils and Foundations,* Vol. 23, No. 1, 1983, pp. 91–101.
[15] Ohta, H. and Nishihara, A., "Anisotropy of Undrained Shear Strength of Clays Under Axi-Symmetric Loading Conditions," *Soils and Foundations,* Vol. 25, No. 2, 1985, pp. 73–86.
[16] Aas, G., "Vane Tests for Investigation of Anisotropy of Undrained Shear Strength of Clays," *Proceedings, Geotechnical Conference,* Oslo, Norway, Vol. 1, 1967, pp. 3–8.
[17] Wiesel, C. E., "Some Factors Influencing In Situ Test Results," *Proceedings, 8th International Conference on Soil Mechanics and Foundation Engineering,* Moscow, U.S.S.R., Vol. 2, 1973, pp. 475–479.
[18] Donald, I. B., Jordan, D. O., and Parker, R. E., "Interpretation of Vane Testing in Clays," *Proceedings, 5th Symposium on In Situ Testing for Design Parameters,* Melbourne, Australia, 1975, pp. 1–25.
[19] Donald, I. B. Jordon, D. O., Parker, R. J., and Toh, C. T., "The Vane Test-a Critical Appraisal," *Proceedings of the 9th International Conference on Soil Mechanics and Foundation Engineering,* Tokyo, Vol. 1, 1977, pp. 81–88.
[20] Richardson, A. M., Brand, E. W., and Memon, A., "In-Situ Determination of Anisotropy of a Soft Clay," *Proceedings, ASTM Specialty Conference on In-Situ Measurement of Soil Properties,* Raleigh, NC, Vol. 1, 1975, pp. 336–349.
[21] Lemasson, H., "Une nouvelle méthode pour la mesure en place l'anisotropie des argiles," *Bulletin de Liaison,* Laboratoire des Ponts et Chaussées, Special III, 1976, pp. 107–116.
[22] Menzies, B. K. and Merrifield, C. M., "Measurements of Shear Stress Distribution on the Edges of a Shear Vane Blade," *Géotechnique,* Vol. 30, 1980, pp. 314–318.
[23] Hill, R., *The Mathematical Theory of Plasticity,* Oxford University Press, London, 1950.
[24] Davis, E. H. and Christian, J. T., "Bearing Capacity of Anisotropic Cohesive Soils," *Journal of Soil Mechanics and Foundation Engineering Division,* Vol. 98, No. SM1, 1971, pp. 126–132.
[25] Yong, R. N. and Silvestri, V., "Anisotropic Behavior of a Sensitive Clay," *Canadian Geotechnical Journal,* Vol. 16, No. 2, 1979, pp. 335–350.
[26] Mitchell, J. K., *Foundamentals of Soil Behavior,* John Wiley and Sons, New York, 1976.
[27] Bishop, A. W., "The Strength of Soils as Engineering Materials," *Géotechnique,* Vol. 16, No. 2, 1966, pp. 91–128.
[28] Law, K. T. and Lo, K. Y., "Analysis of Shear Induced Anisotropy in Leda Clay," *Proceedings,*

Engineering Conference on Numerical Methods in Geomechanics, Blacksburg, VA, Vol. 1, 1976, pp. 329–344.

[29] Ladd, C. C. and Foott, R., "New Design Procedure for Stability of Soft Clays," *ASCE Journal of Geotechnical Engineering Division,* Vol. 100, No. GT7, 1974, pp. 763–786.

[30] Casagrande, A. and Carillo, N., "Shear Failure of Anisotropic Materials," *Journal of the Boston Society for Civil Engineers, Contributions to Soil Mechanics 1941–1953,* 1944, pp. 122–135.

[31] Duncan, J. M. and Seed, H. B., "Anisotropy and Stress Reorientation in Clay," *ASCE Journal of Soil Mechanics and Foundation Engineering Division,* Vol. 92, No. SM5, 1966, pp. 21–50.

[32] Soydemir, C., "Strength Anisotropy Observed Through Simple Shear Tests," *Lauritz Bjerrum Memorial Volume,* Oslo, Norway, 1976, pp. 99–113.

[33] Kenney, T. C. and Landva, H., "Vane Triaxial Apparatus," *Proceedings, 6th International Conference on Soil Mechanics and Foundation Engineering,* Montreal, Quebec, Canada, Vol. 1, pp. 269–272.

[34] Prevost, J. H., "Undrained Shear Tests on Clay," *Journal of Geotechnical Engineering,* Vol. GT1, 1979, pp. 49–64.

[35] Timoshenko, S., *Theory of Elasticity,* McGraw Hill, New York, 1934.

[36] Murf, J. D., "Vane Shear Testing of Anisotropic, Cohesive Soils," *International Journal for Numerical and Analytical Methods in Geomechanics,* Vol. 4, 1980, pp. 285–289.

[37] Aubertin, M., "Etude de l'anisotropie de la résistance au cisaillement non drainé de l'argile de Louiseville à l'aide du scissomètre de chantier," Master thesis, Ecole Polytechnique de Montréal, Montreal, Quebec, Canada, April 1982, 92 pp.

[38] Silvestri, V. and Aubertin, M., "L'anisotropie en place d'une argile sensible du Québec," *Proceedings, International Symposium on In-Situ Testing,* Paris, Vol. 2, 1983, pp. 391–395.

[39] Leroueil, S., Tavenas, F., and Le Bihan, J.-P., "Propriétés caractéristiques des argiles de l'est du Canada," *Canadian Geotechnical Journal,* Vol. 20, No. 4, 1983, pp. 681–705.

[40] Tavenas, F. and Leblond, P., "Etude de tassement d'un remblai par conmult," Rapport d'étape No. GCT 8007, Université Laval, Quebec, Canada, 1980.

[41] Schmertmann, J. H., "Measurements of In Situ Shear Strength," *Proceedings, ASCE Specialty Conference on In-Situ Measurement of Soil Properties,* Raleigh, NC, Vol. 2, 1975, pp. 57–138.

[42] Lavallée, J. G., "Validation d'un profil de résistance d'une argile sensible," *Proceedings, International Symposium on In-Situ Testing,* Paris, Vol. 2, pp. 321–326.

[43] Ladd, C. C., Foott, R., Ishihara, K., Schlosser, F., and Poulos, H. G., "Stress-Deformation and Strength Characteristics," *Proceedings, 9th International Conference on Soil Mechanics and Foundation Engineering,* Tokyo, Japan, Vol. 2, 1977, pp. 421–494.

[44] Lee, I. K., White, W., and Ingles, O. G., *Geotechnical Engineering,* Pitman Publishing Inc., Melbourne, Australia, 1983, p. 120.

[45] Cassan, M., "Réalisation et interprétation," *Les essais in situ en mécanique des sols,* Vol. 1, Eyrolles, Paris, 1984, p. 458.

[46] Menzies, B. K. and Mailey, L. K., "Some Measurements of Strength Anisotropy in Soft Clays Using Diamond-Shaped Shear Vanes," *Géotechnique,* Vol. 26, No. 3, pp. 535–538.

[47] Kimura, T. and Saito, K., "Effect of Disturbance Due to Insertion on Vane Shear Strength of Normally Consolidated Cohesive Soils," *Soils and Foundations,* Vol. 23, No. 2, 1983, pp. 113–124.

[48] Torstensson, B. A., "Time-Dependent Effects in the Field Vane Test," *Proceedings, International Symposium on Soft Clay,* Bangkok, Thailand, 1977, pp. 387–397.

[49] Matsui, T. and Abe, N., "Shear Mechanism of Vane Test in Soft Clays," *Soils and Foundations,* Vol. 21, No. 4, 1981, pp. 69–80.

[50] Hanzawa, H., "Undrained Strength Characteristics of an Alluvial Marine Clay in the Tokyo Bay," *Soils and Foundations,* Vol. 19, No. 4, 1979, pp. 87–103.

[51] Amar, S., Baguelin, F., Jezequel, J. F., and Lemehaute, A., "In-Situ Shear Resistance of Clays," *Proceedings, Specialty Conference on In-Situ Measurement of Soil Properties,* Raleigh, NC, Vol. 1, 1975, pp. 21–44.

[52] Arman, A, Poplin, J. K., and Ahmad, N., "Study of the Vane Shear," *Proceedings, Specialty Conference on In Situ Measurement of Soil Properties,* Raleigh, NC, Vol. 1, 1975, pp. 93–120.

[53] Graham, J., Crooks, J. H. A., and Bell, A. L., "Time Effects on the Stress-Strain Behavior of Natural Soft Clays," *Géotechnique,* Vol. 33, No. 3, 1983, pp. 327–340.

[54] Janbu, N., "Slopes and Excavations in Normally and Lightly Overconsolidated Clays," State-of-the-art Report on Slopes and Excavations, *Proceedings, 9th International Conference on Soil Mechanics and Foundation Engineering,* Tokyo, Japan, Vol. 2, pp. 549–566.

[55] Hanzawa, H., "Three Case Studies for Short Term Stability of Soft Clay Deposits," *Soils and Foundations,* Vol. 23, No. 2, 1983, pp. 140–154.

[56] Murthy, M. K., Sridharan, A., and Nagaraj, T. S., "Prediction of Undrained Strength of Overconsolidated Clays," *Soils and Foundations,* Vol. 22, No. 1, 1982, pp. 78–81.
[57] Skempton, A. W., "The Post-Glacial Clays of the Thames Estuary at Tilburry and Shellhaven," *Proceedings, 3rd International Conference on Soil Mechanics and Foundation Engineering,* Zurich, Switzerland, Vol. 1, 1953, pp. 302–308.
[58] Nakase, A. and Kamei, T., "Influence of Strain Rate an Undrained Shear Characteristics of K_0-Consolidated Cohesive Soils," *Soils and Foundations,* Vol. 26, No. 1, 1986, pp. 85–95.
[59] Mayne, P., "A Review of Undrained Strength in Direct Simple Shear," *Soils and Foundation,* Vol. 25, No. 3, 1985, pp. 64–72.
[60] Brand, E. W., "Discussion of 'Time effects on the stress-strain behavior of natural soft clays' by Graham et al., 1983," *Géotechnique,* Vol. 34, 1984, pp. 435–438.
[61] Kirkpatrick, W. M. and Khan, A. J., "The Influence of Stress Relief on the Vane Strength of Clays," *Géotechnique,* Vol. 34, No. 3, 1984a, pp. 428–432.
[62] Kirkpatrick, W. M. and Khan, A. J., "The Reaction of Clays to Sampling Stress Relief," Géotechnique, Vol. 34, No. 1, 1984b, pp. 29–42.
[63] Morris, D. V., "In Situ Testing of Sensitive Soils," *Proceedings, International Symposium on In-Situ Testing,* Paris, Vol. 2, 1983, pp. 359–362.

J. A. R. Ortigão[1] *and Haroldo B. Collet*[2]

Errors Caused by Friction in Field Vane Tests

REFERENCE: Ortigão, J. A. R. and Collet, H. B., **"Errors Caused by Friction in Field Vane Tests,"** *Vane Shear Strength Testing Soils: Field and Laboratory Studies, ASTM STP 1014,* A. F. Richards, Ed., American Society for Testing and Materials, Philadelphia, 1988, pp. 104–116.

ABSTRACT: Field vane tests have been traditionally employed with reported success in design on soft soils. In Brazil, most of the available vane test equipment requires preboring. With these devices soil-rod friction influences the data and should be accounted for through calibration tests. An investigation was carried out on Rio de Janeiro clay, and the results showed that soil-rod friction tests can provide unreliable results. Therefore, friction should be eliminated through a better mechanical design, as accomplished in the vaneborer apparatus, with which field vane (FV) tests have shown less data scatter. The FV strength profile was evaluated in a stability analysis of an embankment failure on Rio de Janeiro clay, and a factor of safety close to one was obtained in this highly plastic clay. This led to the conclusion that recommended FV correction factors are too conservative for the Rio de Janeiro clay.

KEY WORDS: field vane tests, Rio de Janeiro clays, calibration tests, soil-rod friction, correction factors

Early attempts to obtain the undrained strength of soft clays through field vane tests (FVT) go back as far as the 1920s and 1930s, in Sweden and Germany. Many years later, the test was introduced in Brazil, circa 1948, but only after a decade a field vane apparatus became commercially available through a local manufacturer in Rio de Janeiro. Since then, this test gained widespread use in Brazil.

A FV testing program in the Rio de Janeiro soft clay using the commercially available apparatus has led to difficulties for data evaluation and correction for soil-rod friction and to excessive data scatter. Therefore, alterations in the mechanical design of the equipment were introduced in order to reduce friction to a negligible value.

This paper presents the results of this campaign, comparing the data from two different FV apparatuses, and finally the strength profile of the Rio de Janeiro soft clay and FV correction factors are evaluated.

FV Equipment Types

According to the installation method, FV equipment can be classified into four different categories as shown in Fig. 1.

Unprotected Rods and Vane, Through a Prebored Hole (Fig. 1a)

Despite being the earliest equipment model, this device is still in use today. The measured torque includes soil friction on the rods just above the vane, and the results have to

[1] Associate Professor, School of Engineering, Federal University of Rio de Janeiro, Rio de Janeiro, Brazil.

[2] Professor, Department of Civil Engineering, Fluminense Federal University, Niterói, Brazil.

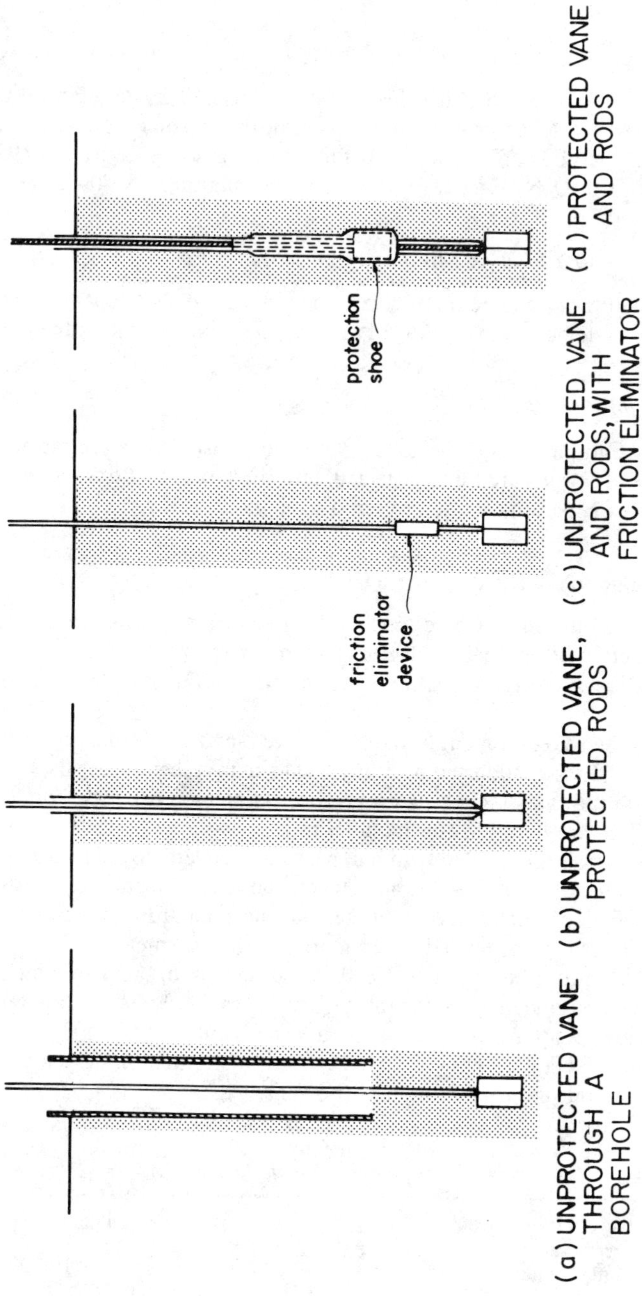

FIG. 1—*Classification of FV equipment according to the method of installation.*

be corrected. Calibration tests without the vane are suggested for this correction (for example, ASTM Method for Field Vane Shear Test in Cohesive Soils [D 2573]).

*Protected Rods and Unprotected Vane (Fig. 1*b*)*

An example of an equipment with this feature is the LCPC (Laboratoire Central des Ponts et Chaussées) [*1*]. A rod protection pipe is required by some standards, such as British Standard (BS) Methods of Test for Soils for Civil Engineering Purposes (BS 1377), or the German Standard (DIN) Vane testing (English translation DIN 4096, 1980).

*Unprotected Vane with a Friction Eliminator Device (Fig. 1*c*)*

This is the case of the vaneborer manufactured by the Swedish company Nilcon AB. The friction elimination device allows a rod rotation of 45° before actuating the vane.

*Protected Rods and Vane (Fig. 1*d*)*

As designed by Cadling and Odenstad [*2*], the vane is installed in a retracted position in a protection shoe. The rods are fully protected, and therefore, friction can be totally eliminated. This type of apparatus is often referred to as the vaneborer.

The FVT Campaign in the Rio de Janeiro Clay

FVT were carried out in the Rio de Janeiro clay during a major research program which involved construction of full-scale instrumented embankments [*3,4*].

At the site of the tests the clay stratum is 11 m thick and its properties are as shown in Table 1.

Figure 2 shows the FV equipment initially employed in this investigation. Manufactured in Brazil for about 15 years and used by many local site investigation firms, FV equipment comprised a 100-mm-diameter casing and 41-mm-diameter AX rods connected to the vane through a 750-mm-long 12-mm-diameter rod.

At least two rod spacers provided with ball bearings were employed in the investigation, one at the top and the other at the bottom. The torque measuring unit comprised a 7200:1 reduction gear, which enabled rotating the vane at the standard rate of 6°/min. A system of spring and deflection gage enabled accurate torque measurements.

The vane was 90 mm in diameter and 180 mm in height, in order to match the torque values to the more accurate range of the torque-spring system. Vanes of different sizes were also used in a previous attempt to investigate the clay's anisotrophy [*5*].

Some tests with this equipment employed a friction elimination device allowing a 45° free rotation of the rods before actuation of the vane.

TABLE 1—*Properties of Rio de Janeiro clay.*

Property	Value
Liquid limit, %	160 to 120
Plastic limit, %	80 to 60
Water content, %	180 to 120
Mean plasticity index, %	80
Total unit weight, kN/m^3	13

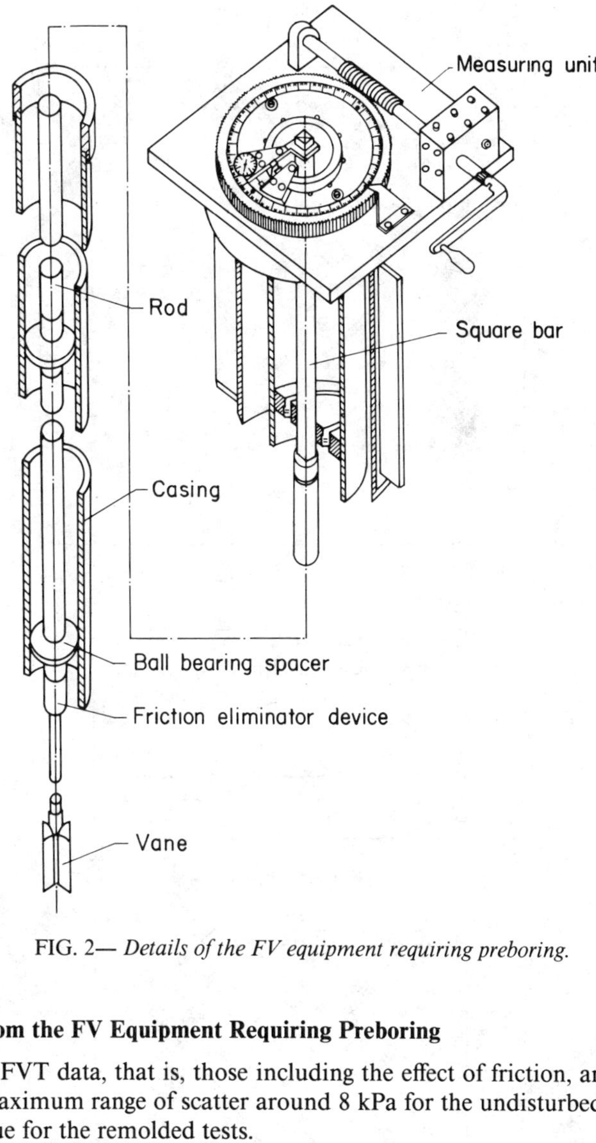

FIG. 2— *Details of the FV equipment requiring preboring.*

Test Results from the FV Equipment Requiring Preboring

Uncorrected FVT data, that is, those including the effect of friction, are plotted in Fig. 3, showing a maximum range of scatter around 8 kPa for the undisturbed tests and about half of this value for the remolded tests.

Calibration or dummy tests without the vane were then carried out to evaluate the effect of friction. Typical test curves are shown in Fig. 4, and peak values, that is, the maximum measured torques on the vane rod only, are plotted in Fig. 5. There is excessive scattering in these data, and maximum values are about 20 Nm, which is equivalent to an undrained strength of 8 kPa, if a 90 by 180 mm (diameter by height) vane was in place. This value is inconsistent with the data in Fig. 3, about twice the mean value given by the tests in remolded clay.

Therefore, it was concluded that the excessive scattering and the high values of friction test data were caused rather by internal friction of the instrument than by soil-rod friction. This excessive amount of internal friction was not present when the vane was in use, as the data of Fig. 3 show.

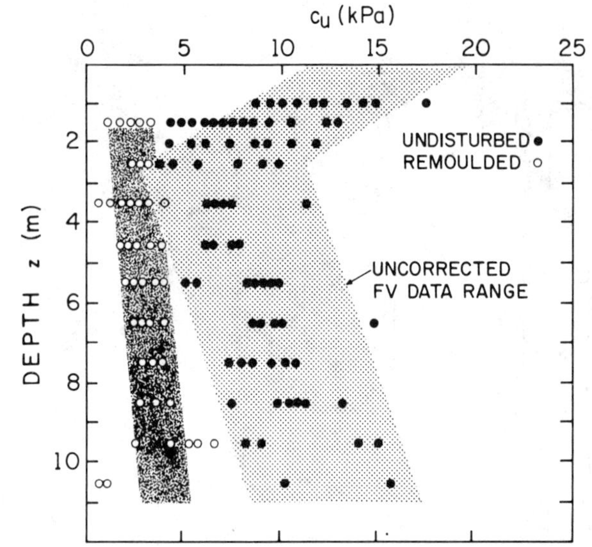

FIG. 3—*FVT data, uncorrected for friction.*

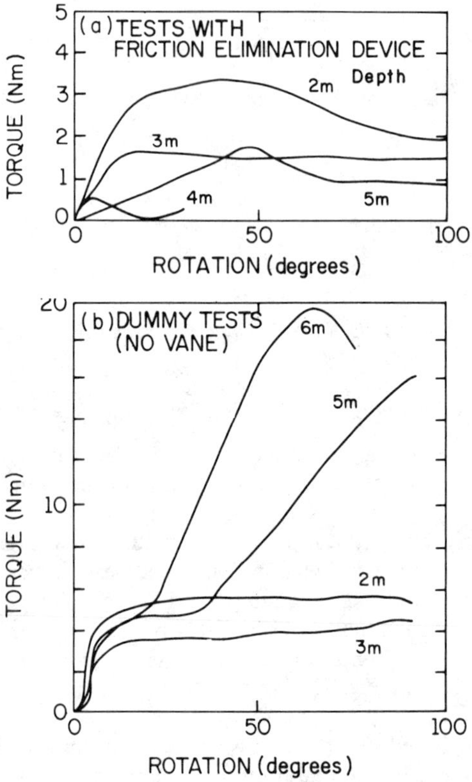

FIG. 4—*Calibration test results for the evaluation of friction: with the* (a) *friction elimination device and* (b) *dummy tests without the vane.*

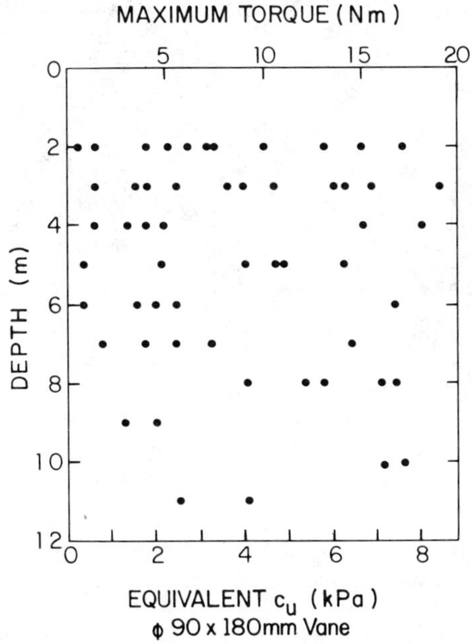

FIG. 5—*Data scatter in friction calibration tests.*

Attempts to minimize this problem by employing well aligned rods and by changing the rod spacers were unsuccessful.

It is interesting to note that similar difficulties with dummy tests were reported some 30 years ago in the United States [*6,7*] with the same type of FV apparatus.

In order to overcome these difficulties, Collet [*8*] proposed an acceptance criterion for the friction test data. Only the torque-rotation curves, which appeared not to be significantly influenced by internal friction, were taken into account. Therefore, returning to Fig. 4*b,* the curves corresponding depths of 5 and 6 m were disregarded because their pattern presented a sudden increase in torque caused by internal frictions. However, those corresponding to depths of 2 and 3 m presented an acceptable pattern and were taken into account.

After all test curves had been filtered by this process, torque peak values from the selected curves were gathered in Fig. 6. As should be expected, linear correlations of these data indicated an increase of friction with depth. Mean friction values were about 5 Nm, which corresponds to an equivalent undrained strength, that is, if measured with a 90 by 180 mm (diameter by height) vane of about 2 kPa. This is a reasonable value, and it is compatible with uncorrected FV data shown in Fig. 3.

Finally, the correlation shown in Fig. 6*a* was selected to represent the effect of friction versus test depth. The resulting corrected FV data are plotted in Fig. 7. Excessive data scatter still remains making it difficult to select a design on profile.

The Vaneborer

In 1980, the aforementioned difficulties and errors led the writers to a complete redesign of the FV equipment, in order to eliminate or reduce friction to negligible values. Figure

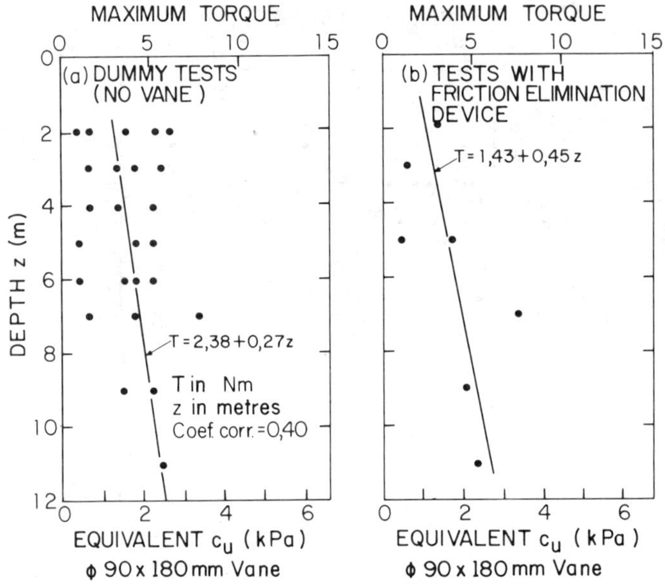

FIG. 6—*Maximum torques measured in friction calibration after data filtering* (a) *without the vane and* (b) *with friction eliminator device.*

8 presents the apparatus that was built according to the original conception of Cadling and Odenstad [*2*].

Torque was transmitted by 20-mm-diameter rods in sections of 1 m, operating inside a 38-mm-diameter protective casing. In order to avoid internal friction, special steel bars and pipes were selected and were carefully machined, in order to keep the axial alignment.

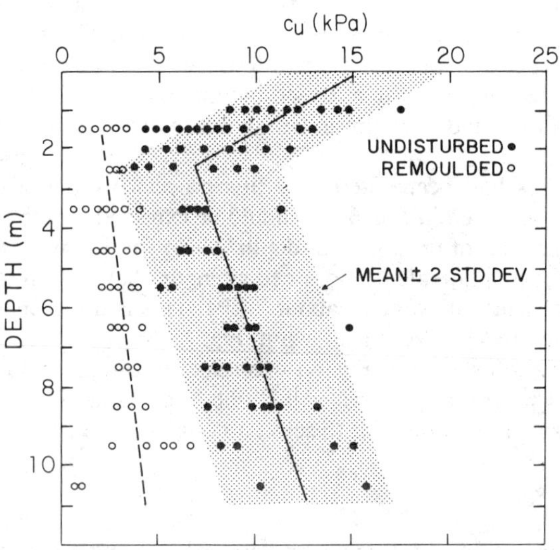

FIG. 7—*FV results from tests with the equipment that requires preboring.*

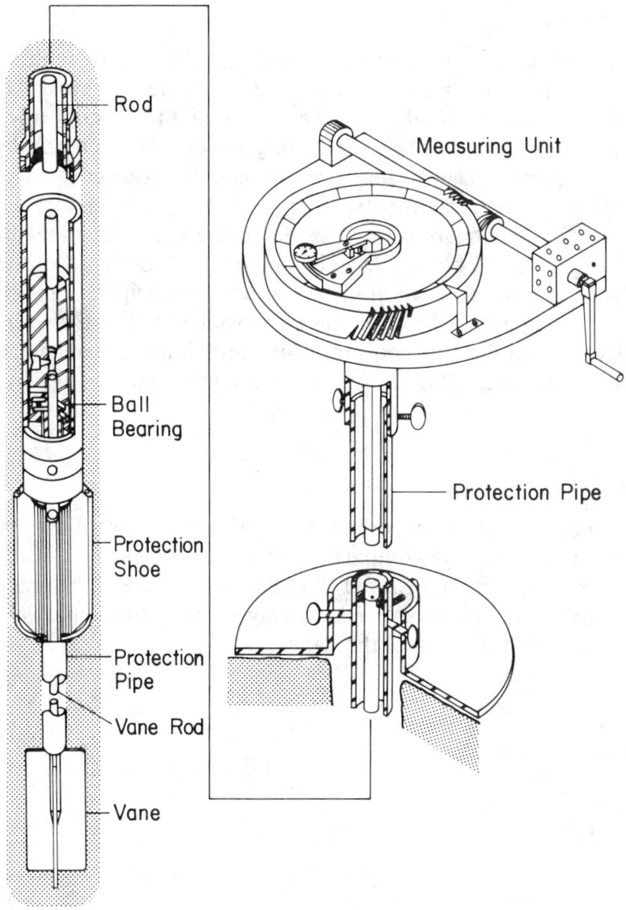

FIG. 8—*The vaneborer.*

Additionally, ball bearing spacers were provided to maintain clearance between the pipes and rods.

The lower part of the apparatus contained a protective shoe that accommodated a vane up to 130 mm long and 65 mm in diameter, protecting it during insertion by pushing the equipment into the soft ground.

A male-female cone-type clamp, apparently more effective than the one used by Cadling and Odenstad [*2*], was located in the lower part of the apparatus. This device enabled the rods to be kept in the upward position until the equipment reached the desired depth, when the rods were unlocked by a quick blow on their top and the vane was inserted 0.50 m into the clay. A 20-mm-diameter protective pipe was used to avoid any soil-rod friction with the 10-mm-diameter rod connected to the vane.

The readout unit was modified in order to sit directly on and to be attached to the protective pipe. A 300-mm-diameter footing provided with a central hole was attached to the protective pipe so as to keep the readout unit stable during the tests. Additionally, a more sensitive torque spring and a 1-μm deflection gage were fitted.

Vaneborer Calibrations

Calibrations carried out on the vaneborer consisted of field tests with and without the vane until it could be assured that the internal frictions were negligible. The readout unit was calibrated with several load-unload torque cycles, and the results, after simple statistical treatment, are summarized in Fig. 9, enabling a choice between a 50 by 100-mm or a 65 by 130-mm (diameter by height) vane, to select the calibration constant and to estimate the accuracy of the torque measurements.

For example, reading this figure with a c_u value of 20 kPa, a torque value of 20 Nm is obtained for the larger vane. This vane size is preferred because it led to a smaller error of less than 2% of the torques, as shown in the upper part of the figure. This chart also shows that for torques less than 7 Nm the error can be considerable and the spring element presents nonlinearity in its calibration constant. Therefore, it can be concluded that the minimum c_u that can be measured with a reasonable accuracy with the 65- by 130-mm vane is about 5 kPa.

Vaneborer Test Results

FVT data obtained with the vaneborer at 0.5-m-depth increments in six boreholes are plotted in Fig. 10. These data are compared with the previous series in Fig. 11 and have their statistical correlations presented in Tables 2 and 3. Figure 12 compares the sensitivities of the clay, that is, the ratio between the clay strength in the undisturbed state and after remolding with some 20 turns of the vane.

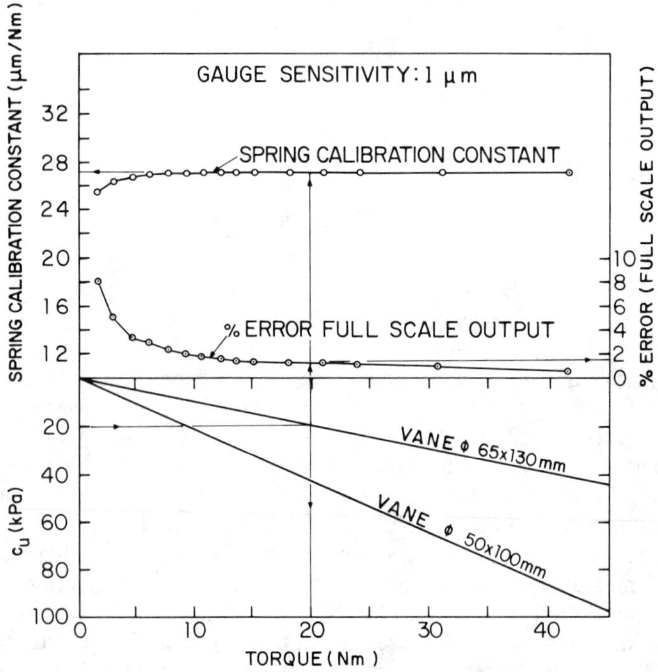

FIG. 9—*Readout unit calibration results.*

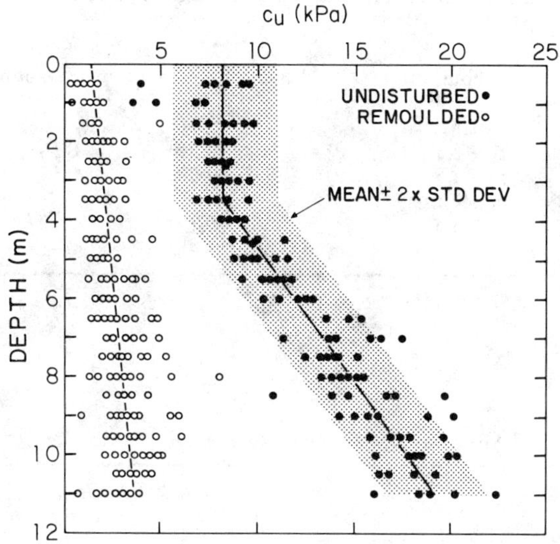

FIG. 10—*FVT results with the vaneborer.*

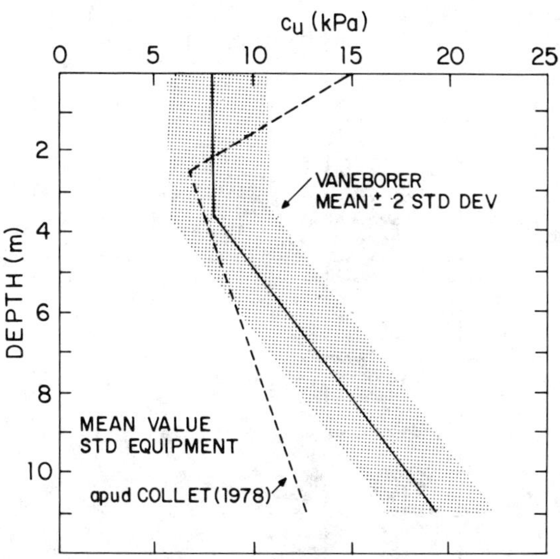

FIG. 11—*Comparison between FV profiles obtained with the vaneborer and with the preboring equipment.*

TABLE 2—*Correlations with the undisturbed test data.*

Equipment Type	Depth, m	Equation[a]	Coefficient of Correlation	Standard Error of Estimate, kPa
Requiring preboring	upper crust $z \leqq 2.5$	$c_u = 15.3 + 4.0z$	0.58	2.46
	$z > 2.5$	$c_u = 4.97 + 0.69z$	0.59	2.15
Vaneborer	upper crust $z \leqq 3.5$	$c_u = 7.36 + 0.19z$	0.107	1.35
	$z > 3.5$	$c_u = 3.00 + 1.48z$	0.913	1.53

[a] c_u = undrained strength, kPa, and z = depth, m.

TABLE 3—*Correlations with the remolded test data.*

Equipment Type	Equation[a]	Coefficient of Correlation	Standard Error of Estimate, kPa
Requiring preboring	$c_{ur} = 0.6 + 0.44z$	0.65	. . .
Vaneborer	$c_{ur} = 1.38 + 0.22z$	0.57	0.992

[a] c_{ur} = remolded undrained strength, kPa, and z = depth, m.

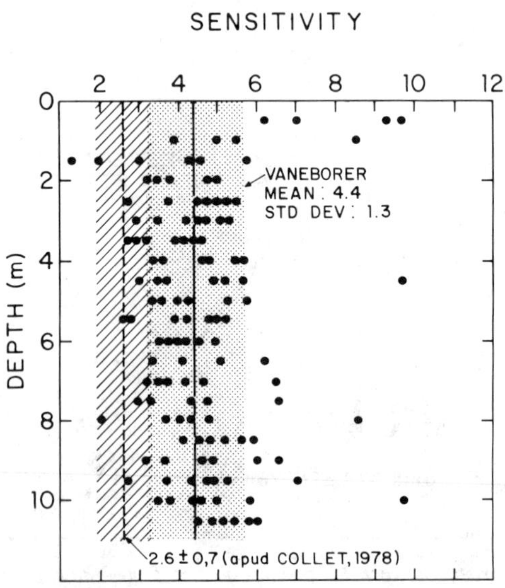

FIG. 12—*Comparison between sensitivity data obtained with the vaneborer and with the preboring equipment.*

In relation to the previous FVT series, the vaneborer data show:

- an approximately constant undrained strength at the upper part of the clay ($z \leqq 3.5$ m), the mean value being 8.3 kPa;
- a decrease in data scattering for the undisturbed tests;
- a standard error of estimate about half of the previous values and a greater coefficient of correlation below the clay crust; and
- an increase in clay sensitivity mean values from 2.6 to 4.4.

Evaluation of the Vaneborer Test Results in the Rio de Janeiro Clay

The mean and uncorrected strength profile obtained with the vaneborer was evaluated through a total stress stability analysis of a test embankment built up to failure in the same testing area [*3*].

A summary of the stability analyses, which have been described in detail in Refs *3* and *13,* is shown in Fig. 13, showing that a safety factor very close to one was obtained for failure conditions. Therefore, no correction seems to be necessary for this clay.

This conclusion is not in agreement with the correction factors suggested by Bjerrum [*9*], Azzouz et al. [*10*], and recently by Aas et al. [*11*].

Indeed, according to these authors, the vane strength should be reduced by about 40% for a clay having a plasticity index of 80% and a normalized vane strength ratio (c_u(FV)/σ'_{vo}) ranging from 0.7 to 0.95.

The reasons for these corrections not to be applicable to this clay are yet not clear.

In a recent discussion Ortigão et al. [*12*] pointed out two more cases of embankments on highly plastic clays in Brazil in which vane corrections also did not apply. It was then argued[3] that this might be related to the clay's organic matter content. However, in two out

[3] Leroueil, S., personal communication, 1987.

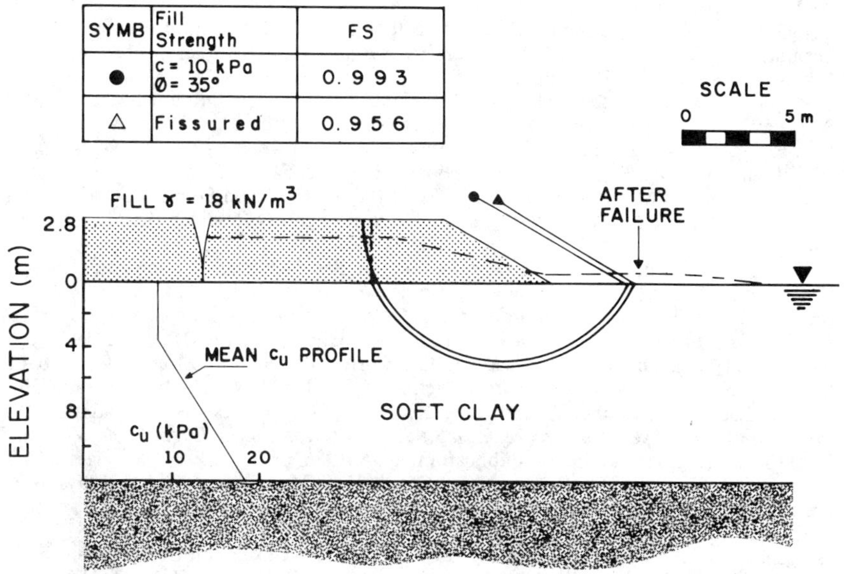

FIG. 13.—*Stability analysis of an embankment on Rio de Janeiro clay.*

of these three cases the organic content was low, around 5%, therefore not supporting this suggestion.

Conclusions

The FV equipment using an unprotected vane through a prebored hole is still used in Brazil and certainly in other countries. Results of the investigation carried out with it in the Rio de Janeiro clay showed that the soil-rod and internal frictions affected significantly the results and are difficult to eliminate.

A frictionless vaneborer device was built according to Cadling and Odenstad [2], leading to results showing less scattering of data. A new FV mean profile was obtained and evaluated in an analysis of an embankment failure in the same test site, yielding a safety factor very close to one at failure, therefore contradicting widely accepted FV strength corrections. The reasons for this are still being investigated.

Acknowledgments

The authors acknowledge the financial support given by the Brazilian Road Research Institute during the research program on the behavior of Brazilian soft clays. Thanks are also due to Dr. E. M. Palmeira who revised the text and made helpful comments.

References

[1] Lemasson, H., "Ensemble carottier à piston stationnaire, scissométre," *Remblais sur sols compressibles,* spécial T, Laboratoire des Ponts et Chaussées, 1973, pp. 276–281.

[2] Cadling, L. and Odenstad, S., "The Vaneborer: an Apparatus for Determining the Shear Strength of Soils Directly in the Ground," *Proceedings of the Royal Swedish Geotechnical Institute,* No. 2, 1950, 87 pp.

[3] Ortigão, J. A. R., Werneck, M. L. G., and Lacerda, W. A., "Embankment Failure on Clay Near Rio de Janeiro," *Journal of Geotechnical Engineering, Transactions of the ASCE,* Vol. 109, No. 2, 1983, pp. 1460–1479.

[4] Ortigão, J. A. R. and Palmeira, E. M., "Geotextile Performance at an Access Road on Soft Ground Near Rio de Janeiro," *Proc. 2nd International Conference on Geotextiles,* Las Vegas, NV, Vol. 1, Section 3B, 1982, pp. 353–358.

[5] Costa-Filho, L. M., Werneck, M. L. G., and Collet, H. B., "The Undrained Strength of a Very Soft Clay," *Proceedings of the 9th ICSMFE,* Tokyo, Japan, Vol. 1, 1977, pp. 79–82.

[6] Vey, E. and Schlessinger, L., "Soil Shear Tests by Means of Rotating Vanes," *Proceedings of the 29th Annual Meeting,* Highway Research Board, 1949, pp. 544–553.

[7] Bennett, G. B. and Mecham, J. G., "The Use of the Vaneborer on Foundation Investigation of Fill," *Proceedings of the 3rd Record Annual Meeting,* Highway Research Board, 1953, pp. 486–496.

[8] Collet, H. B., "Ensaios de palheta de campo em argilas da Baixada Fluminense," MSc thesis, Universidade Federal do Rio de Janeiro, Rio de Janeiro, Brazil, 1978, 243 pp.

[9] Bjerrum, L., "Embankments on Soft Ground," *ASCE Specialty Conference on Performance of Earth and Earth Supported Structures,* Purdue University, Lafayette, IN, Vol. 2, 1972, pp. 1–54.

[10] Azzouz, A. M., Baligh, M. M., and Ladd, C. C., "Corrected Field Vane Strength for Embankment Design," *ASCE Journal of Geotechnical Engineering,* Vol. 109, No. 5, 1983, pp. 730–733.

[11] Aas, G., Lacasse, S., Lunne, T., and Høeg, K., "Use of In Situ Tests For Foundation Design," *Proceedings of the ASCE Conference on In Situ Tests in Geotechnical Engineering,* Virginia Tech, Blacksburg, Geotechnical Special Publication No. 6, 1986, pp. 1–30.

[12] Ortigão, J. A. R., Coutinho, R. Q., and Sant'anna, L. A. M., "Failures on Soft Clays in Brazil," Discussion submitted to the *International Symposium on Geotechnical Engineering of Soft Soils,* Mexico, to be published in Vol. 2), 1987.

[13] Ortigão, J. A. R., Werneck, M. L. G., and Lacerda, W. A., Closure on Embankment Failure on Clay near Rio de Janeiro, *ASCE Journal of Geotechnical Engineering,* Vol. 111, No. 2, pp. 262–264.

Marius Roy[1] *and André Leblanc*[2]

Factors Affecting the Measurements and Interpretation of the Vane Strength in Soft Sensitive Clays

REFERENCE: Roy, M. and Leblanc, A., "**Factors Affecting the Measurements and Interpretation of the Vane Strength in Soft Sensitive Clays,**" *Vane Shear Strength Testing in Soils: Field and Laboratory Studies, ASTM STP 1014,* A. F. Richards, Ed., American Society for Testing and Materials, Philadelphia, 1988, pp. 117–128.

ABSTRACT: Vane tests have been carried out in the field and in the laboratory to investigate some factors affecting the measurements and interpretation of the test, specially the disturbance caused by the insertion of the vane into the clay and the time and shear velocity effects on the strength. Strength rotation curves and failure diagrams have been investigated in the laboratory. The results have shown that the insertion of the vane produces disturbance and generates pore pressure so that the vane strength measured is reduced while time effect leads to an increase in shear strength. Study of the failure diagram shows that the actual shear surface is larger than the cylindrical surface defined by the diameter of the blades.

KEY WORDS: vane, shear strength, clays, field tests, laboratory tests

The vane shear test has been developed in the late 1940s, and a review of its history is presented by Flodin and Broms [*1*]. Since that time, it has been the most commonly used apparatus for in-situ measurement of undrained shear strength of soft soils.

Studies carried out by Osterberg [*2*], Eden and Hamilton [*3*], Eden [*4*], Flaate [*5*], La Rochelle et al. [*6*], and Wiesel [*7*] have shown that many factors can influence the undrained shear strength measured by the vane apparatus, and some of these factors are associated with the interpretation of the test results. Principally, the following hypotheses govern the interpretation:

1. The intrusion of the vane into the clay causes essentially negligible disturbance, meaning that the undrained shear strength is not affected.
2. No consolidation occurs before or during application of the torsional torque.
3. The failure surface is defined by the cylindrical movement of the vane blade edges.
4. The shear strength is mobilized simultaneously over the entire surface of the cylinder and is the same in the vertical and horizontal directions.

With a view to assess the applicability or the limitations of these hypotheses, several series of tests were conducted at two sites in Quebec. This paper presents and discusses the results of these tests. In addition, the X-ray failure diagram taken during laboratory tests was examined, and a certain form of progressive failure was observed.

[1] Professor, Laval University, Department of Civil Engineering, Quebec G1K 7P4, Canada.
[2] Engineer, Laboratoire de Construction de Québec Inc., Quebec GIN 2E6, Canada.

Test Program

In order to study the main factors having a potential influence on measurement of undrained shear strength in clays, a program of in-situ tests was carried out. The following points were considered:

(1) the influence of the vane blade thickness,
(2) the influence of time between intrusion of the vane and shearing, and
(3) the influence of shear rate.

In addition, a laboratory program was undertaken to study the form of the stress-strain curve and the failure surface that develops during testing.

In-situ Tests

In each of the two sites studied, the program involved an equal number of soundings, and identical procedures were followed.

To measure the influence of vane blade thickness or disturbance, we carried out six soundings; one using a standard Nilcon vane and the other ones using home-made vane blade thicknesses of 0.38, 0.78, 1.02, 1.57, and 2.34 mm. Each profile was obtained by performing tests at 0.5-m intervals. For close monitoring of the measurements by means of the standard vane, one test in three was carried out with the standard vane in each sounding.

The influence of consolidation time was studied with the standard vane. Tests in each of the five soundings were conducted at 0.5-m intervals and at depths between 7 and 15 m. The standard test and modified test incorporating a waiting period were regularly alternated. The waiting period ranged from 15 to 10 000 min (7 days); the selected time intervals being 15, 60, 1440, 2880, 5000, 7200, and 10 000 min. In the standard vane the delay time was of 1 min.

The portion of study dealing with shear rate was also carried out with the standard vane. The vane was introduced by pushing, and the shearing was applied at rates of 0.0066, 0.028, 0.042, 0.220, and 0.501 degrees per second. Each sounding at depth between 7 and 11 m involved tests at 0.5-m intervals. While shear rate was changed for each test, the program provided for one test in three to be performed at the selected standard rate of 0.22°/s. Four to five soundings were conducted at each test site to provide a proper coverage of the shear rate aspect.

Laboratory Tests

The laboratory program was designed to examine two specific points: (1) the form of the stress-strain curve was observed by measuring the torque with a torque cell connected to an X-Y recorder, and (2) the development of the failure mode around the vane was examined using X-rays during insertion and shearing of the vane in Arvida clay, deposit of Champlain clay, and in an artificial material simulating Champlain clay [*8*].

Description of Test Sites

All field results were obtained by conducting the same series of tests at the Saint-Louis de Bonsecours site in Richelieu County and the Saint-Alban site in Portneuf County. The soil properties of the sites are briefly described below.

Saint-Louis de Bonsecours is located approximately 80 km east of Montreal along the Yamaska River, in the immediate vicinity of landslides that occurred in 1945 and 1968

[9]. The stratigraphic profile is made up of highly organic topsoil 1.65 m thick over medium sand approximately 1.20 m thick. Below that is a layer more than 15 m deep of silty clay that is fairly homogeneous and grayish-blue in color.

The Saint-Alban site is at the east end of the village of the same name, which is located approximately 70 km west of Quebec City on the north shore of the St. Lawrence River. The stratigraphic profile consists of 1.20 m of very fine sand, overlying more than 15 m of silty clay. Examination of samples taken at various depths indicates that the clay is highly organic and contains shells.

The properties of the clays under study are presented in Table 1, while Fig. 1 shows the related standard vane shear strength profiles at depths between 7 and 15 m. These profiles show that strength increases with depth. However, very apparent breaks at depths of 11 and 12.5 m indicate that the Saint-Alban deposit is not homogeneous and includes layers with slightly different properties.

Description of Measurement Apparatus

The entire field program was carried out with the Nilcon vane developed at Chalmers University of Technology in Sweden. For the purposes of the study, the Nilcon unit was modified by the addition of a Francon electric motor with a torque rating of 1150 kg·cm and a gear transmission permitting adjustment of vane rotation rate to values between 0.01° and 2°/s.

To study the importance of the disturbance of the clay during positioning of the vane, a series of five different vanes were manufactured from stainless steel. To eliminate the volume and size effects [6], the vane should have the same volume as the 130- by 65-mm vane manufactured by Nilcon and used as a standard in this study. Accordingly, rectangular vanes were produced with blades of different thicknesses (Fig. 2) and shaped in such a manner that their horizontal cross-sectional area, or volume, remains constant.

The laboratory program was carried out using a Wykeham-Farrance vane apparatus modified to yield accurate measurements of vane rotation during testing. In all tests, the 20- by 15-mm vane was retained as our standard with a delay time of less than 1 min and a rate of rotation of 17°/min. Many other rate of rotations were available from a gear box and an electric motor.

TABLE 1—*Typical values of the properties of the clays studied.*

Soil Properties	Saint-Alban Site, Depth 6 m	Saint-Alban Site, Depth 12 m	Saint-Louis de Bonsecours Site, Depth 6 m	Saint-Louis de Bonsecours Site, Depth 12 m
Water content, %	35	39	69	62
Liquid limit, %	22	39	42	42
Plastic limit, %	16	21	22	28
Plasticity index	6	18	19	13
Liquidity index	2.6	1.7	2.0	2.5
Clay content, %	65	60	80	69
Silt content, %	32	37	20	30
Sand content, %	3	3	0	1
Unit weight, kN/m^3	15.7	16.5	16.2	17
Sensitivity (laboratory vane)	4	14	8	8
Salt content, g/L	3	6	0.2	0.4

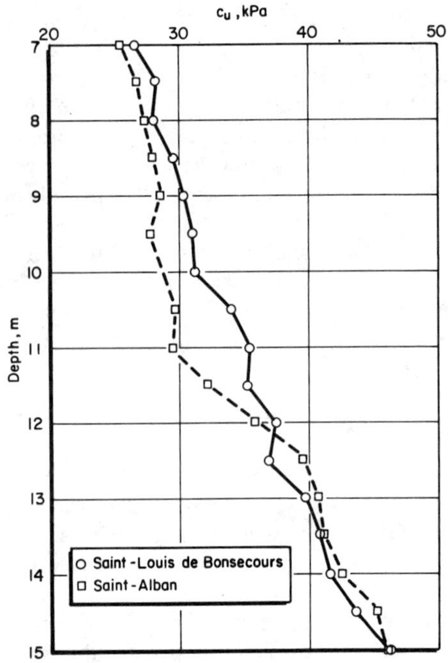

FIG. 1—*Results of vane shear strength.*

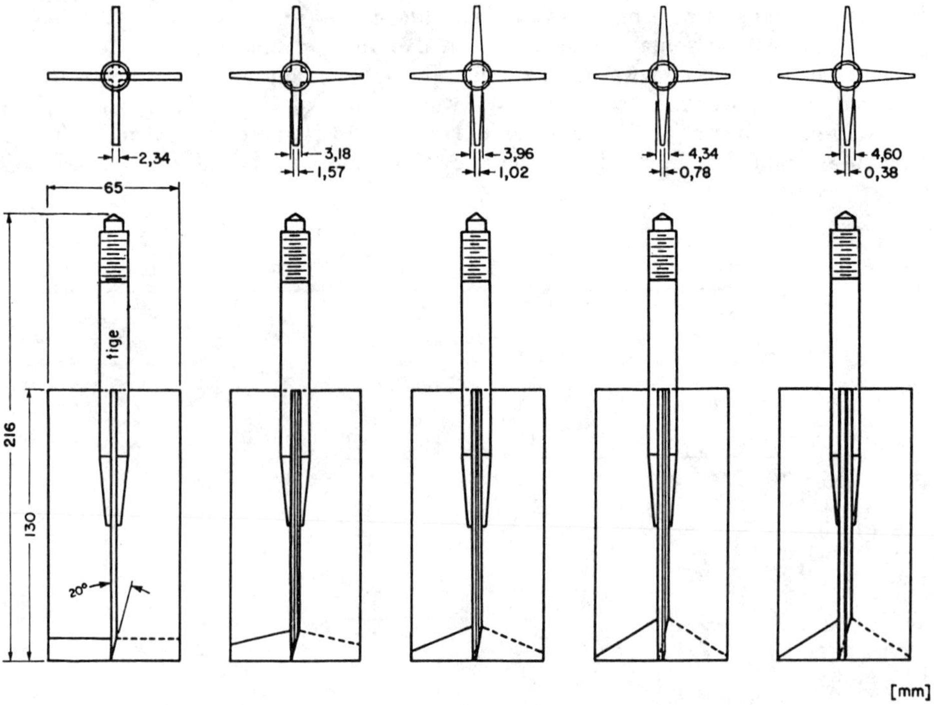

FIG. 2—*Vane dimensions at constant volume.*

Influence of Disturbance

The intrusion of the vane into the clay will produce disturbance and changes, over a given volume, in the natural properties and structure in the clay. According to Cadling and Odenstad [10], the disturbance around the vane blades can be assessed by means of the following expression:

$$\alpha = 4e/\pi d$$

where α is the perimeter ratio in relation to blade thickness e over the circumference of the cylindrical vane blade edges of diameter d.

For all the vane shown in Fig. 2, the test results are presented in Fig. 3 in the form of a relation between α and c_u/c_{uo} where c_u is the shear strength value associated with a given blade thickness and c_{uo} is the undrained shear strength measured with the standard vane.

Each point represents the mean value of five or six measurements. The linear form of the relation clearly indicates that disturbance varies with blade thickness. It can also be seen that disturbance varies with clay type, since the results are different for each of the two sites studied. According to these results, the shear strength extrapolated to zero disturbance, or zero blade thickness, would be approximately 9 and 6.5% higher than the standard vane strength for the clays of Saint-Louis and Saint-Alban, respectively.

Examination of the presented results clearly shows that the insertion of the vane created a certain amount of disturbance of the clay and that the measured shear strength is affected differently depending on the properties of the clays encountered and the thickness of the vane blades used. Earlier study by La Rochelle et al. [6] on the Saint-Louis de Bonsecours site has shown a 15% increase in undrained shear strength extrapolated for a zero blade thickness. In their study, as the blade increase was also associated to an increase of the horizontal cross-sectional area, there is more disturbance by the intrusion of the vane into the soil mass.

These results show clearly that disturbance can be reduced by a proper design of the vane blades without compromising on its sturdiness. Some efforts should be made to sensitize the manufacturers and adjust standards to such observations.

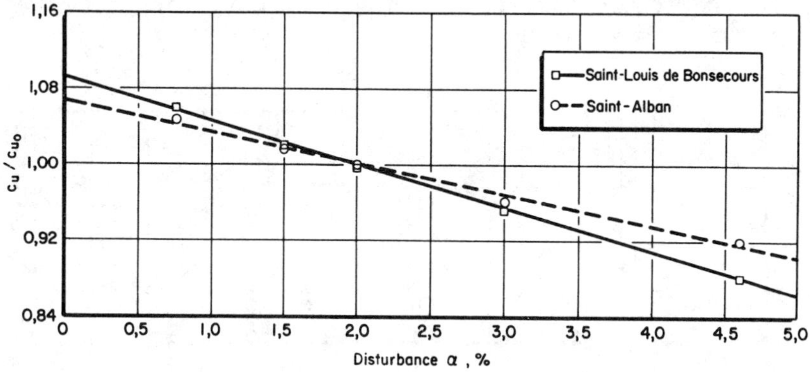

FIG. 3—*Extrapolation of vane strength for zero blade thickness.*

Consolidation

The influence of consolidation around the vane blades was measured by extending the waiting period between intrusion of the vane and the beginning of shearing. As the waiting period, which is normally of 1 min in the standard test, was prolonged to up to 10 000 min, the results are presented in semilogarithmic form (c_u/c_{uo} as a function of log t). The values of c_{uo} were obtained by the standard procedure. All of these tests were made with the standard vane blades, but with waiting periods ranging from 15 to 10 000 min.

Figure 4 shows the results obtained for each of the sites. Each point on the diagram represents the mean value of three to six tests conducted for the selected waiting times. It can be noted that the waiting period has a very significant effect on shear strength and that the magnitude of the time effect is not the same for different types of clay. Saint-Louis de Bonsecours clay exhibits a shear strength increase of approximately 20% after 1 h of waiting. Thereafter, shear strength continues to rise slowly to a maximum of 21.5%. Saint-Alban clay, on the other hand, shows a less pronounced shear strength increase that takes longer to develop. It would appear that a minimum waiting period of 4 to 5 hours is required to obtain the full benefit of the time effect.

Introduction of a waiting period before shearing involves a nonnegligible time effect or consolidation, as it is demonstrated by the results in Fig. 4. Torstensson [*11*] also measured this time effect, on two sites in Sweden, and showed an increase comparable to ours, 17 and 20%. Similarly, Flaate [5] noted an increase of approximately 19% for Norwegian clay.

The variation in shear strength with time is undoubtedly associated with (1) the excess pore pressures generated at the time of insertion and (2) the complex consolidation mechanism taking place in the clay around the vane. The results show that the vane test is conducted in a unreconsolidated soil, partially disturbed by the insertion of the vane. The consolidation effect introduces a change in the effective stresses in the soil around the vane and a substantial increase in the undrained shear strength.

Shear Rate

In this series of tests, the applied shear rate was modified to cover a range between 0.01 and 2°/s. All tests were conducted with a standard vane using the standard delay time of one minute.

The results are shown in Fig. 5 and are presented in the form, log v versus c_u/c_{uo}, where c_{uo} is the conventional measured shear strength. It can be seen that for shear rates below the standard rate of 0.22°/s the ratio c_u/c_{uo} increases significantly. However, for rates above

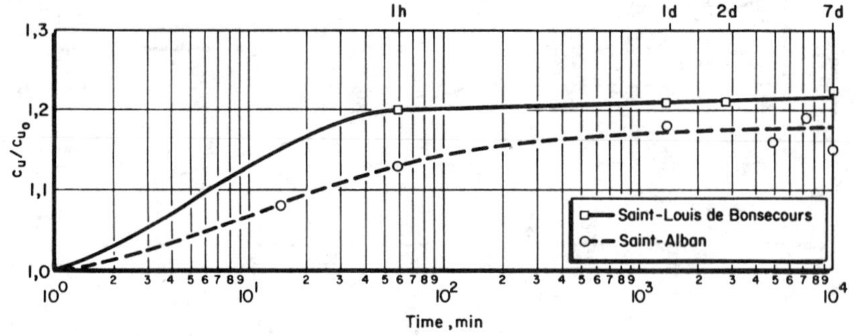

FIG. 4—*Consolidation effect on the undrained shear strength.*

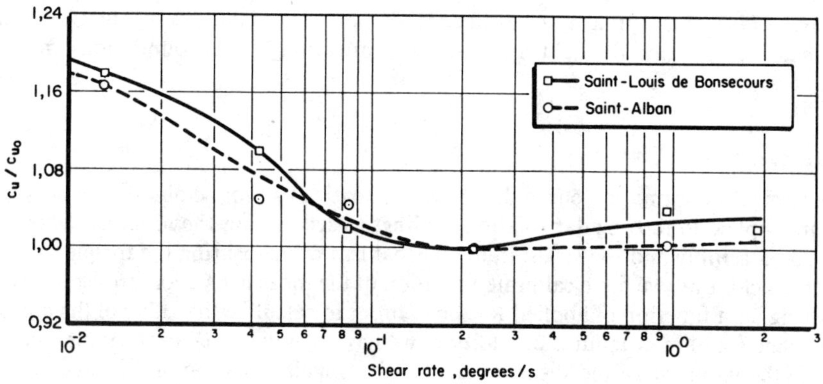

FIG. 5—*Shear rate effect on the undrained shear strength (Canadian clays).*

0.22°/s, the increase is not significant for Saint-Alban clay and approximately 3% for Saint-Louis clay.

The results presented in Fig. 5 have been obtained without allowing reconsolidation after the insertion of the vane in the clay. Because of this procedure we have probably measured undrained strength over the shear rate of 0.2°/s and partially drained strength for slower shear rates.

Above 0.2°/s, our results show a very small increase of the shear strength with the increase of the shear rate (viscous effect). This increase is smaller than the values obtained by Skempton [*12*], Cadling and Odenstad [*10*], and Aas [*13*] in similar tests. After a consolidation period of 15 h, Wiesel [*7*] has conducted consolidated vane tests that clearly show the rate effect on shear strength as shown in Fig. 6.

In the range of partially drained tests, the consolidation effect is so important that we are not able to measure the shear rate effect. This finding is supported by the results of Fig. 4 and indicates that there is a change in effective stresses on the failure plane during the test.

These results show the importance of the shear rate and the consolidation effects on the shear strength. The critical shear stress is time-dependent (Torstensson [*11*]) and according

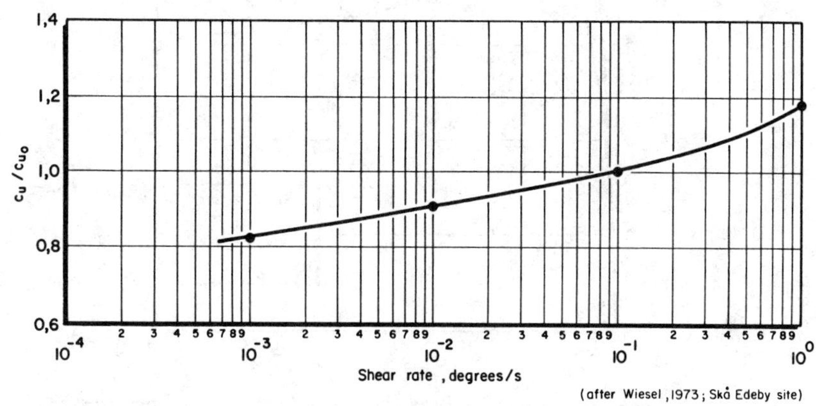

FIG. 6—*Shear rate effect on the undrained shear strength (Swedish clay).*

to Bjerrum [*14*], this primarily rate effect explains a part of the discrepancy between the vane shear strength and the field shear strengths prevailing under foundations and in natural slopes.

Resistance-Rotation Curves

The strength-deformation curve during vane shearing cannot be obtained unless corrections are applied to the angular rotation readings to account for the torsional deformation of the rod assembly and the strain in the mechanism for measuring the torque.

For the field tests, we have calibrated rotation of the measuring apparatus (rod assembly and spring) as a function of applied torque in order to calculate the value of the correction for the vane rotation at failure. In addition, we have conducted laboratory measurements to define the variation of the shear resistance with angular rotation of the vane.

For all our field results, the mean rotation to the failure was found to be 5° for the Saint-Louis site and 3.5° for the Saint-Alban site, reflecting the difference in the nature of the soil.

The series of laboratory tests on Saint-Alban clay was conducted following the usual standard procedure. The results presented in Fig. 7 clearly show the increase of the undrained shear resistance, which peaks between 3° and 4° and drops towards a lower value at large deformation.

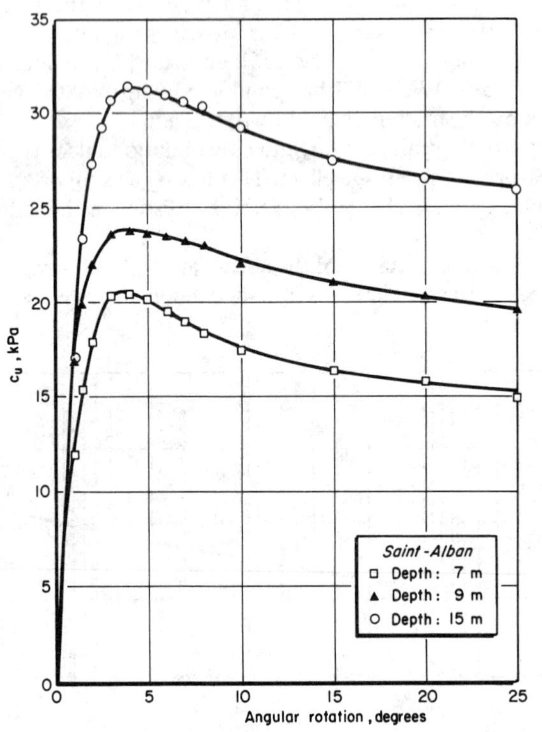

FIG. 7—*Strength-angular rotation curves measured from the laboratory vane test.*

The results of the vane tests conducted in the laboratory provide good confirmation for the in-situ tests: calculated rotation at the point of failure falls into a narrow range between 3 and 6°. During these tests, we examined the failure mode as discussed below.

Failure Mode

The failure mode was observed by means of a series of X-ray photographs taken before insertion of the vane into the specimen, after insertion, and at various vane rotation angles during shear.

Development of this technique required extensive tests, which will not be reviewed in this paper. For our immediate purposes, we will limit ourselves to presenting typical results from which it is possible to follow the development of the failure diagram during shear.

To facilitate the presentation, the results obtained with the artificial material are grouped in Fig. 8. Figure 8*a* shows the vane positioned over the specimen; in Fig. 8*b*, the same vane has been introduced into the specimen. In both cases, the results are the same; insertion of the vane unit does not seem to have caused any particular effect. In Fig. 8*c*, the rotation is 4°, which according to the results discussed above, just precedes the peak. At this point, there is yet no sign of the onset of shearing; obviously, the same can be said for lower rotation values. In Fig. 8*d*, which corresponds to a rotation of 12°, a whitish area can be

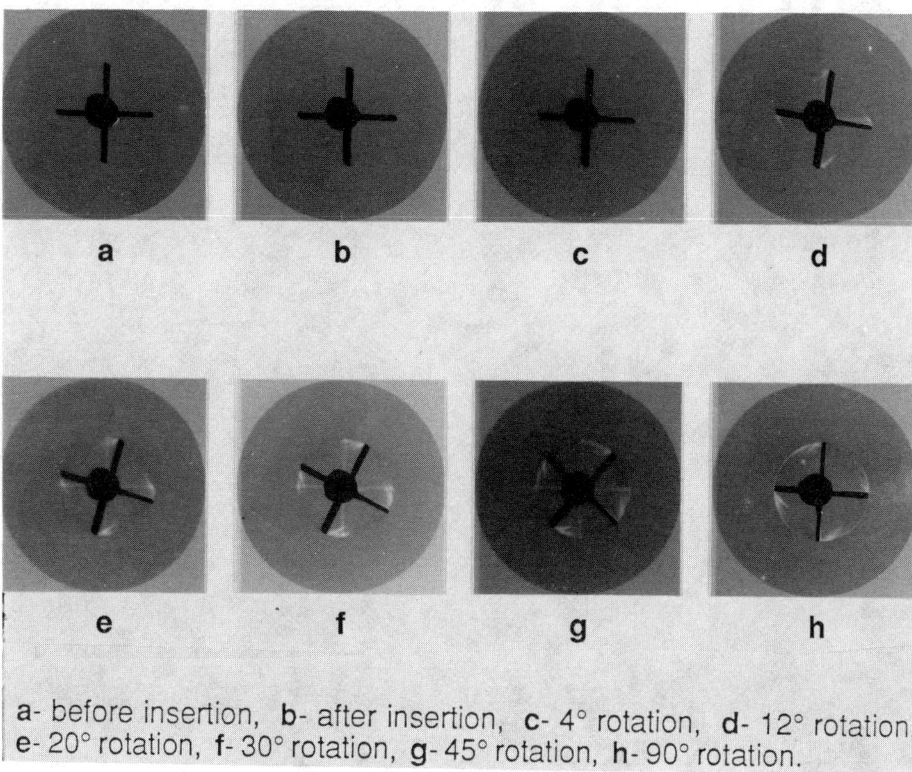

FIG. 8—*Development of the shear failure observed by X-rays, artificial material.*

seen behind each of the vane blades, which are turning in a clockwise direction. This indicates slight separation of the clay behind the blades. In the next two views (*e* and *f*), corresponding to 20° and 30°, the light colored areas are expanding and the beginnings of a failure shape, which is not yet circular, can be seen. In *g*, at 45° rotation, these areas extend over the entire sheared zone and clearly show that a circular failure surface defined by the ends of the blades has developed. The failure is completely defined when the rotation reaches 90° (Fig. 8*h*).

These results suggest that failure develops near the edges of the blades and at the same time highly disturbed soil appeared in the back of the blade. As the disturbed zone created around the blades at the time of the insertion increases with the thickness of the blades and extended during shearing, the highly disturbed soil is developed more rapidly and the strength measured is then reduced. In all cases, the simplify failure surface should be defined by a diameter slightly larger than the diameter of the blades.

The same study was carried out on sample of Arvida clay, which is highly sensitive and strongly overconsolidated; the results are shown in Fig. 9. During the insertion (Fig. 9*b*), fissures starting on the sides and ends of the blades and crossing the entire specimen were noted (Fig. 9*b*). The subsequent phases are clearly influenced by the presence of the fissures but are otherwise similar to those described above for the artificial material. New fissures appeared in the specimen as shearing progressed.

These results are believed to be representative of the development of the failure mode

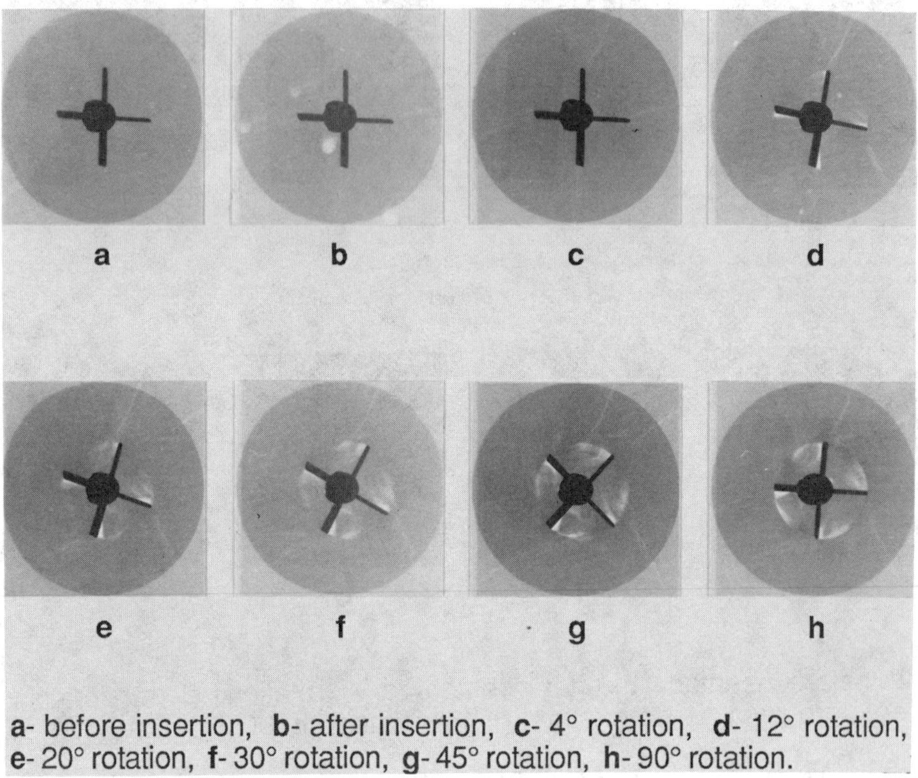

FIG. 9—*Development of the shear failure observed by X-rays, natural stiff clay.*

during shear at large strains. However, the X-ray pictures have only a remote relation with the failure conditions derived from the shear resistance-rotation curves, which show that failure is developing at small angular strains of the order of 4°. This may be due to the fact that this technique does not detect small strains in the soil.

The results observed using X-rays and presented in Figs. 8 and 9 show that the failure surface develops in the course of shearing and that a large zone of highly disturbed soil appeared in the same time. This failure mode gives support to a simplify shear surface of a cylindrical form with a diameter slightly larger than that of the vane. It would not seem unreasonable to suggest that the diameter of the cylindrical shear surface is approximately 5% larger than that defined by the diameter of the vane.

Arman et al. [*15*] also studied the failure diagram by carrying out X-ray examinations of section taken from soil subjected to vane shearing. They report that the shear surface can be identified by a thin disturbed layer outside the surface circumscribed by the blade ends. Thus, the effective diameter of the shear surface can be expressed as Bd, where d is the diameter of the vane and B is a coefficient to be determined experimentally. Skempton [*12*] studied this problem and suggested a value of $B = 1.05$. The use of such a coefficient in the calculation of the shear strength leads directly to a decrease in shear strength of the order of 16%. With this in mind, it must be remembered that the failure mode that develops during vane testing is of major importance for the correct interpretation of the tests.

Examination of the above results also shows that the shear strength measured by means of the vane test can be underestimated because of the disturbance and pore pressures generated by the insertion of the vane, and overestimated because of the assumption that the shear surface is defined strictly by the diameter of the vane.

Conclusion

The examination of the hypotheses at the basis of the interpretation of vane tests leads us to formulate the following conclusions:

- The insertion of the vane creates a certain amount of disturbance, which is linked to blade thickness and to the type of clay under study. Use of a better designed vane should therefore permit further reduction of disturbance, with a 5 to 10% increase in measured shear strength.
- The gain in shear strength obtained with the time delay between insertion of the vane and the beginning of shearing supports the fact that displacement of soil by the blades will set up important pore pressure in some cases.
- The time delay to obtain full consolidation is an essential condition to succeed in the measurement of the effect of time to failure in field tests on the shear strength (undrained and drained conditions).
- The rate effect (time to failure) measured after reconsolidation of the soil around the blades is very important and leads to the identification of a correction that must be introduced to ensure the stability of foundations and natural slopes.
- The maximum delay time of 1 min must be the value to introduce in the standard.
- I believe that the failure mode is not yet well known and that other investigations are required to clarify the interpretation of the vane test.

Acknowledgments

Dr. P. La Rochelle, Université Laval, is acknowledged for his valuable discussion with the authors. The present study was carried out at the Geotechnical Section of Université

Laval under operating grants and research grants from the ministère de l'Education, Gouvernement du Québec, and the Natural Sciences and Engineering Research Council of Canada.

References

[1] Flodin, N. and Broms, B., "Historical Development of Civil Engineering in Soft Clays," *Soft Clay Engineering,* Brand and Brenner, Eds., 1981, Elsevier, pp. 25–156.

[2] Osterberg, J. D., *Symposium on Vane Shear Testing of Soils, STP 193,* American Society for Testing and Materials, Philadelphia, 1956, pp. 1–7.

[3] Eden, W. J. and Hamilton, J. J., "The Use of a Field Vane Apparatus in Sensitive Clay," *Symposium on Vane Shear Testing of Soils, STP 193,* American Society from Testing and Materials, Philadelphia, 1956, pp. 41–53.

[4] Eden, W. J., "An Evaluation of the Field Vane Test in Sensitive Clay," *Vane Shear and Cone Penetration Resistance of In-Situ Soils, STP 399,* American Society for Testing and Materials, Philadelphia, 1966, pp. 8–17.

[5] Flaate, K., "Factors Influencing the Results of Vane Tests," *Canadian Geotechnical Journal,* Vol. 3, 1966, pp. 18–31.

[6] La Rochelle, P., Roy, M., and Tavenas, F., "Field Measurements of Cohesion in Champlain Clays," *Proceedings of the 8th International Conference on Soil Mechanics and Foundation Engineering,* Moscow, Vol. 1.1, 1973, pp. 229–239.

[7] Wiesel, C. E., "Some Factors Influencing In Situ Vane Test Results," *Proceeding of the 8th International Conference on Soil Mechanics and Foundation Engineering,* Moscow, Vol. 1.2, 1973, pp. 475–480.

[8] Tavenas, F., Roy, M., and La Rochelle, P., "An Artificial Material for Simulating Champlain Clays," *Canadian Geotechnical Journal,* Vol. 10, No. 3, 1973, pp. 489–503.

[9] Lefebvre, G. and La Rochele, P., "The Analysis of Two Slope Failures in Cemented Champlain Clays," *Canadian Geotechnical Journal,* Vol. 11, No. 1, 1974, pp. 89–109.

[10] Cadling, L. and Odenstad, S., "The Vane Borer, an Apparatus for Determining the Shear Strength of Clay Soils Directly in the Ground," *Royal Swedish Geotechnical Institute, Proceedings,* Vol. 2, 1950, pp. 1–87.

[11] Torstensson, B. A., "Time Dependent Effects in the Field Vane Test," *Proceedings of the International Symposium on Soft Clays,* Bangkok, 1977, pp. 387–397.

[12] Skempton, A. W., "Vane Test in the Alluvial Plains of River Firth near Grangemouth," *Géotechnique,* Vol. 1, No. 2, 1948, pp. 111–125.

[13] Aas, G., "A Study of the Effect of Vane Shape and Rate of Strain on the Measured Values of in Situ Shear Strength of Clays," *Proceedings of the 6th International Conference on Soil Mechanics and Foundation Engineering,* Montréal, Vol. 1, 1965, pp. 141–145.

[14] Bjerrum, J., "Embankments on Soft Ground," *Proceedings of the Specialty Conference of Earth and Earth Supported Structures, ASCE,* Vol. 2, 1972, pp. 1–54.

[15] Arman, A., Poplin, J. K., and Ahmad, N., "Study of the Vane Shear," *Proceedings of Specialty Conference on In Situ Measurement of Soil Properties,* Raleigh, Vol. 1, 1975, pp. 93–120.

Part III: Laboratory Vane Theory and Interpretation

Daizo Karube,[1] Satoru Shibuya,[2] Takao Baba,[3] and Yasuhiro Kotera[4]

Analysis of a Vane Test Based on Effective Stress

REFERENCE: Karube, D., Shibuya, S., Baba, T., and Kotera, Y., **"Analysis of a Vane Test Based on Effective Stress,"** *Vane Shear Strength Testing in Soils: Field and Laboratory Studies, ASTM STP 1014,* A. F. Richards, Ed., American Society for Testing and Materials, Philadelphia, 1988, pp. 131–149.

ABSTRACT: A laboratory investigation has been made into rate effect on undrained strengths of a reconstituted clay. A "cylinder shear" apparatus has been developed recently to examine shear characteristics of the vane test in a simplified manner. A suite of laboratory tests has been carried out with this apparatus that included pore-pressure measurements on the shear surface. A series of laboratory vane tests has also been performed to be compared with those results. These test results allowed an interpretation to be made of the vane test rate effect in terms of effective stress and within the framework of the original Cam clay model. The undrained strength anisotropy of the clay was also examined in tests in which specimens from the same block sample, but cut out with different angles relative to the plane of sedimentation, were sheared in a direct shear device.

KEY WORDS: laboratory vane shear tests, direct shear tests, effective stress, constitutive equation

It is widely acknowledged that the undrained shear strength of soil S_u is significantly influenced by shearing mode (for example, plane strain or axisymmetric shear) and the rate of straining. Bjerrum [*1*], for instance, has developed empirical factors to correct the strengths obtained from vane tests, $(S_u)_{\text{vane}}$, to give values appropriate for the analysis of stability problems with embankment on soft ground. Bjerrum proposed that S_u mobilized in field is

$$(S_u)_{\text{field}} = (S_u)_{\text{vane}}\,\mu = (S_u)_{\text{vane}}\,\mu_R\,\mu_A \tag{1}$$

where μ_R and μ_A are factors correcting for the rate effect and for soil anisotropy, respectively. According to Bjerrum, these correction factors are closely related to the plasticity index (PI) of the soil.

Considering μ_R, the in-situ test results by Torstensson [*2*] support Bjerrum's postulate that the rate effect is more significant as PI increases in value. La Rochelle et al. [*3*] have observed that the residual strengths determined in triaxial (UU) tests are scarcely affected by the rate of shearing (that is, $\mu_R \approx 1$), irrespective of plasticity index of the soil. This matches the field vane test results reported by Torstensson, which also showed the residual strength to be free from time effects.

[1] Associate professor, Department of Civil Engineering, Kobe University, Kobe, Japan.
[2] Research assistant, Institute of Industrial Science, University of Tokyo, formerly Imperial College of Science and Technology, London, England.
[3] Engineer, Shimizu Construction Company, Ltd., Tokyo, Japan.
[4] Government officer, Osaku Local Government, Osaka, Japan.

While it is true that S_u in saturated soil is not dependent on changes in applied total stress, effective stress analyses with pore-pressure measurements are more meaningful than total stress analyses in many field problems. When analyzing the problem in a systematic way, there are advantages in using effective stress strength parameters that are considerably more stable than the total stress undrained shear strengths. For example, the angle of shearing resistance ϕ' is little influenced by drainage conditions and the rate of shearing, though it varies slightly with the shearing mode. Effective stress analyses are thus often used when back analyzing such occurrences as slope failures. However the effective stress approach loses appeal when considering field problems where the magnitudes of pore pressures at failure have to be predicted before construction. It is recognized at the outset that reliable vane test data for field design problems can only be derived through the use of a sophisticated and versatile constitutive soil model.

The objective of this paper is to analyze the vane test using an effective stress method. It has been experimentally demonstrated that the shear resistance mobilized in the vane test is approximately the same as that determined when a cylinder with a rough surface, mounted vertically in a clay specimen, is rotated along the long axis of the cylinder. An analysis in such a test is made to predict the shear resistance mobilized around the cylinder with respect to the angle of rotation of the cylinder, and this is compared to experimentally observed data. The paper also describes the rate effects that have been examined in tests with varying speed of rotation of the cylinder.

Soil Used

The soil used for all types of tests was reconstituted alluvial marine clay that had originally been sampled from the north of Osaka Bay in Japan. The clay was mixed thoroughly in batches with distilled water to give a water content of about 120%. That was then deaired using a vacuum pump.

The index properties are summarized as $G_s = 2.72$, $w_L = 107\%$, and $I_p = 70\%$, and the clay fraction (<5 μm) equal to 74%.

Tests Performed

Four series of "cylinder shear test," vane shear test, direct shear test, and triaxial compression tests were carried out in the laboratory.

In vane tests, it is usually assumed that the torque measured can be attributed to the shear resistance mobilized on the right cylindrical shear surface whose shape coincides with the dimensions of the vane blades. In order to examine the soil behavior on this postulated surface, some attempts have been made in the laboratory to record the pore-pressure response of the failing soil using a pore-pressure probe mounted at the tip of the vane blade. However the measurement of pore pressure in such a test is likely to be strongly affected by the time lag associated with a pressure probe of very small surface area. Theoretical analyses have also been made using the finite-element method, but these demonstrate the difficulties of studying the mechanisms of failure when the shear deformation in the surrounding soil is highly complicated.

In a standard vane with a blade aspect ratio H/D of 2 the area of vertical component of the shear surface in the right cylinder is four times larger than that of the horizontal component. The main part of the overall torque is therefore provided by shear resistance mobilized on vertical section of the shear surface. The basic idea of the cylinder shear test is to examine this behavior in a simplified way using a cylindrical shear rod, which is mounted vertically in the soil specimen and rotated around the long axis. In such a test, the stress distribution in the surrounding soil is far simpler than that in vane tests, since no interaction between vertical and the horizontal shear mechanism exists. Additionally, the pore pressures developed in the test can be readily monitored if the central section of the shear rod is instrumented with a porous stone.

Cylinder Shear Test

The cylinder shear test consists of two stages of consolidation and shear. Figure 1 shows a schematic diagram of specimen setup during consolidation of the cylinder shear test. The consolidation ring, which is 60 mm in diameter and 70 mm high, has a rectangular outside shape. As can be seen in the figure, the consolidation ring is floated by four coil springs at the corners. The ring is therefore allowed to move vertically by adjusting the screw placed on top of each spring. At the central length of the ring, a circular porous stone is buried in the inner wall. This porous stone is connected to a drainage pipe and a pressure gage (Fig. 2). Total lateral pressure acting on the wall can be measured using a pressure gage instrumented in the close vicinity of the circular porous stone.

A shear rod of 15 mm diameter is located in the center of the floating consolidation ring through the top loading plate and the base. The perspex shear rod is instrumented on its central section with a porous stone, which is indicated as pore-pressure probe in Fig. 1. A pressure gage is connected to it so as to measure the pore-pressure response on surface of the shear rod. A total of 16 V-shaped grooves are engraved around the shear rod including

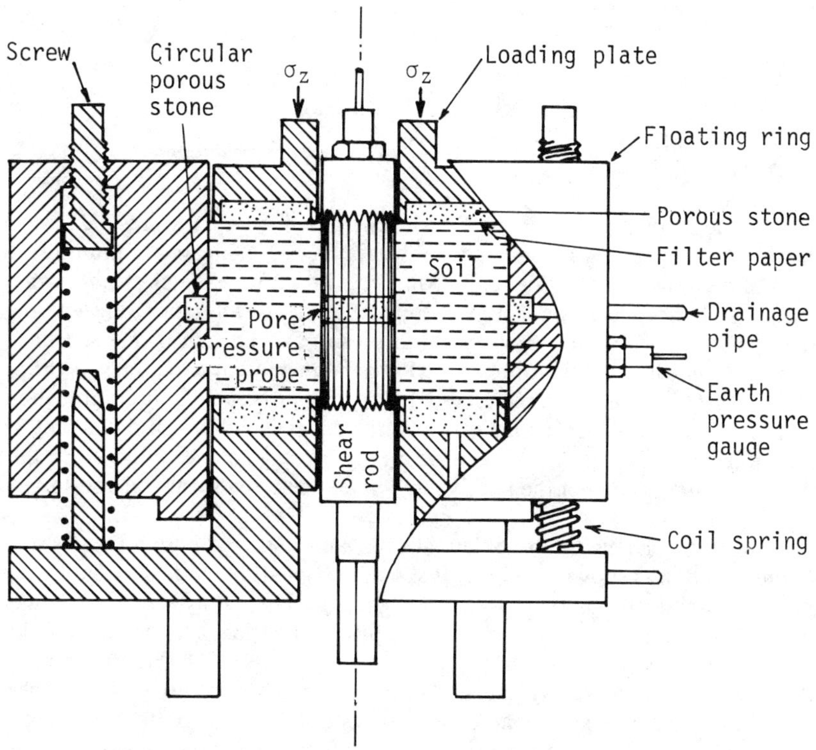

FIG. 1—*Cylinder shear apparatus.*

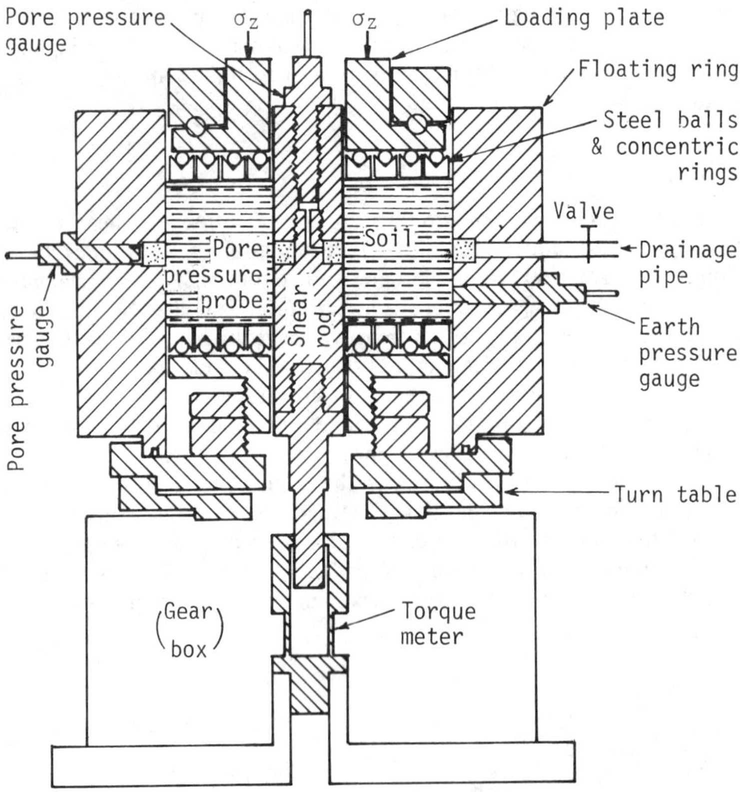

FIG. 2—*A schematic diagram of the cylinder shear testing.*

the porous stone section in the center. The deadweight of the shear rod is balanced during consolidation using a spring hanging above the apparatus. A square aluminum section of lower part of the shear rod is designed to connect it into torque meter during shear.

For each test the soil slurry was gently spooned into the consolidation ring with the shear rod set in position. Filter papers were placed on top and bottom of the specimen. The specimen was consolidated one-dimensionally using incremental loadings of the vertical stress from 5 to 74 kPa, using an incremental ratio of approximate unity. Throughout consolidation, the center of the specimen was maintained level with the mid-depth of the consolidation ring by adjusting the coil springs and the spring balance attached to the shear rod. This procedure decreased the relative movement between shear rod and the soil during consolidation.

A typical record of changes in total lateral stress and shear rod pore pressures measured during consolidation is shown in Fig. 3. It should be noted that almost equal increases in magnitude of total lateral pressure were observed on each loading, and it took about 8 h for excess pore-water pressure to achieve 90% consolidation for each increment of loading.

After the completion of consolidation at the vertical stress of 74 kPa, the specimen was unloaded instantly to zero vertical stress. The filter papers were removed to be replaced by thin rubber membranes. These membranes placed on the top and bottom of the specimen assured undrained conditions during shear. The loading plate and base were also replaced by those for shear.

A schematic diagram of the cylinder shear testing arrangements is shown in Fig. 2. In

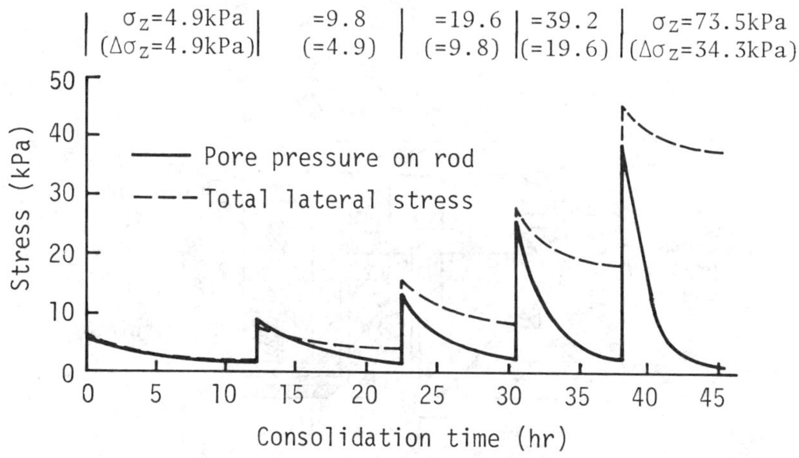

FIG. 3—*Change in pore pressure and the total lateral stress measured in a cylinder shear test.*

order to reduce the friction acting on the top and bottom of the specimen, four concentric rings containing steel thrust bearings were used between rubber membrane and the end platen. The specimen in the consolidation ring, together with the shear rod in the center, was fixed to a turntable, and the shear rod was connected to a torque meter. A vertical stress of 74 kPa, which was equal to the final consolidation pressure, was then applied to the specimen. When pore pressure in the shear rod reached to equilibrium, the drainage valve was closed, and the specimen was subject to shearing.

Torque developed when the shear rod that was fixed to the torque meter and remained stationary during shear was subjected to shearing resistance mobilized from the surrounding soil at the commencement of shear. The rate of rotation of the turntable can be varied in steps from 0.001 to 1°/s. Torque, angle of rotation of the turntable, and the vertical displacement measured at the top of the loading plate were continuously recorded during shear. The test was terminated when the angle of rotation reached 60°.

Vane Shear Test

A view of consolidation arrangements for the vane test is shown in Fig. 4. Like the cylinder shear test, the vane blade was positioned in the center of the consolidation ring before spooning the soil slurry into the ring. The self-weight of the vane was supported using a wire during the first loading of consolidation and by the surrounding soil for the rest of the consolidation stage. The vane blades made from 1-mm-thick stainless steel and three different blade configurations were used with H/D of 20/10, 15/15, and 15/30 in mm.

The specimens were consolidated in the ring, together with the vane in it, using an identical stress history to the cylinder shear procedure. After completing the consolidation, the specimen was unloaded and set onto the turntable. The vane rod was connected to the torque meter, and the specimen was subject to shear after the vertical stress of 74 kPa had been applied. The speed of rotation of the turntable was fixed at a standard rate of 0.1°/s. The apparatus used for the vane shear testing was generally similar to that for the cylinder shear tests, although the concentric rings with steel thrust bearings were not used for the vane tests.

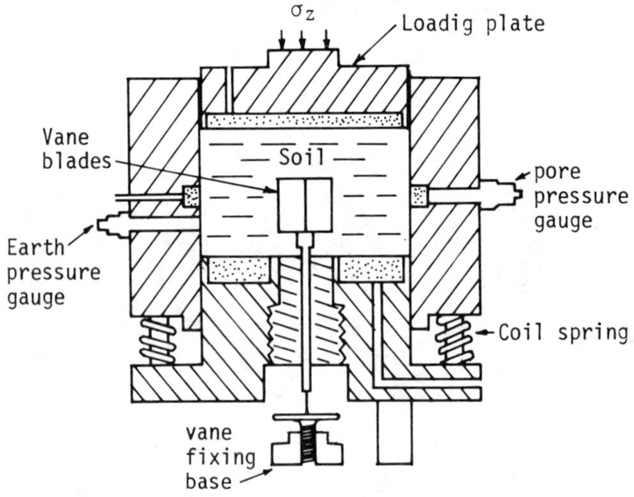

FIG. 4—*A view of consolidation in the vane test.*

Direct Shear Test

A series of direct shear tests was performed in order to examine undrained strength anisotropy of the sample.

The soil slurry was poured into a cylindrical consolidation cell, which was 140 mm in diameter and 250 mm high, and then this was consolidated using incremental loadings of the vertical stress from 12 to 74 kPa. Several disk-shaped specimens, which were 60 mm in diameter and 15 mm thick, were cut out of the same block sample with different cutting angles θ relative to the horizontal (that is, the plane of sedimentation). Angles θ of 0°, 30°, 45°, 60°, and 90° were selected for the tests.

The specimens were subject to undrained shear during which a vertical stress of 34 kPa was applied. The rate of shearing was 1 mm/min.

Triaxial Compression Tests

Two undrained triaxial compression tests were carried out so as to obtain the angle of shearing resistance of the specimen. Two cylindrical specimens, 35 mm in diameter and 80 mm high, were trimmed out of the block sample that provided the specimens for the direct shear tests. These triaxial specimens were vertically cut out (that is, $\theta = 0°$).

The specimens were isotropically consolidated in a triaxial cell to effective confining pressures of 98 and 196 kPa and then sheared at a constant axial strain rate of 0.1%/min. Pore pressures during shear were measured at the bottom of the specimen.

Test Results

Cylinder Shear Tests

Results of the cylinder shear tests are summarized in Table 1. The symbols of ω and $\dot{\omega}$ denote angular rotation of the consolidation ring relative to the shear rod and the rate of rotation, respectively. The shear stress mobilized on the shear surface around the shear rod

TABLE 1—*Summary of cylinder shear tests.*

		At Failure						
Test	$\dot{\omega}$°/s	τ, kPa	u, kPa	σ_z, kPa	$\frac{\tau}{\sigma_z}$	ω, deg	w, %	K_o
C-1	1.0	25.4	0.8	73.5	0.345	2.0	64.3	0.49
C-2	0.1	19.5	2.1	73.5	0.265	1.7	65.0	0.50
C-2a	0.1	19.6	3.8	73.5	0.267	1.6	65.8	0.51
C-3	0.033	21.8	4.1	73.5	0.296	1.9	64.9	0.47
C-4	0.01	18.3	5.2	73.5	0.249	2.2	64.0	0.45
C-5	0.001	18.6	2.6	73.5	0.235	2.9	62.6	0.47
C-6	0.1	38.5	1.6	147	0.269	2.2	58.6	0.50

τ_0 is calculated using

$$\tau_0 = \frac{2T}{\pi D^2 H} \tag{2}$$

where

T = torque measured,
D = diameter of shear rod (15 mm), and
H = specimen thickness.

The values of σ_{vo}, K_o, and w are vertical stress, coefficient of earth pressure at rest, and the water content of the specimen measured at the end of consolidation, respectively. Pore pressures measured on the shear surface are denoted using a symbol of u. The subscript f implies the failure, which is defined for these tests and the vane tests to mean the instant when the torque reaches the maximum.

The pore pressures, shear stress, and total lateral stress are plotted against ω in Fig. 5. The value of u' stands for pore pressure measured on inner wall of the consolidation ring. The residual strength was scarcely influenced by the rate of shearing.

The shear strength and pore pressure at failure are examined in relation to the rate of shearing, which is shown in Fig. 6. The relationship between shear strength and the elapsed time at failure is shown in Fig. 7 where $(\tau_f)_{\dot{\omega}=0.1}$ and $(t_f)_{\dot{\omega}=0.1}$ denote shear strength and the elapsed time at failure obtained from Test C2, respectively. As can be seen in Fig. 7, the shear strength increased in magnitude as t_f decreased. This matches the experimental findings reported by Torstensson [*2*] and Bjerrum [*4*].

Vane Shear Test

In vane tests, it is usually assumed that the torque originates from the shear stresses mobilized on the right cylindrical surface whose shape coincides with the dimensions of the vane blades, that is

$$T = (\pi D^2 H \tau_v)/2 + (\pi D^3 \alpha \cdot \tau_h)/2 \tag{3}$$

where

T = torque,
τ_v, τ_h = shear stress mobilized on the vertical and the horizontal shear surface, respectively, and

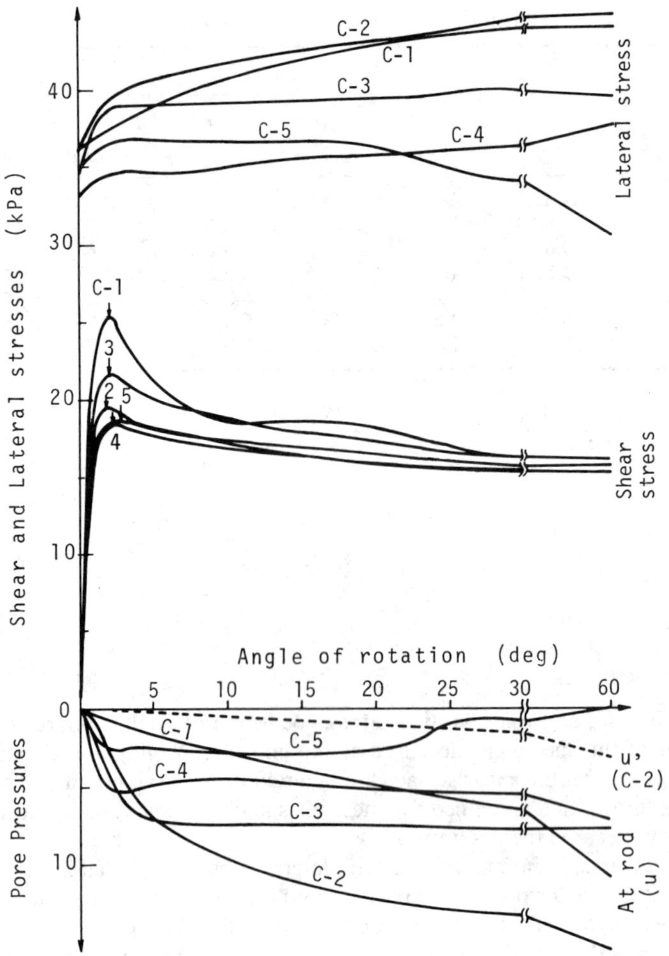

FIG. 5—*Cylinder shear test results.*

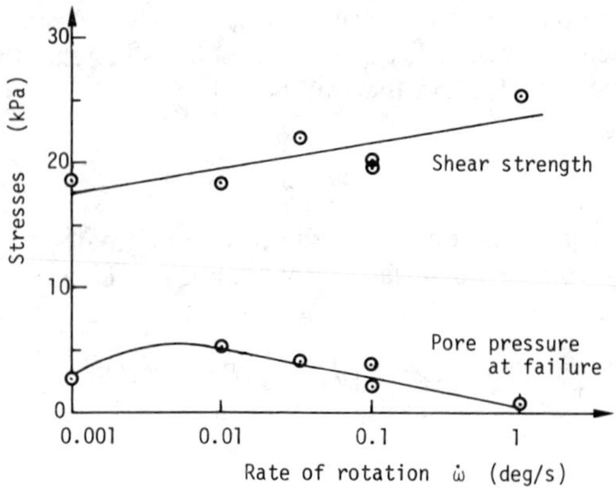

FIG. 6—*Pore pressures and shear stresses developed at the maximum torque in the cylinder shear test.*

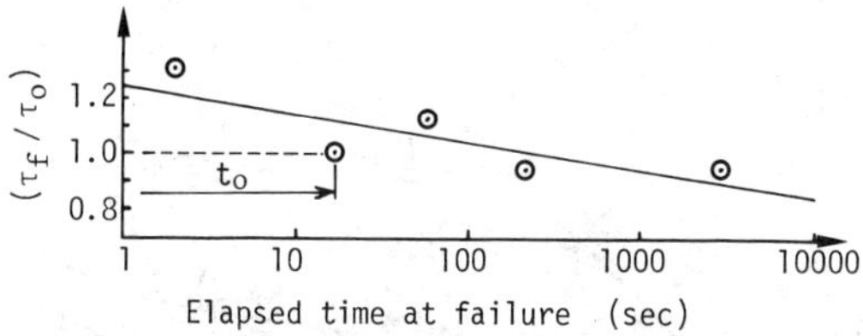

FIG. 7—*Relationship between shear strength and the elapsed time at failure in the cylinder shear test.*

α = a coefficient that depends on distribution of τ_h along the radius of the vane blade (for example, α = ⅓ for a uniform distribution and α = ¼ for a triangular distribution).

Results of the vane shear tests are summarized in Table 2. The shear stress τ in the vane test is derived by assuming $\tau = \tau_v = \tau_h$ and α = ⅓, that is

$$\tau = \frac{1}{1 + \frac{1}{3} \cdot \frac{D}{H}} \frac{2T}{\pi D^2 H} \tag{4}$$

The shear strengths, together with the changes in total lateral stress, are examined for the vane tests with respect to ω, that is, shown in Fig. 8. Results of the cylinder test C2, in which the rate of shearing was the same as that of the vane tests, are included for comparison. The amount of increases in σ_r showed a tendency to be larger as the diameter of the vane increased. Figure 9 shows the relationship between ω_f and the value of D/H of the vanes. The trend is that the angular rotation at failure increased in magnitude as D/H increased.

Direct Shear Test

Figure 10 shows the variation of the shear strength of the direct shear tests with respect to the cutting angle θ where shear strengths are normalized using the shear strength of 18.4 kPa obtained from the horizontally sheared test at $\theta = 0°$. It is well-known that the rotation of principal stress axes occurs in this kind of test with the angle of rotation being 45° at failure [5]. The shear strength of test at $\theta = 45°$ showed the highest value; however, the shear strength of test at $\theta = 135°$ whose major principal stress direction rotated from the

TABLE 2—*Summary of vane shear tests.*[a]

		At Failure				
Test	D/H, mm/mm	τ, kPa	u, kPa	ω, deg	w, %	K_o
V-1	10/20	23.9	0.1	5.9	65.2	0.45
V-2	15/15	20.3	0.4	6.0	64.9	0.45
V-3	30/15	22.5	0.1	11.9	64.3	0.48
C-2	15/∞	19.5	0.0	1.7	65.4	0.51

[a] $\dot{\omega}$ = 0.1°/s, and σ_z = 73.5 kPa for all tests.

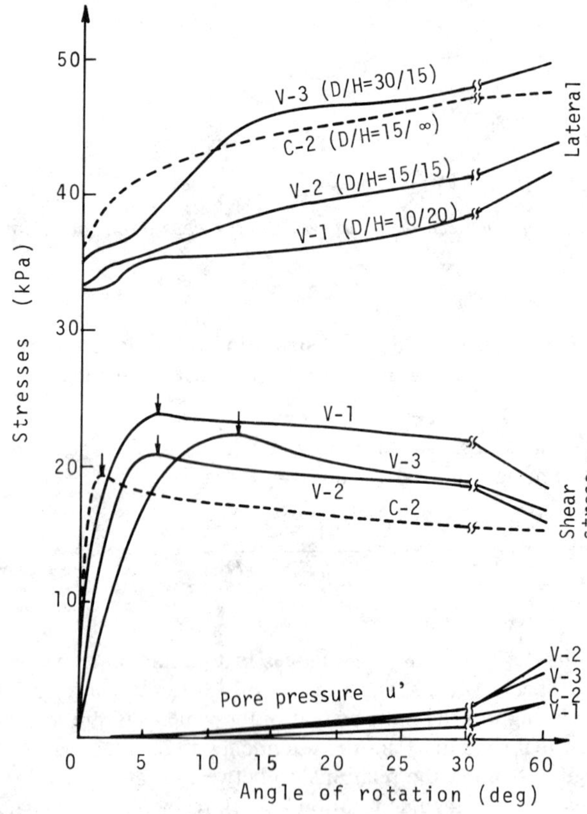

FIG. 8—*Results of the vane shear tests.*

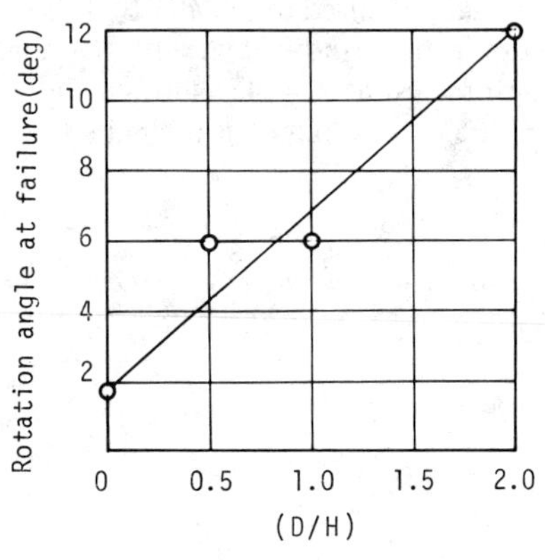

FIG. 9—*Angle of rotation at failure with respect to the ratio of* D/H *observed in the vane tests.*

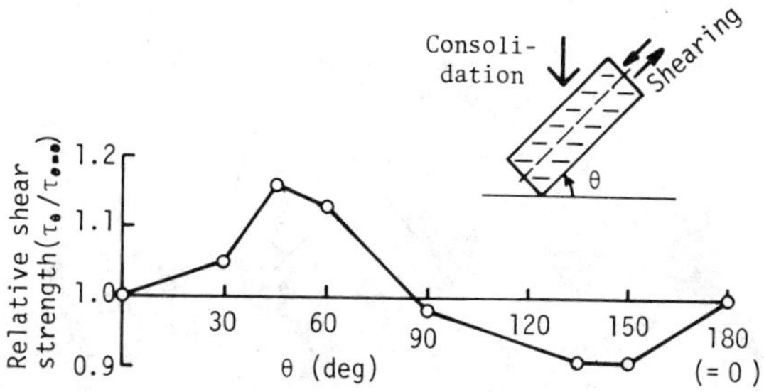

FIG. 10—*Undrained strength anisotropy of the soil observed in direct shear tests.*

vertical almost 90° at failure was the lowest. This matches the experimental finding reported by Shibuya and Hight [*6*].

In direct shear tests, the shear strength of test at $\theta = 0°$ is equivalent to that mobilized on horizontal shear surface of the vane tests. Figure 11 shows a specimen sheared in a direct shear device whose shearing mode corresponds to that of the vertical shear surface in the vane tests, whose direction of major principal stress rotates 90°. This particular type of shear was simulated in another shear test at $\theta = 90°$. The value of τ_v/τ_h obtained from the direct shear tests was approximately unity. It indicates that above explanation of Fig. 10 can not be applied to this particular shearing mode.

Triaxial Compression Test

The undrained effective stress paths are shown in Fig. 12 using a dotted line. The angle of shearing resistance ϕ' was 30.2° for the soil.

Interpretation of Test Results

Comparison of Test Results Between Vane Shear and Cylinder Shear

The specimens of three vane tests (V1, V2, and V3) and a cylinder shear test (C2) were prepared by using an identical consolidation history and sheared with a standard rate of $\dot{\omega}$ equal to 0.1°/s. As can be seen in Fig. 8, Test V2, in which the diameter of the vane was

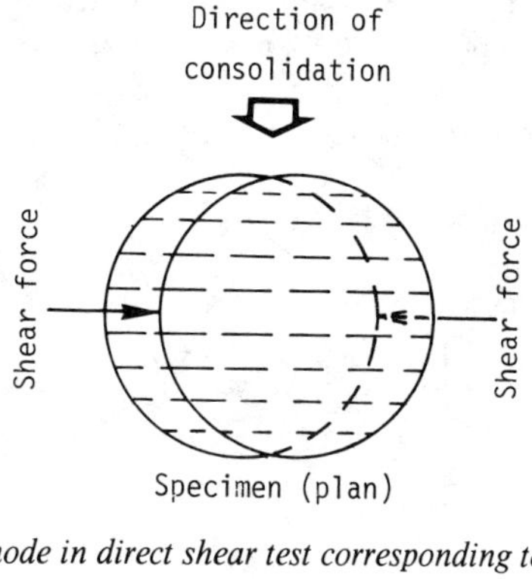

FIG. 11—*A shearing mode in direct shear test corresponding to the vertical shear surface of the vane test.*

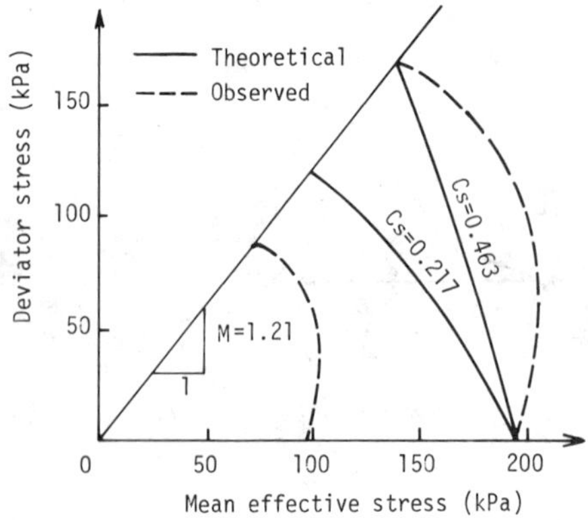

FIG. 12—*Undrained effective stress paths of triaxial tests.*

equal to that of the shear rod used for the cylinder shear tests, gave the closest shear strength to that of Test C2. In a comparison of ω_f between the two tests of V2 and C2, the value observed from Test V2 was larger than that of Test C2. This may refer to the difference in value of D/H between the two tests, that is, the vane test results show that the value of ω_f increased as D/H increased (Fig. 9).

Provided that the shear stresses of τ_v and τ_h in vane tests are independent upon the value of D/H, the two shear stresses can be estimated in a series of tests with various blade aspect ratios. Equation 3 is modified as

$$\frac{2T}{\pi D^2 H} = \tau_v + \alpha\tau_h(D/H) \tag{3a}$$

Figure 13 shows the vane test results, together with the result of Test C2, that are examined in relation to Eq 3*a*. The vertical axis intercept and slope of the solid line correspond to the values of τ_v and $\alpha\tau_h$, and those were 21.6 and 7.6 kPa, respectively. Thus the observed value of τ_v/τ_h could be in the range between 0.68 for $\alpha = 1/4$ and 0.91 for $\alpha = 1/3$. The observed value of τ_v/τ_h was approximately unity in the direct shear tests (Fig. 10). Therefore, it may be inferred that the difference in values between τ_v and τ_h, observed in the vane tests, may well be attributed to the corresponding magnitudes of effective normal stresses acting on the vertical and horizontal shear surfaces of σ_h' and σ_v', respectively; that is, the value of (σ_h'/σ_v') in the vane tests could have been in the range between 0.68 and 0.91. It should also be noted that the value of τ_v observed from Test C2 was 19.6 kPa, which was close to τ_v obtained from the vane shear tests. This may suggest that the shearing

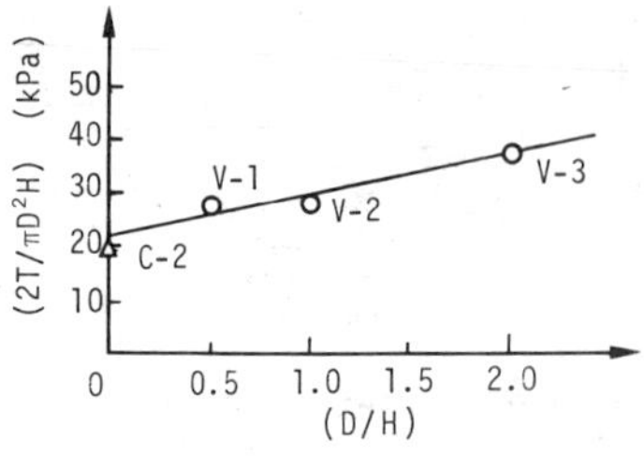

FIG. 13—*Shear strength anisotropy measured in the vane tests.*

mechanism of the cylinder shear tests was similar to that achieved on the vertical shear surface in the vane shear tests.

Analysis of Cylinder Shear Test Results

Figure 14 shows a three-dimensional coordinate (z,r,θ), used for analyzing the cylinder shear tests. Radius of the shear rod and the specimen thickness are represented using r_o and H, respectively.

At the end of consolidating the specimen, the major and minor principal stresses, σ_1 and σ_3, were σ_z' and $\sigma_r'(\sigma_\theta')$, respectively. The coefficient of earth pressure at rest K_o is equal to (σ_r'/σ_z'). When the specimen was subject to shear, there developed little shear stresses on the horizontal plane, since the concentric rings with steel bearings were placed on the top and bottom of the specimen. The vertical stress, therefore, remained to be one of the principal stresses throughout shear.

The magnitude of the vertical stress applied to the specimen during shear was equal to the magnitude at the end of consolidation. Thus the mean total stress P_T is

$$P_T = P_{T0} = (1 + 2K_0)\sigma_z/3 \tag{5}$$

where P_{T0} is the mean total stress at the end of consolidation. The value of P_T may be assumed to have been stayed constant throughout shear.

Assuming undrained conditions in the specimen, the mean effective stress P is

$$P = (\sigma_1' + \sigma_2' + \sigma_3')/3 = P_{T0} - u \tag{6}$$

where u denotes pore pressure at any element in the specimen.

Considering stress equilibrium in the circumferential direction, shear stress $\tau_{r\theta}$, that is, at a distance r from the center of the shear rod, can be derived

$$\tau_{r\theta} = \frac{1}{2\pi H}\frac{T}{r^2} \tag{7}$$

Provided that the plane with the maximum effective stress ratio (τ/σ') is in the circumferential direction, $\tau_{r\theta}$ can be expressed in terms of principal stresses, that is

$$\tau_{r\theta} = \frac{(\sigma_1' - \sigma_3')}{(\sigma_1' + \sigma_3')}\sqrt{\sigma_1'\sigma_3'} \tag{8}$$

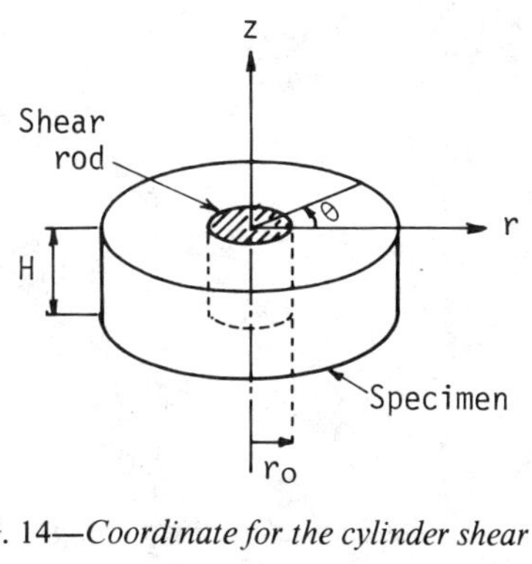

FIG. 14—*Coordinate for the cylinder shear test.*

Combining Eqs 7 and 8

$$\frac{(\sigma_1' - \sigma_3')}{(\sigma_1' + \sigma_3')} \sqrt{\sigma_1'\sigma_3'} = \frac{1}{2\pi H} \frac{T}{r^2} \tag{9}$$

Strains in the vertical, radial, and circumferential directions are expressed using symbols of ϵ_z, ϵ_r, and ϵ_θ, respectively. Shear strains are generally represented by γ in this paper. If undrained conditions are satisfied in the specimen, those normal strains are all zero. A component of shear strain, $\gamma_{zr}(\gamma_{z\theta})$, is also zero since the vertical stress is a principal stress.

As shown in Fig. 15*a*, the soil specimen is divided into a number of concentric "hypothetical rings" for each with the thickness of Δr. Figure 15*b* shows a segment of a hypothetical ring, A-B-C-D, which is at a distance of r from the center of the shear rod when subject to shear. An idealized deformation characteristic is postulated such that A and B move to A′ and B′ relative to C and D, respectively. Assuming $\widehat{AA'} = \widehat{BB'}$, a component of shear strain $\gamma_{r\theta}$ may be derived, which is

$$2\gamma_{r\theta} = \frac{\widehat{BB'}}{\Delta r} = \frac{\widehat{AA'}}{\Delta r} \tag{10}$$

The component of shear strain can be expressed in terms of the major and minor principal strains, ϵ_1 and ϵ_3, which is

$$2\gamma_{r\theta} = (\epsilon_1 - \epsilon_3) \tag{11}$$

The vertical strain ϵ_z is now the intermediate principal strain, which remains zero.

An angular rotation of $\Delta\omega$, shown in Fig. 15*b*, for any hypothetical ring is obtained using Eqs 10 and 11, that is

$$\Delta\omega(\text{deg}) = \frac{360}{2\pi} \frac{\widehat{BB'}}{r} = \frac{360}{2\pi} (\epsilon_1 - \epsilon_3) \frac{\Delta r}{r} \tag{12}$$

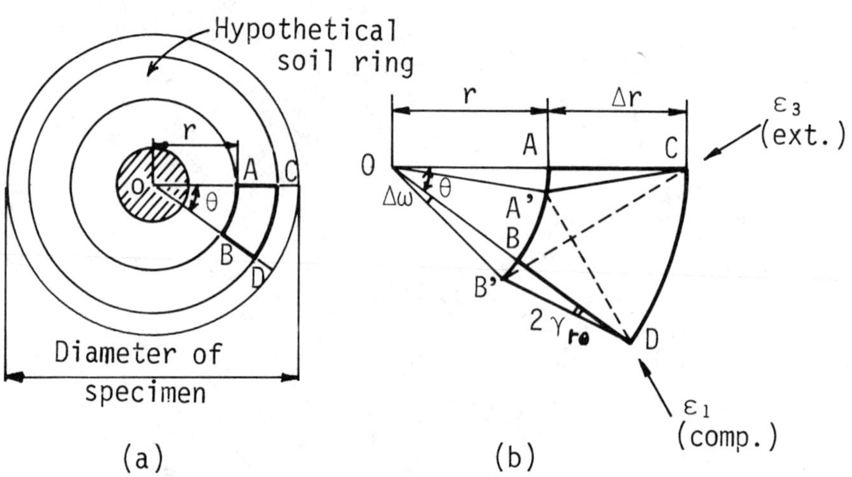

FIG. 15—*Deformation mechanism of cylinder shear tests.*

The angular rotation of the shear rod ω may be derived by integrating $\Delta\omega$ of all hypothetical soil rings along the radial direction.

Summarizing, when a soil specimen is subject to shear in the cylinder shear apparatus, the relationship between principal stresses and the mobilized torque of the shear rod has been given in Eq 9 for an element of soil that is at a distance of r from the center of the shear rod. The corresponding principal strains have also been given in Eq 12. Therefore it is now possible for the resultant strains that develop in the soil when sheared to be related to the mobilized torque of the shear rod through an appropriate soil model.

Figure 16 shows a three-dimensional principal stress space viewed from the space diagonal. An outer triangle represents a plane of constant P_{TO}, which is normal to the space diagonal. At the end of consolidation of the cylinder shear tests, the stress state of the specimen was at point C. Total stress path, when the specimen was subject to shear, traveled along a path from C to F staying on the P_{TO}-plane since the vertical stress applied to the specimen was maintained to be constant throughout shear. A dashed curve $\widehat{CF'}$ is the corresponding effective stress path projected onto the P_{TO}-plane from the origin.

In the cylinder shear tests, the direction of the major principal stress rotates 90° from the vertical to the horizontal. However, the difference had no influence with undrained shear strength of the specimen. To the authors' knowledge, no constitutive model for practical use is available to simulate the effects of continuous principal stress rotation. Accordingly an alternative stress path is considered such that the stress state of the specimen suddenly jumps from C to the isotropic stress state of point O at the commencement of shear, and follows an effective stress path $\widehat{OF'}$. This postulation appears to be reasonable when considering the experimental observation made by Shibuya and Hight [6]; that is, most of the principal stress rotation in the simple shear test occurs at the beginning of the test involving very small shear strains. An angle δ, shown in Fig. 16, is defined using a

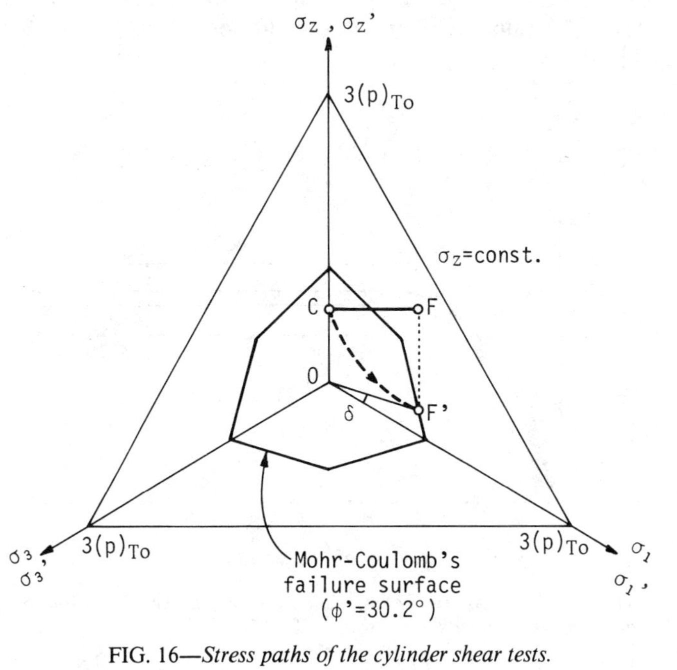

FIG. 16—*Stress paths of the cylinder shear tests.*

coefficient of intermediate principal stress $b = (\sigma_2 - \sigma_3)/(\sigma_1 - \sigma_3)$, that is

$$\delta = \arctan\,[\sqrt{3}b/(2 - b)]$$

The value of δ is equal to 30° ($b = 0.5$) for an isotropic elastic material, however around 17° ($b = 0.3$) for most of soils when sheared under plane strain conditions.

The original Cam clay model, proposed by Roscoe et al. [7], has been used for analyzing the soil behavior of the cylinder shear tests under the simplified stress path without principal stress rotation. The undrained effective stress path and shear strain, $\epsilon = 2(\epsilon_1 - \epsilon_3)/3$, are

$$q = \frac{MP}{(1 - C_s/C_c)} \ln\left(\frac{P_o}{P}\right) \text{ where } P_o = P_{To} \tag{13}$$

$$\epsilon = -\frac{(C_c - C_s)C_s}{2.3M(1 + e)C_c} \ln\left\{1 - \frac{1}{M}\frac{q}{p}\right\} \tag{14}$$

where e, C_c, C_s, and q are void ratio, compression index, swell index of soil, and deviatoric stress, respectively. The effective stress ratio at failure M has been defined using Mohr-Coulomb's failure criterion, that is

$$M = \frac{2 \sin \phi'}{1 - (1 - 2b) \sin \phi'/3} \tag{15}$$

Figure 17 shows predicted relationships between torque and the angular rotation of cylinder shear test using the original Cam clay model. The result of Test C4 is shown to be compared to these relations with different b values. The reason is that the rate of rotation of $\dot{\omega} = 0.01°/\text{s}$ of Test C4 corresponds approximately to the rate of straining of 0.1%/min of the triaxial tests. The values of C_c and C_s used for the calculations are 0.703 and 0.217,

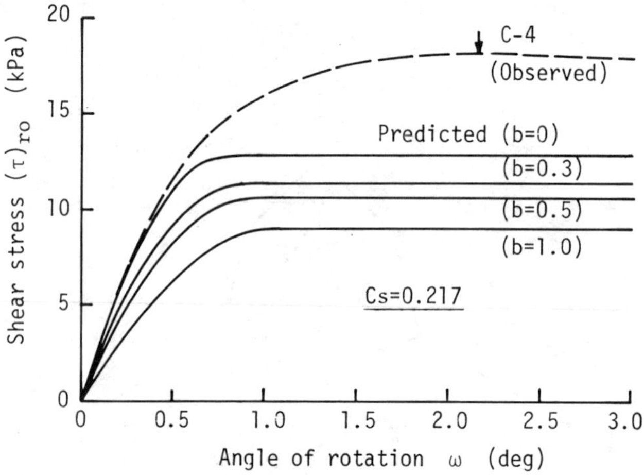

FIG. 17—*Predicted and observed relationships between angle of rotation and the mobilized shear stress in the cylinder shear test* ($C_s = 0.217$).

respectively. As can be seen in the figure, the predicted shear strength for b equal to 0.3 is about a half of the observed value.

It is generally difficult to determine the value of swell index of soils since the rebound curve in a plot of e versus log p is usually not straight. As shown in Fig. 12, the predicted undrained effective stress path of the triaxial test using C_s = 0.217 is far away from the observed path. Therefore a value of C_s = 0.463 has been obtained so as that the predicted undrained strength matches that of the observed value. The results using the value of C_s = 0.463 is shown in Fig. 18. It should be noted that the result at b = 0 is the most consistent with the observed result.

Predictions of pore pressures measured on the surface of the shear rod have been made for different b values, that is, shown in Fig. 19. The predicted magnitude of pore pressures is approximately three times larger than that of Test C4, irrespective of the b value. In an attempt to examine the significant difference in magnitude between predicted and the observed pore pressure response, the distribution of pore pressures in the surrounding soil of the shear rod has been calculated using the original Cam clay model, which is shown in Fig. 20. This shows an extremely high gradient of pore-pressure distribution in the surrounding soil close to the surface of the shear rod. It may be presumed that some partial dissipation of pore pressures occurred in the sample associated with shear. This may have triggered a certain amount of increase in magnitude of the effective normal stress acting on the surface of the shear rod. As a result, the observed shear strength may be much larger than that of the predicted value.

Concluding Remarks

An attempt has been made to analyze the vane test using a cylinder shear test apparatus that has been developed by the authors. The conclusions are summarized as follows:

1. In the vane shear tests, the shear strength mobilized on the horizontal shear surface was larger than that of the vertical shear surface. This difference in magnitude of the shear

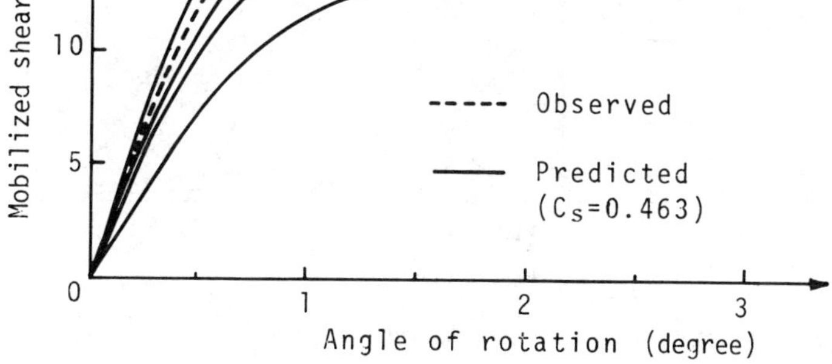

FIG. 18—*The same relationships as Fig. 17* (C_s = 0.463).

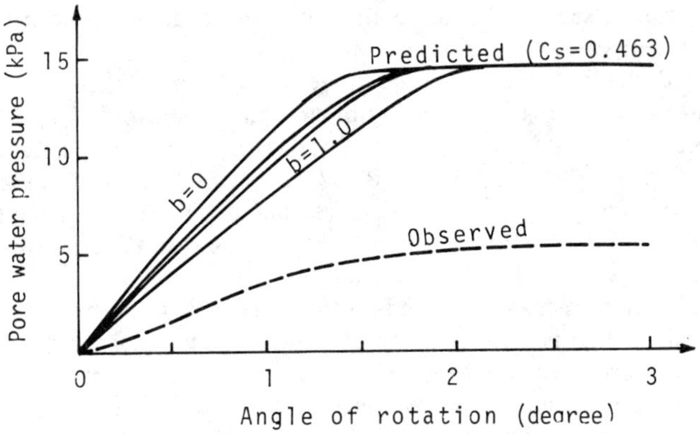

FIG. 19—*Predicted and the observed pore pressure on the shear rod during cylinder shear test.*

strengths may simply be attributed to the different magnitudes of the effective normal stress acting on the two shear surfaces.

2. The residual strengths of the reconstituted clay obtained from a series of cylinder shear tests were scarcely influenced by the rate of shearing over the range of the rotation speed between 1 and 0.001°/s.

3. The shearing mechanism of the cylinder shear test is compatible with that observed on the vertical shear surface of the vane tests.

4. The stress-strain and strength characteristics of soil in the cylinder test has been analyzed using the original Cam clay model. The theoretical prediction was consistent with the observed behavior of the soil in terms of the mobilized shear stress, however, it was not able to simulate the partial dissipation of pore pressure that occurred in the specimen.

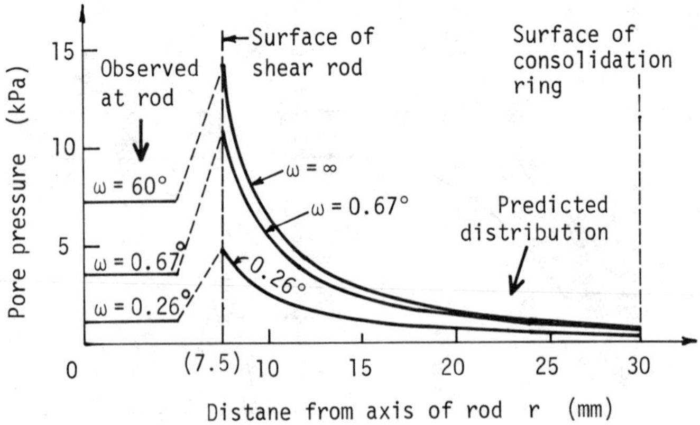

FIG. 20—*Predicted pore pressure distribution in the surrounding soil of the cylinder shear test.*

References

[1] Bjerrum, L., "Embankment on Soft Ground," *Proceedings of the Conference on Performance of Earth and Earth-Supported Structures,* Vol. 2, American Society of Civil Engineers, 1972, pp. 1–54.

[2] Torstensson, B. A., "Time-Dependent Effects in the Field Vane Test," *Proceedings of the Symposium on Soft Clay,* Bangkok, Thailand, 1977, pp. 387–397.

[3] La Rochelle, P., Trak, B., Tavenas, F., and Roy, M., et al., *Canadian Geotechnical Journal,* Vol. 11, No. 1, 1974, pp. 142–164.

[4] Bjerrum, L., "Problems of Soil Mechanics and Construction on Soft Clays and Structurally Unstable Soils," *Proceeding of the 8-ICSMFE,* Vol. 3, 1973, pp. 111–153.

[5] Oda, M. and Konishi, J., "Rotation of Principal Stresses in Granular Materials During Simple Shear," *Soils and Foundations,* Vol. 14, No. 4, pp. 39–53.

[6] Shibuya, S. and Hight, D. W., "A Bounding Surface for Granular Materials," *Soils and Foundations,* Vol. 27, No. 4, pp. 123–136.

[7] Shibuya, S. and Hight, D. W., "On the Stress Path in Simple Shear," *Geotechnique,* Vol. 37, No. 4, pp. 511–515.

[8] Roscoe, K. H., Schofield, A. N., and Thurairajah, A., *Géotechnique,* Vol. 13, No. 2, 1963, pp. 211–240.

Julio A. De Alencar,[1] Dave H. Chan,[1] and Norbert R. Morgenstern[1]

Progressive Failure in the Vane Test

REFERENCE: De Alencar, J. A., Chan, D. H., and Morgenstern, N. R., **"Progressive Failure in the Vane Test,"** *Vane Shear Strength Testing in Soils: Field and Laboratory Studies, ASTM STP 1014,* A. F. Richards, Ed., American Society for Testing and Materials, Philadelphia, 1988, pp. 150–165.

ABSTRACT: A finite-element analysis of the undrained vane test has been performed using an elasto-plastic constitutive relation with strain-softening behavior. A hyperbolic strain softening model has been developed to model the post peak behavior of the soil. This constitutive relationship assumes an elastic behavior up to peak strength, and the material subsequently softens according to a hyperbolic relationship. As a result one can model different rates of post peak softening in a compact manner while maintaining the same peak and residual strength. The results of the analysis show that the peak torque measured in a vane test depends not only on the peak strength and the residual strength of the material, but also that the rate of post peak softening has a dramatic effect on the peak torque and the torque-rotation relationship of the test. If the material possesses moderate rates of softening behavior, the maximum torque is close to the value calculated based on peak strength. In other words, the results of the vane test can be used to deduce the peak strength of the material. At high rates of softening, the progressive development of the failure zone results from stress concentration effects at the blades of the vane, and the strength is mobilized in a nonuniform manner around the vane at maximum torque. Therefore the distribution of stresses around the perimeter of the vane is not known and the assumption of uniform mobilization of shear strength will lead to an incorrect evaluation of the peak strength of the material. This effect is especially pronounced in very strain sensitive soils. The finite-element results also reveal a localized failure zone around the vane when the maximum torque is reached. Comparisons between the finite-element results and existing experimental studies of soil behavior around a vane are in agreement for the failure mechanism developed during a vane test.

KEY WORDS: vane tests, progressive failure, strain softening, finite-element analysis, plasticity

The vane shear test is one of the most frequently used tests to estimate the in-situ undrained shear strength of a cohesive soil in saturated soft clays. Although the vane shear test is a simple and effective test to obtain such information for a soil, it is well recognized that the determination of undrained shear strength using the results from a vane test is not a simple task. The results of a vane test are affected by many factors such as the rate of rotation of the vane, the time elapse between the insertion of the vane and the beginning of the test, the height to diameter ratio of the vane blade, the drainage conditions around the vane, strength anisotropy, and progressive failure of the soil around the vane. Many studies have been conducted to evaluate the significance of these factors [*1–8*], and standards have been set in an attempt to obtain consistent results from the vane test. Among these factors, the influence of progressive failure has been identified, but it has not been studied extensively. In deducing the undrained shear strength from the vane test, it is commonly assumed that the rectangular blades of the vane shear the soil along a circumscribing

[1] Ph.D. student, assistant professor, and professor, respectively, Department of Civil Engineering, The University of Alberta, Edmonton, Alberta T6G 2G7, Canada.

cylinder and the mobilized shear strength is uniform over the surface of rotation at maximum torque. Because of the stress concentration effect near the tip of the blade, the mobilized strength at any stage in a strain hardening material will be higher near the blade than some distance away. Since there is no decrease in strength after peak for such materials, the mobilized strength is essentially uniform at peak torque. The assumption of uniform mobilized shear strength to determine the undrained shear strength for the soil is therefore a good approximation. However, for a strain softening soil, the decrease in strength after peak will result in nonuniform strength distribution around the vane and therefore the determination of the undrained shear strength becomes difficult. In order to deduce the undrained shear strength of the soil, the distribution of the mobilization of shear strength must be known around the blade. Attempts have been made to study the stress distribution induced by the rotation of the blades using numerical models based on linear elasticity [*9,10*] and an elasto-plastic strain hardening model [*11*]. To study the problem of progressive failure, a strain softening model must be used. The present study shows the results of the stress and strength distribution around the vane for a strain softening soil.

Strain Softening Finite-Element Model

To study the effect of the rate of post peak softening on progressive failure analysis, an elastic-plastic model with a variable rate of softening has been adopted. Lo [*12*] used a strain softening model in which the post peak softening part of the stress-strain curve is approximated by a hyperbola. The stress-strain relationship for the material is given by

$$S' = \frac{\epsilon'}{a + b\epsilon'} \tag{1}$$

where

$S' = S/S_f$,
$S =$ nominal shear stress,
$S_f =$ shear strength,
$\epsilon' = \epsilon/\epsilon_f$,
$\epsilon =$ shear strain, and
$\epsilon_f =$ shear strain at peak strength.

The physical significance of a and b has been discussed by Lo and will be discussed later in relation to the plasticity formulation of the model.

The idea of approximating the post peak part of the stress-strain curve by a hyperbola makes it possible to vary the rate of softening in a compact manner. The calculations are performed by means of the finite-element method. Details of the basic finite-element formulation can be found in many texts [*13*] and will not be repeated here, but the part of the formulation related to the development of the hyperbolic strain softening model is presented.

Considering the principle of virtual displacement, the basic incremental finite-element equation is given by (assuming small straining)

$$[\mathbf{K}]\{\Delta\delta\} = \{\Delta\mathbf{R}\} \tag{2a}$$

where

$[\mathbf{K}] = \int_v [\mathbf{B}]^T[\mathbf{C}][\mathbf{B}]\, dv$

and where

$\{\Delta\delta\}$ = increments of displacement,
$[\mathbf{B}]$ = strain displacement matrix,
$[\ ]^T$ = transpose of a matrix,
$\{\ \}$ = vector,
$\langle\ \rangle$ = transpose of a vector,
$\{\Delta\sigma\}$ = increment of total stress vector,
= $\langle \Delta\sigma_{xx}, \Delta\sigma_{yy}, \Delta\sigma_{xy}, \Delta\sigma_{zz}, \Delta\sigma_{yz}, \Delta\sigma_{xz}\rangle^T$;
$\{\Delta \boldsymbol{R}\}$ = increment of external applied load,
v = entire volume of the body.

The incremental load vector $[\Delta \boldsymbol{R}]$ includes the load caused by the body force and surface traction. The load vector $\{\Delta \boldsymbol{R}\}$ is given by

$$\{\Delta \boldsymbol{R}\} = \int_v [\mathbf{N}]^T\{\gamma\}\, dv + \int_{s_t}[\mathbf{N}]^T[\mathbf{N}]\{\mathbf{P}\}\, ds \tag{3}$$

where

$[\mathbf{N}]$ = interpolation function matrix,
$\{\gamma\}$ = body force vector,
$\{\mathbf{P}\}$ = nodal surface traction vector, and
s_t = surface subjected to external traction.

Making the usual assumption in the theory of plasticity, the constitutive matrix can be expressed as

$$\{\mathbf{d}\sigma\} = [\mathbf{C}]\{\mathbf{d}\epsilon\} \tag{4a}$$

where

$$[\mathbf{C}] = [\mathbf{C}^{\mathrm{E}}] - \frac{[\mathbf{C}^{\mathrm{E}}]\left\{\dfrac{\partial Q}{\partial \sigma}\right\}\left\langle\dfrac{\partial F}{\partial \sigma}\right\rangle[\mathbf{C}^{\mathrm{E}}]}{\left\langle\dfrac{\partial F}{\partial \sigma}\right\rangle[\mathbf{C}^{\mathrm{E}}]\left\{\dfrac{\partial Q}{\partial \sigma}\right\} - \left\langle\dfrac{\partial F}{\partial \epsilon^P}\right\rangle\left\{\dfrac{\partial Q}{\partial \sigma}\right\}} \tag{4b}$$

where F and Q are the yield function and plastic potential, respectively, and $[\mathbf{C}^{\mathrm{E}}]$ is the elastic constitutive matrix, which can be determined by Hooke's law. For simplicity only undrained analysis will be considered.

Note that the first term in the denominator represents perfectly plastic deformation, and the second term represents strain hardening and softening deformation. The following three conditions may occur

1. $\left\langle\dfrac{\partial F}{\partial \epsilon^P}\right\rangle\left\{\dfrac{\partial Q}{\partial \sigma}\right\} < 0$ for strain softening deformation;
2. $\left\langle\dfrac{\partial F}{\partial \epsilon^P}\right\rangle\left\{\dfrac{\partial Q}{\partial \sigma}\right\} = 0$ for perfectly plastic deformation; and
3. $\left\langle\dfrac{\partial F}{\partial \epsilon^P}\right\rangle\left\{\dfrac{\partial Q}{\partial \sigma}\right\} > 0$ for strain hardening deformation.

In this hyperbolic strain softening model, the yield function is taken to be the same as the peak strength of the material, which is given by

$$F = q - \kappa = 0 \tag{5}$$

where

$$q = \sqrt{3J_2}$$

and

$\kappa = \kappa_p =$ uniaxial compressive peak strength and
$J_2 =$ second stress invariant of the stress deviation tensor.

The strength of the material decreases gradually after peak, and the yield function in the post peak region is defined as

$$F = q - \kappa = 0 \tag{6a}$$

where

$$\kappa = \kappa_p\left(1 - \frac{\bar{\epsilon}^p}{a + b\bar{\epsilon}^p}\right) \tag{6b}$$

and

$$\begin{aligned} \bar{\epsilon}^P &= d\bar{\epsilon}^P = \text{equivalent plastic strain} \\ d\bar{\epsilon}^P &= (\tfrac{2}{3} de^P_{ij} de^P_{ij})^{1/2} \\ de^P_{ij} &= d\epsilon^P_{ij} - d\epsilon_{\kappa\kappa}\delta_{ij}/3 \\ d\epsilon^P_{ij} &= \text{increment of plastic strain tensors} \end{aligned} \tag{6c}$$

At peak strength, $\bar{\epsilon}^p = 0$, Eq 6b reduces to Eq 5. At very large strain, $\bar{\epsilon}^p = \infty$, the shear strength should approximate the residual strength of the material. Substitute $\bar{\epsilon}^p = \infty$ into Eq 6b; the yield function becomes

$$F = q - \kappa_r = 0 \tag{7}$$

where

$$\kappa_r = \kappa_p(1 - 1/b)$$

Therefore

$$b = 1/(1 - \kappa_r/\kappa_p) \tag{8}$$

The brittleness index I_B is defined as [*14*]

$$I_B = (\kappa_p - \kappa_r)/\kappa_p = 1 - \kappa_r/\kappa_p \tag{9}$$

thus the b parameter is simply the reciprocal of the brittleness index

$$b = 1/I_B \tag{10}$$

The b parameter depends only upon the amount of softening of the material. The significance of the a parameter can be found by differentiating κ with respect to $\bar{\epsilon}^p$ in Eq 6b. One obtains

$$\frac{\partial \kappa}{\partial \bar{\epsilon}^p} = -a\kappa_p/(a + b\bar{\epsilon}^p)^2 \tag{11}$$

Let $\bar{\epsilon}^p = 0$

$$\frac{\partial \kappa}{\partial \bar{\epsilon}^p} = -\left(\frac{\kappa}{a}\right) \tag{12}$$

In other words κ_p/a is the tangent of the initial slope of the post peak stress strain relationship of the material. A typical stress-strain relationship of this model and the meaning of the a and b parameters are illustrated in Fig. 1.

To determine an expression for the constitutive relationship given by Eq 6, the following four situations are being considered

$$F(\sigma_{ij}, \epsilon_{ij}^P) < 0,\ F(\sigma_{ij} + \Delta\sigma_{ij}, \epsilon_{ij}^P + \Delta\epsilon_{ij}^P) < 0 \tag{13a}$$

for elastic deformation

$$F(\sigma_{ij}, \epsilon_{ij}^P) = 0,\ F(\sigma_{ij} + \Delta\sigma_{ij}, \epsilon_{ij}^P + \Delta\epsilon_{ij}^P) < 0, \text{ and } \left\langle \frac{\partial F}{\partial \sigma_{ij}} \right\rangle \{d\sigma_{ij}\} < 0 \tag{13b}$$

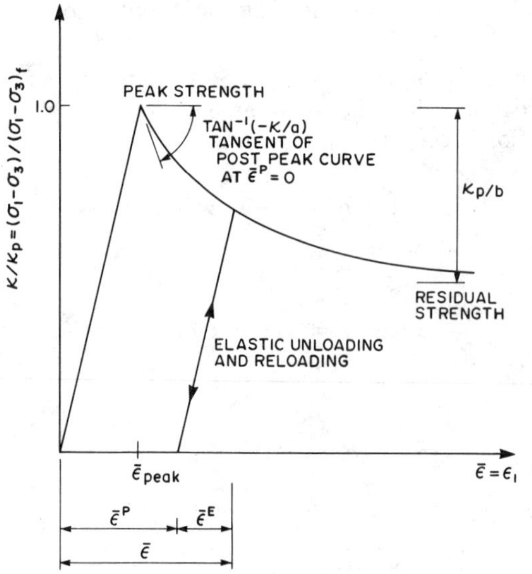

FIG. 1—*Stress-strain relationship for the hyperbolic strain softening model.*

for elastic unloading

$$F(\sigma_{ij}, \epsilon_{ij}^P) = 0,\ F(\sigma_{ij} + \Delta\sigma_{ij}, \epsilon_{ij}^P + \Delta\epsilon_{ij}^P) = 0,\ \text{and}\ \left\langle \frac{\partial F}{\partial \sigma_{ij}} \right\rangle \{d\sigma_{ij}\} > 0 \tag{13c}$$

for plastic deformation and strain hardening

$$F(\sigma_{ij}, \epsilon_{ij}^P) = 0,\ F(\sigma_{ij} + \Delta\sigma_{ij}, \epsilon_{ij}^P + \Delta\epsilon_{ij}^P) = 0,\ \text{and}\ \left\langle \frac{\partial F}{\partial \sigma_{ij}} \right\rangle \{d\sigma_{ij}\} < 0 \tag{13d}$$

for plastic deformation and strain softening.

The consistency condition is usually expressed as

$$\begin{aligned} &F(\sigma_{ij}, \epsilon_{ij}^P) > 0 \text{ is not permissible and} \\ &F(\sigma_{ij}, \epsilon_{ij}^P) = 0 \text{ for plastic deformation} \end{aligned} \tag{14}$$

This can be rewritten in terms of equivalent plastic strain as

$$dF = \left\langle \frac{\partial F}{\partial \sigma} \right\rangle \{d\sigma\} + \frac{\partial F}{\partial \kappa}\frac{\partial \kappa}{\partial \bar{\epsilon}^P} d\bar{\epsilon}^P = 0 \tag{15}$$

where

$$d\bar{\epsilon}^P = \lambda \overline{Q};$$

$$\overline{Q} = \sqrt{\frac{2}{3}} \left(\frac{\partial Q}{\partial \sigma_{ij}} \frac{\partial Q}{\partial \sigma_{ij}} - \frac{1}{3} \frac{\partial Q}{\partial \sigma_{mm}} \frac{\partial Q}{\partial \sigma_{nn}} \right)^{1/2}$$

$Q =$ plastic potential $= F$ for associated flow rule.

From Eq 4b, the hardening term in the elasto-plastic matrix can be replaced by

$$\left\langle \frac{\partial F}{\partial \bar{\epsilon}^P} \right\rangle \left\{ \frac{\partial Q}{\partial \sigma} \right\} = \frac{\partial F}{\partial \kappa} \frac{\partial \kappa}{\partial \bar{\epsilon}^P} \overline{Q} \tag{16}$$

From Eq 5

$$\frac{\partial F}{\partial \kappa} = -1 \tag{17}$$

and

$$\frac{\partial \kappa}{\partial \bar{\epsilon}^P} = -a\kappa_P/(a + b\bar{\epsilon}^P)^2 \tag{18}$$

It is possible to determine the a and b parameters from the triaxial test results assuming homogeneous deformation. In order to fit the hyperbola in the post peak range, Eq 6b can be rewritten in a linear form as

$$\left(\frac{\bar{\epsilon}^P}{1 - \kappa/\kappa_P} \right) = a + b\,\bar{\epsilon}^P \tag{19}$$

By plotting ($\bar{\epsilon}^P/1 - \kappa/\kappa_P$) as the ordinates and $\bar{\epsilon}^P$ as the abscissae, the y intercept will yield the a parameter, and the slope of the straight line will be equal to the b parameter. Detail of the determination of these parameters can be found in Chan [*15*]. Typical ranges of values for parameters a and b for a clay near Edmonton, Alberta, are 3 to 5 and 1 to 3, respectively.

Finite-Element Model for the Vane Test

The finite-element mesh for the vane test is shown in Fig. 2. Two-dimensional 8-node quadrilateral and 6-node triangular isoparametric elements were used with a total of 200 elements and 615 nodes. The mesh is an idealization of the section at the mid-height of the vane. Although it is possible to perform three-dimensional analysis using the computer code SAFE (Soil Analysis by Finite Element), which is used in this study, the present work assumes plane strain conditions, neglecting, therefore, the end effect existing in a real vane test of finite length. Previous publications [*8,10,*] indicate that the shear stress distribution in the vertical direction is nonuniform.

Despite its limitations for some cases, the two-dimensional model is able to take into account the necessary conditions (see Bjerrum [*14*]) for progressive failure to occur and is sufficient to illustrate the influence of this mechanism when performing the vane test in strain softening materials.

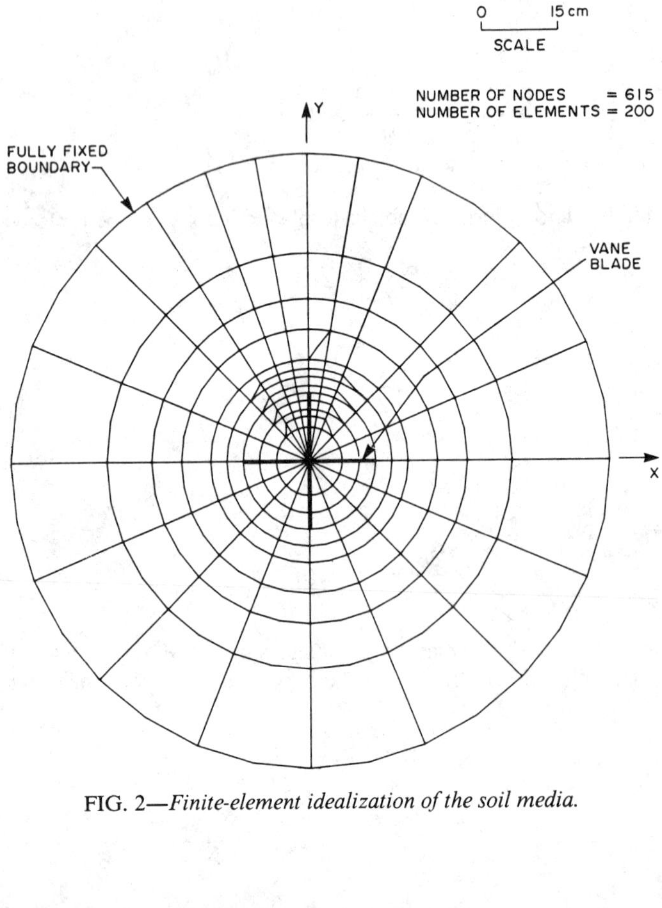

FIG. 2—*Finite-element idealization of the soil media.*

Since there is no symmetry, the entire mesh will include all four blades with the diameter of the blades equal to 9 cm. In principle the stress distribution will be the same at each quadrant of the vane; therefore mesh refinement is performed at the tip of only one blade to obtain better resolution of stresses at this location. To model an infinite media, the boundary of the mesh is taken to be 20.25 cm from the axis of rotation of the vane. During rotation of the blades, the soil will be in compression on one side of the blade, and tension is developed on the other side. In reality, a small separation will occur in this tension zone. To model the separation condition, a small gap is introduced between the blade and the soil.

The blades were rotated incrementally in the counter-clockwise direction by prescribing displacements at the nodes of the elements adjacent to the blades. It is expected that the torque will decrease after the peak for the strain softening soil; therefore the rotation of the blades is prescribed instead of prescribing the torque in order to obtain the post-peak behavior of the test. Since displacements along the blades are prescribed directly at the nodes of the soil elements, the blades are therefore modeled as rigid and infinitesimally thin.

Results of the Finite-Element Analysis

The four stress-strain relationships that have been used are shown in Fig. 3. Materials 1 and 4 are modelled as elastic perfectly plastic materials with strength equal to the peak and residual strengths, respectively. Materials 2 and 3 have the same peak strength as Model 1 and same residual strength as Model 4. Materials 2 and 3 have different rates of post-peak softening. An elastic modulus of 6000 kPa and peak and residual undrained shear strengths of 50 and 30 kPa with Poisson's ratio of 0.45 were used in the analysis.

Figure 4 shows the torque-rotation relationship for all four analyses. It is seen that the

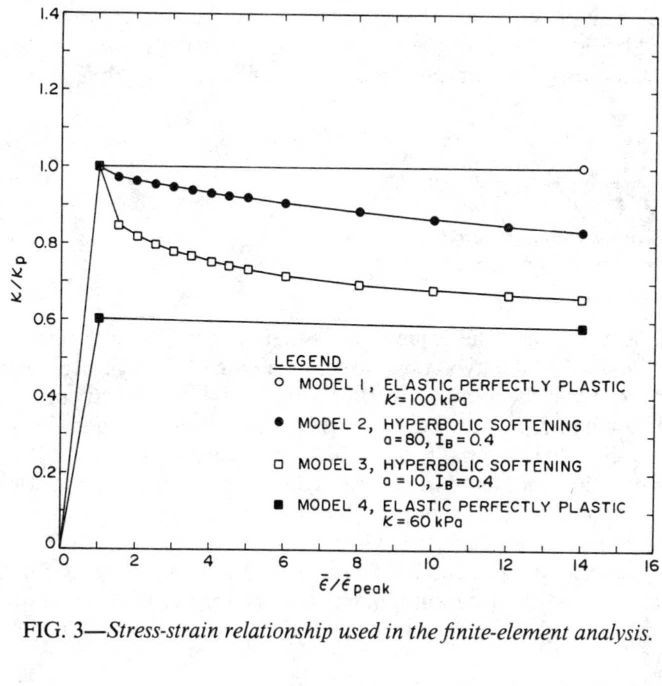

FIG. 3—*Stress-strain relationship used in the finite-element analysis.*

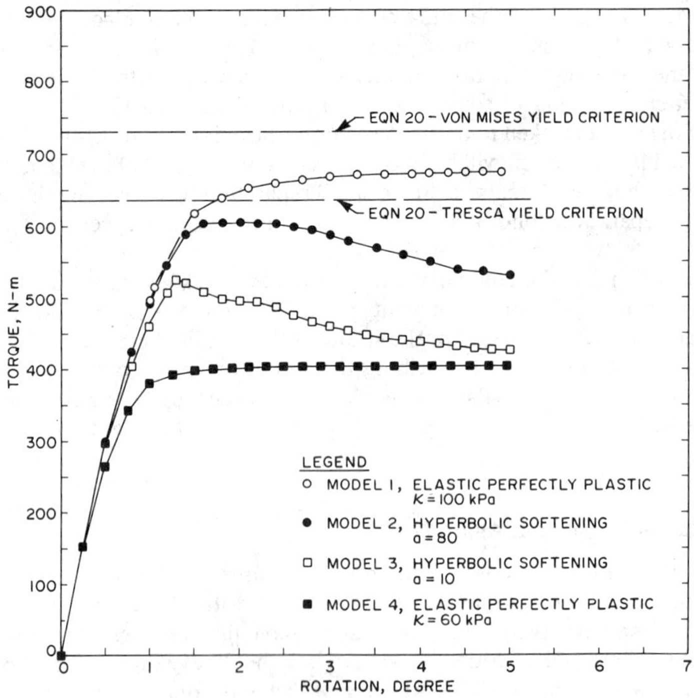

FIG.4—*Torque-rotation relationship of the vane tests.*

maximum torque of the test varies with the rate of softening. For the elastic perfectly plastic material, the rate of softening is zero, therefore the maximum torque is close to the value calculated based on peak strength of the material assuming uniform strength distribution. The calculation of maximum torque based on peak strength is given by

$$\tau_{max} = \pi D^2/2C_u \tag{20}$$

where

C_u = undrained shear strength;
D = diameter of vane, and
τ_{max} = maximum torque.

For C_u = 50 kPa, D = 9 cm, the value of τ_{max} is equal to 636 N·m. Equation 20 assumes Tresca yield criterion. For the hyperbolic softening model, the von-Mises yield criterion is used. Therefore based on the von-Mises criterion assuming pure shear condition, the maximum torque is equal to 735 N·m. The finite-element solution shown in Fig. 4 lies between these two values since the material is not subjected to pure shear around the vane.

The maximum torque measured from the test decreases with increasing rate of softening. Moreover, the rate of decrease in torque after peak is also dependent on the rate of softening. Since the rate of softening of the material is not known without laboratory tests, the calculation of the peak strength based on maximum torque will lead to incorrect results. Figure 4 also shows that the rotation required to reach maximum torque decreases with increasing rate of softening. At higher rates of softening, the propagation of the yield zone is faster, which leads to a lower maximum peak torque and a smaller angle of rotation.

Figures 5 through 7 show the variation of tangential shear stress at an angle 30° from the y axis along the radial distance from the center of the vane. The shear stress increases to its maximum value at the tip of the blade. For the cases of elastic perfectly plastic materials, the shear stress remains practically unchanged as shown in Fig. 5 when rotation increases from 1.5° to 5.0°. In the case of strain softening materials, the shear stress at the tip of the blade starts to decrease after reaching a maximum value as shown in Fig. 6. The decrease in shear stress is due to local post peak deformation of the strain softening material. At higher rates of softening the decrease in shear strength is more pronounced after peak as shown in Fig. 7.

The progressive failure mechanism is best illustrated by Figs. 8 through 10, which show the tangential shear stress around the vane. For the elastic perfectly plastic cases, Fig. 8 indicates that the shear stress reaches its maximum value at around 2.4° of rotation with no decrease in shear stress after peak. No progressive failure develops in this case. In the interpretation of these results, yielding can arise from shearing of the soil in front of the blade or due to the development of tension behind the blade. The model can yield in compression and in tension, and thus yielding propagates from both sides of the blade. For the case of strain softening materials, part of the soil is subjected to pre-peak behavior and part of the soil is at post-peak deformation as shown in Figs. 9 and 10. The point of the maximum strength propagates from the edge of the blade. Post-peak deformation is indicated by a decrease in shear stress after peak, which occurs first near the edge of the blade. The effect is even more pronounced at higher rates of softening as shown in Fig. 10.

The development of the yield zone for the above cases is illustrated in Figs. 11 and 12.

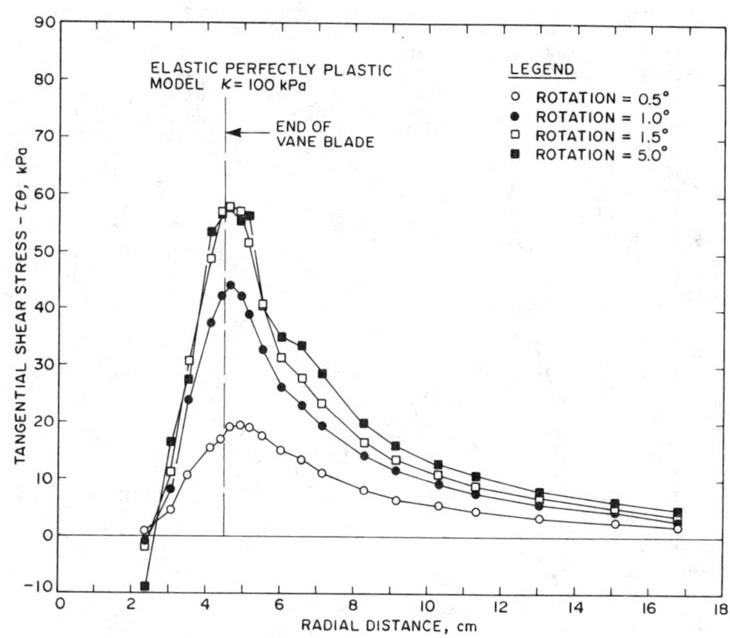

FIG. 5—*Tangential shear stress in the radial direction, Model 1.*

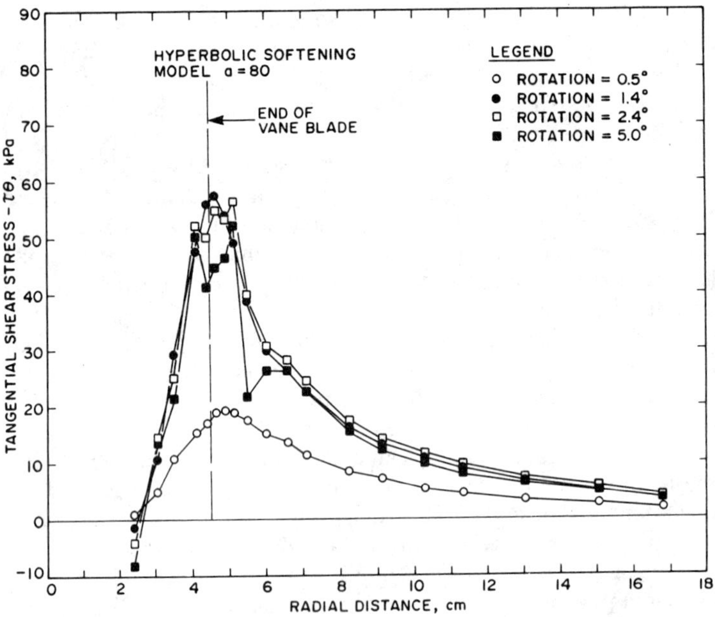

FIG. 6—*Tangential shear stress in the radial direction, Model 2.*

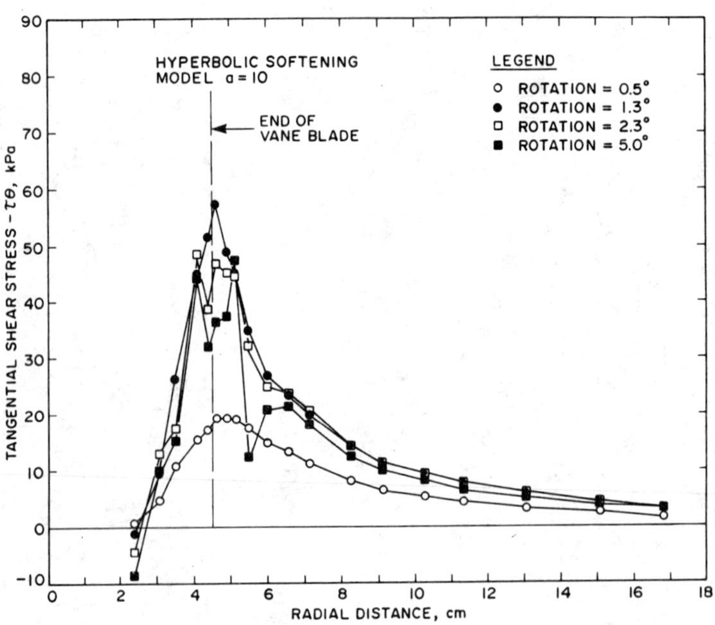

FIG. 7—*Tangential shear stress in the radial direction, Model 3.*

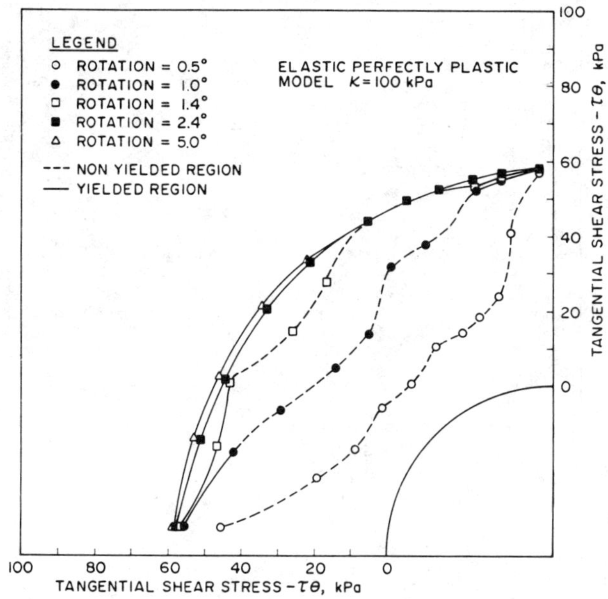

FIG. 8—*Tangential shear stress around the vane, Model 1.*

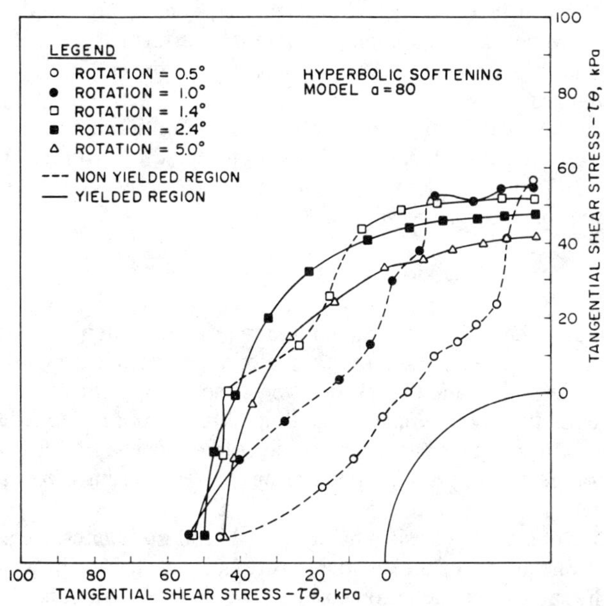

FIG. 9—*Tangential shear stress around the vane, Model 2.*

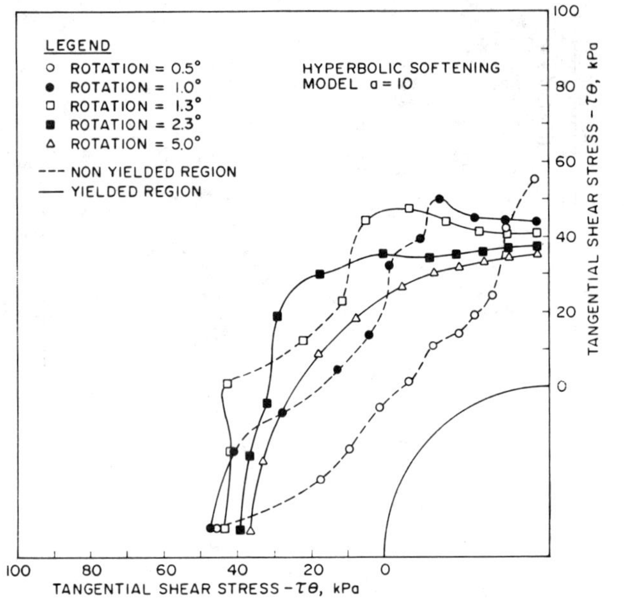

FIG. 10—*Tangential shear stress around the vane, Model 3.*

The zones of yielding are very localized for both elastic perfectly plastic and strain softening materials. Yielding occurs in a relatively narrow region forming a ring around the vane. Yielding starts near the blade and progressively spreads around the vane. Maximum torque is reached when the ring of yielded material has formed. Further rotation leads to a decrease in torque for the strain softening material, but the size of the yield zone remains practically unchanged.

The concentration of shear strains in a relatively thin zone around the vane can be observed in the experimental laboratory results presented by Kimura and Saitoh [*16*], as shown in Fig. 13. The finite-element results presented above show a similar pattern of localized deformation around the vane.

Conclusion

It has been shown that in a strain-softening material, the interpretation of the vane test depends upon the complete stress-strain curve of the material. Progressive failure develops, and the maximum torque observed in the vane test depends upon the rate of post-peak softening exhibited by the material. Both maximum rotation to failure and post-peak torque-rotation are also influenced by the strain-softening behavior. Calculations based on a hyperbolic strain-softening model material have been presented to illustrate these features of the vane test.

The vane test is one of the classical in-situ tests of soil mechanics. Neglect of progressive failure underlies one aspect of the test that has to be corrected for in practice in an empirical manner. The same dependence on complete stress-strain curve is to be anticipated for other in-situ tests involving shear failure. All such tests require knowledge of the complete stress-strain curve to aid in their interpretation. Therefore, except in the case of the simplest soils, laboratory tests on undisturbed samples are essential even where reliance is placed on extensive in-situ testing.

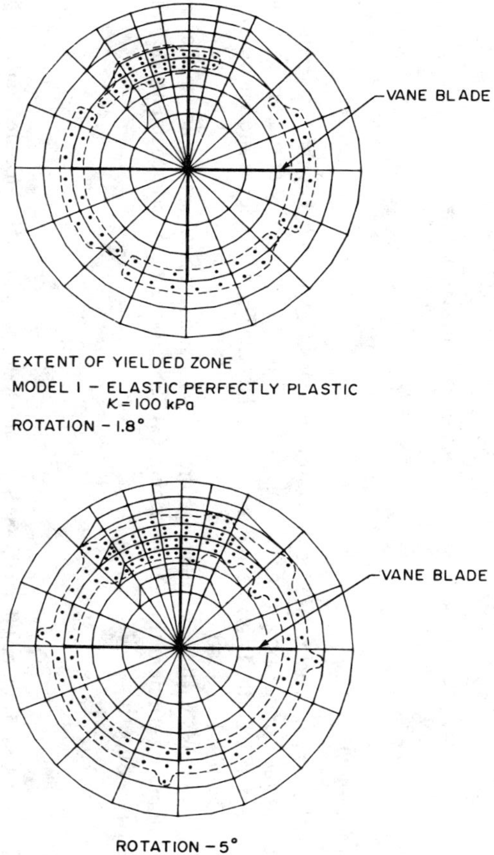

FIG. 11—*Yield zone around the vane, Model 1.*

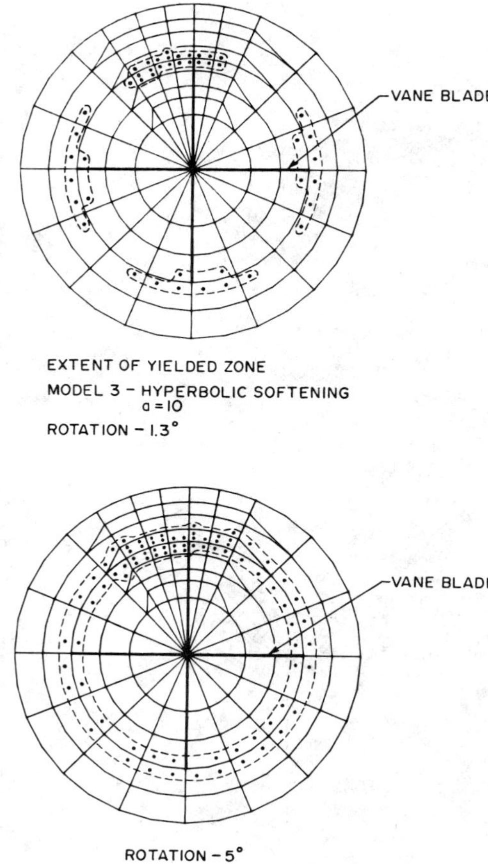

FIG. 12—*Yield zone around the vane, Model 3.*

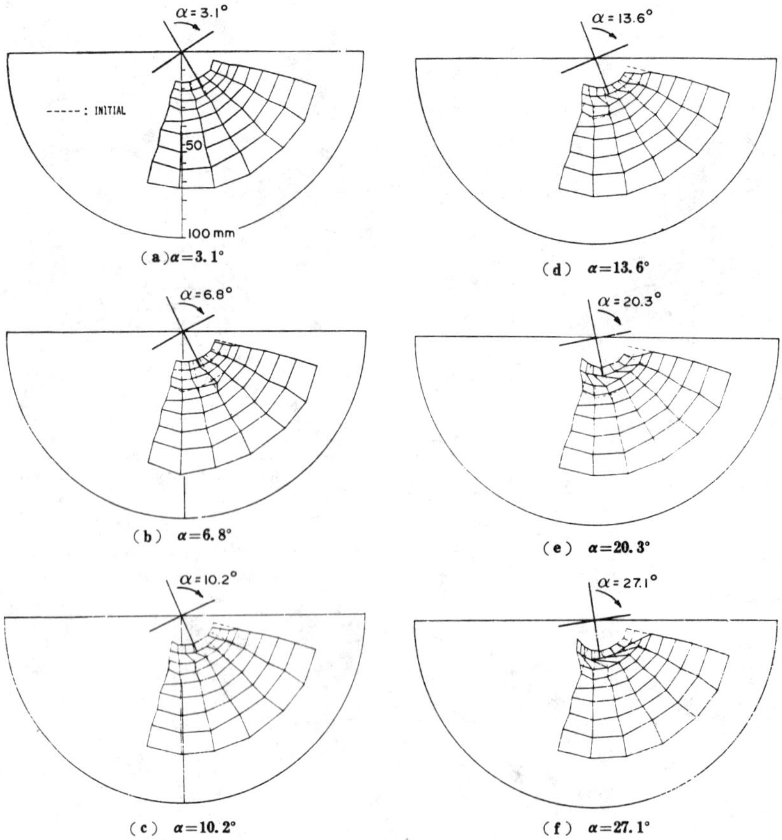

FIG. 13—*Movement of lead shot during the rotation of vane* [16] .

Acknowledgment

The authors wish to acknowledge the computing facilities provided by the University of Alberta and the Super-Computer facilities provided by the University of Calgary and the Government of Alberta. This research is supported by the Natural Science and Engineering Research Council of Canada.

The authors also wish to acknowledge the financial support given to the first author by the Brazilian Government through its agency CAPES (Coord. de Aperfeicoamento de Pessoal de Nivel Superior).

References

[*1*] Aas, G., "A Study of the Effect of Vane Shape and Rate of Strain on the Measured Values of In Situ Shear Strength of Clays," *Proceedings of the 6th International Conference of Foundation Engineering,* Montreal, Quebec, Canada, Vol. 1, 1965, pp. 3–8.

[*2*] Aas, G., "Vane Tests for Investigation of Anisotropy of Undrained Shear Strength of Clays," *Proceedings of the Geotechnical Conference,* Oslo, Norway, Vol. 1, 1967, pp. 3–8.

[*3*] Flaate, K., "Factors Affecting the Results of Vane Tests," *Canadian Geotechnical Journal,* Vol. 3, 1967, pp. 18–31.

[4] La Rochelle, P., Roy, M., and Tavenas, F., "Field Measurements of Cohesion in Chaplain Clays," *Proceedings of 8th International Conference on Soil Mechanics and Foundation Engineering,* Moscow, Vol. 1.1, 1973, pp. 229–236.

[5] Weilsel, C. E., "Some Factors Influencing In Situ Vane Test Results," *Proceedings of the 8th International Conference on Soil Mechanics and Foundation Engineering,* Moscow, Vol. 1.2, 1973, pp. 475–479.

[6] Torstensson, B. A., "Time Dependent Effects in the Field Vane Test," *Proceedings of the International Symposium on Soft Clays,* Bangkok, 1977, pp. 387–397.

[7] Menzies, B. K. and Mailey, L. K., "Some Measurements of Strength Anisotropy in Soft Clays Using Diamond Shaped Shear Vane," *Geotechnique,* Vol. 26, 1976, pp. 535–538.

[8] Menzies, B. K. and Merrifield, C. M., "Measurements of Shear Distribution at the Edges of a Shear Vane Blade," *Geotechnique,* Vol. 30, No. 3, 1980, pp. 314–318.

[9] Cadling, L. and Odenstad, S., "The Vane Borer, an Apparatus for Determining the Shear Strength of Clay Soils Directly in the Ground," *S.G.I., Proceedings No. 2,* 1950.

[10] Donald, I. B., Jordan, D. O., Parker, R. J., and Toh, C. T., "The Vane Test—A Critical Appraisal," *Proceedings of the 9th International Conference on Soil Mechanics and Foundation Engineering,* Tokyo, Vol. 1, 1977, pp. 81–88.

[11] Matsui, T. and Abe, N. "Shear Mechanism of Vane Test," *Soils and Foundations,* Vol. 21, No. 4, 1981, pp. 69–80.

[12] Lo, K. Y., "An Approach to the Problem of Progressive Failure," *Canadian Geotechnical Journal,* Vol. 9, 1972, pp. 407–429.

[13] Bathe, K. J., *Finite Element Procedure in Engineering Analysis,* Prentice Hall Inc., N.J., 1982.

[14] Bjerrum, L., "Progressive Failure in Slopes of Over-consolidated Plastic Clay and Clay Shales," *Proceedings of the American Society of Civil Engineers,* Vol. 93, No. SM5, 1967, pp. 1–49.

[15] Chan, D., "Finite Element Analysis of Strain Softening Materials," unpublished Ph.D thesis, University of Alberta, Edmonton, Canada, 1986.

[16] Kimura, T. and Saitoh, K., "Effect of Disturbance Due to the Insertion on Vane Shear Strength of Normally Consolidated Cohesive Soils," *Soils and Foundations,* Vol. 23, 1983, pp. 113–124.

Ronald C. Chaney[1] *and Gregory N. Richardson*[2]

Measurement of Residual/Remolded Vane Shear Strength of Marine Sediments

REFERENCE: Chaney, R. C. and Richardson, G. N., **"Measurement of Residual–Remolded Vane Shear Strength of Marine Sediments,"** *Vane Shear Strength Testing in Soils: Field and Laboratory Studies, ASTM STP 1014,* A. F. Richards, Ed., American Society for Testing and Materials, Philadelphia, 1988, pp. 166–181.

ABSTRACT: The measurement of both residual and remolded shear strength using the vane shear apparatus is discussed. Residual shear strength is defined to be measured after 90° of vane rotation. In contrast, remolded shear strength is dependent on whether it is measured by vane remolding or hand remolding. To accomplish vane remolding, a minimum of three vane revolutions are required. The relative strengths of the various remolding methods show that (1) field vane remolding gives highest strength, (2) followed by either laboratory vane remolding or hand remolding. Order of the last two appears to depend on soils plasticity. Vane remolding is shown to be influenced by a soils anisotropy while hand remolding is not. Case studies are presented for a DSDP site in the North Pacific and the Mississippi Fan in the Gulf of Mexico. Standardization and measurement procedures to obtain repeatable and comparable results are presented.

KEY WORDS: vane shear, anisotropy, remolding, shear strength, marine, North Pacific, Gulf of Mexico, Mississippi Fan, residual strength, sensitivity

Laboratories typically perform two vane shear tests at each location, one to obtain the undisturbed undrained vane shear strength S_u and another to obtain an undrained shear strength S_{ur} on the same material after destruction of the structure along the surface of sliding. A sensitivity of the soil S_t is defined as S_u/S_{ur}. The term, sensitivity, indicates the effect of remolding on the consistency of a clay, regardless of the physical nature of the causes of the change [*1*]. The vane shear strength test on the disturbed material is performed in the same manner as the original test. Between the two tests the sample's original structure is destroyed, either by physically removing the sample and mixing it by hand with a spatula [*2*] or with a mechanical mixer, or by rapidly rotating the vane within the solid through several revolutions [*3*]. Few comparisons of the two techniques are available, although Richards [*4*] indicated that soil remolding by hand, with subsequent retesting by laboratory vane, yields a somwhat lower value of shearing strength on the disturbed material. Higher sensitivity values thereby result.

The resulting vane shear strength after destruction of the soil structure has been variously called the remolded strength, residual strength, or ultimate strength. The actual term to be used is dependent upon how the soil structure was destroyed. The process of kneading or working a clay is commonly referred to as remolding. Soils that have had their natural structure modified by manipulation in this manner are called remolded soils (ASTM Terms and Symbols Relating to Soil and Rock [D 653]). The softening effect is probably

[1] Professor and director, Humboldt State University Marine Laboratory, P.O. Box 690, Trinidad, CA 95570.

[2] Consultant, Soil and Material Engineers, Raleigh, NC.

due to two different causes: (1) destruction of the orderly arrangement of the molecules in the adsorbed layers and (2) collapse of the structure that the clay acquired during the process of sedimentation. In contrast, the residual or ultimate shear strength was defined by Skempton [5] as the ultimate shearing resistance after large displacements under fully drained conditions. After a surface of sliding forms and extensive slip occurs, the bonds between the soil particles are destroyed and the particles along the surface of sliding assume an orientation favorable to a low resistance to shear along the surface.

In the following sections the variation of shear strength as effected by the following variables will be discussed: (1) remolded or residual vane strength as compared to the undisturbed vane shear strength, (2) effect of anisotropy on the variation of the measured vane shear strength values, (3) effect of the number of revolutions of the vane on the determination of vane remolded shear strength, and (4) the effect of the liquidity index on the remolded shear strength. This will be followed by case studies from the North Pacific and the Gulf of Mexico.

Vane shear strength results presented in this paper were performed using a Wykeham/Farrance device with torsional springs. Tests were conducted normally at 9°/min unless noted otherwise. A test series was conducted using various rates of vane rotation on a pelagic clay material from the North Pacific as shown in Fig. 1. A review of Fig. 1 shows that for the test conditions and material being tested there was no effect of rotation rate on the resulting vane shear strength. A similar test program showed the same effect for the other materials used in this study.

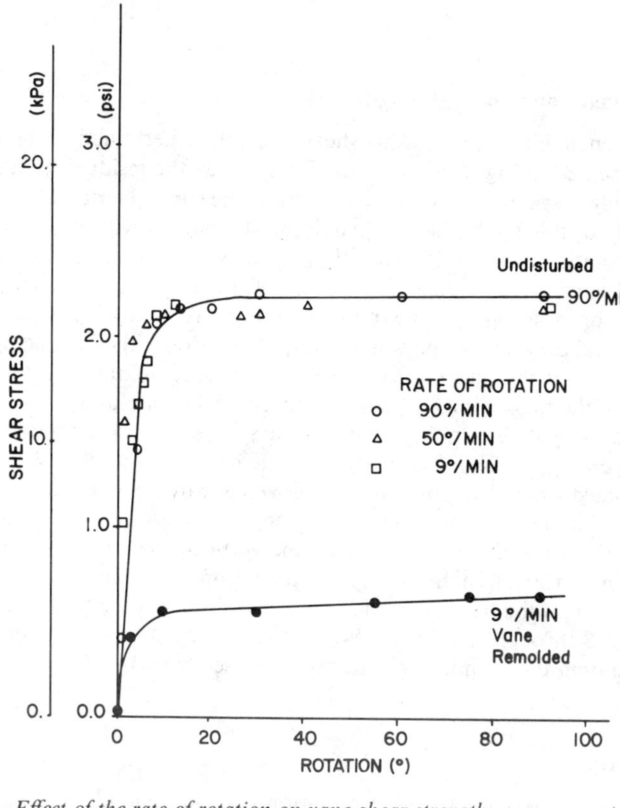

FIG. 1—*Effect of the rate of rotation on vane shear strength measurements using a torsional spring, DSDP Leg 96, Site 617A, Core 4-3A, undisturbed.*

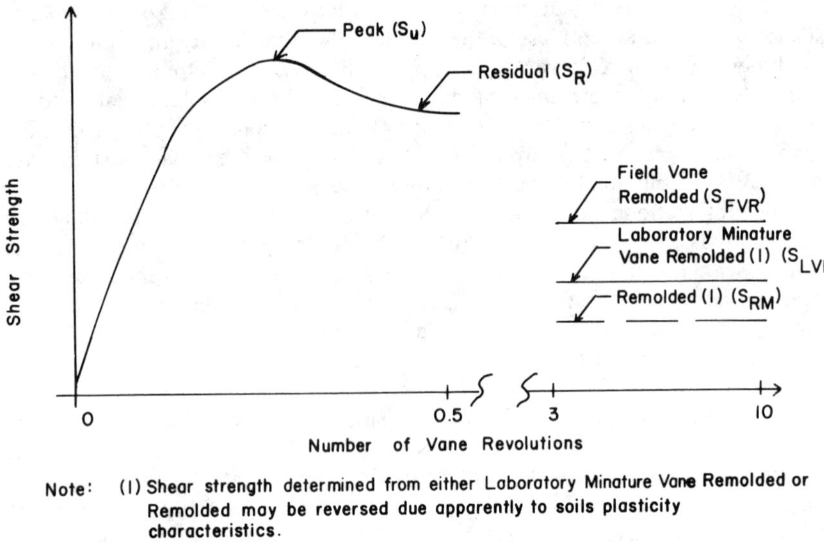

FIG. 2—*Relationship between strengths measurements determined by the vane shear strengths.*

Remolded/Residual Vane Shear Strength

An idealized schematic of the various shear strengths determined using the vane shear apparatus is presented in Fig. 2. A review of Fig. 2 shows the residual strength, also called the ultimate strength as defined by Skempton [*5*], is the strength after a large deformation. For the purposes of this study the residual shear strength S_R will be defined as the vane strength after a rotation of 180° (0.5 revolutions of the vane). In contrast, the determination of the remolded vane shear strength depends on whether the test was conducted using an in-situ vane S_{FVR} or in the laboratory. In the laboratory, remolded vane shear strength will be shown to depend on whether the soil was remolded using either a vane S_{LVR} or by hand manipulation S_{RM}. For the purposes of this paper, vane remolding will be considered an approximation of the more thorough hand or mechanical remolding.

A comparison of relative remolded vane shear strengths as determined by these three methods is $S_{FVR} > S_{LVR}$ or S_{RM}. The reason for the observed variation between field and laboratory remolded vane shear strengths is believed due to the reduction in applied stress, [*6,7*]. Law [*8*] has shown that the vane shear strength increases with increasing horizontal stress and depends to a much lesser degree on the vertical stress. In contrast, the variation between the vane shear strengths as determined from the laboratory miniature vane remolded test S_{LVR} and the hand remolded vane test S_{RM} can vary by a factor of up to 2.5 as shown in Figs. 3 through 6. Whether S_{RM} or S_{LVR} is largest, based on very limited data, appears to depend on the liquidity index of the soil (see Fig. 7).

Effect of Anisotropy

The effect of anisotropy on the determination of vane remolded and hand remolded vane shear strengths was investigated using box cores from the Eel River Fan located off

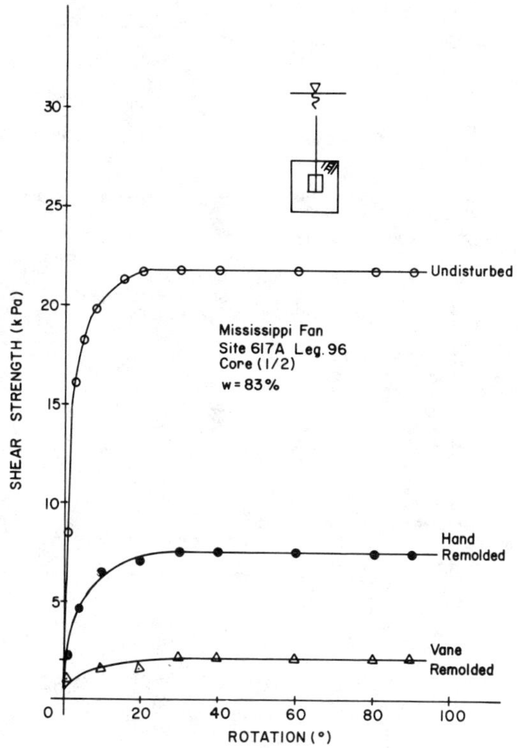

FIG. 3—*Relationship between undisturbed, hand remolded, and vane remolded laboratory miniature vane strengths, Mississippi Fan, DSDP, Leg 95, Site 617A, Core 1-2, and vertical vane orientation.*

the northern California coast near Eureka. The box cores were taken from a water depth of approximately 500 m. The soil recovered in the box core was a silty clay. Vane tests were performed in both the vertical (Fig. 4) and the horizontal (Figs. 5 and 6) directions. The horizontal tests were conducted at depths of 6 and 33 cm in the box core. A comparison between the vertical and horizontal tests indicates that the hand remolded vane tests at the top of the box core are equal (2.5 kPa). The hand remolded vane test at a depth of 33 cm is increased to a value of approximately 3.5 kPa. This behavior would be expected with increasing depth and decreasing water content.

In comparison, the shear strength as determined in the vane remolded test is dependent on the direction in which the test is performed. A review of Fig. 4 for the vertical test indicates a vane remolded shear strength of approximately 6 kPa, while in Fig. 5 for the horizontal test a remolded shear strength of approximately 3.3 kPa is determined. The undisturbed vane shear strengths presented in Figs. 4 through 6 exhibit this same behavior.

Effect of the Number of Revolutions on the Determination of the Residual Vane Shear Strength

The effect of the number of revolutions of the vane on the determination of the laboratory remolded vane shear strength for a marine clay is shown in Fig. 8. A review of Fig. 8 shows that for a case shown the shear strength decreases rapidly during the first three vane

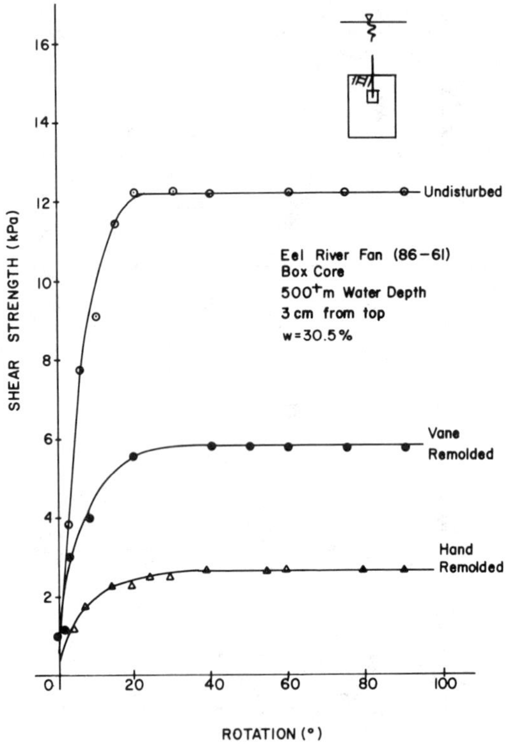

FIG. 4—*Relationship between undisturbed, hand remolded, and vane remolded laboratory miniature vane strengths, Eel River Fan (86-61), 3 cm depth, vertical vane orientation.*

revolutions. After the first three revolutions the shear strength is relatively constant over the interval from 3 to 10 vane revolutions. A summary of selected recommendations for the number of vane revolutions to determine remolded vane shear strengths is presented in Table 1. Based on the above limited information the vane remolded shear strength should be determined only after a minimum of between 3 to 5 revolutions of the vane.

Effect of the Liquidity Index on Remolded Vane Shear Strength

A number of investigators have studied the relationship between remolded shear strength and the liquidity index (LI). Pyles [*11*] showed that the ratio of vane remolded shear strength S_{LVR} to the undisturbed peak vane shear strength S_u is constant for all values of liquidity index (LI) but was dependent on whether vane strengths were determined in the field or laboratory. The vane remolded shear strength in Pyles study was determined after three revolutions. A linear relationship between LI and the logarithm of sensitivity S_t has been shown by Bjerrum [*12*] and later expanded by Eden and Hamilton [*10*]. In this work Bjerrum showed that as the liquidity index increases the sensitivity also increases. A form of this relationship can be presented as the liquidity index (LI) versus the logarithm of the remolded shear strength as shown in Fig. 14. This graph is a modified form of one previously presented by Sullivan et al. [*13*] and incorporates data from this study. The data presented in Fig. 14 were developed using different testing techniques (CU triaxial, vane,

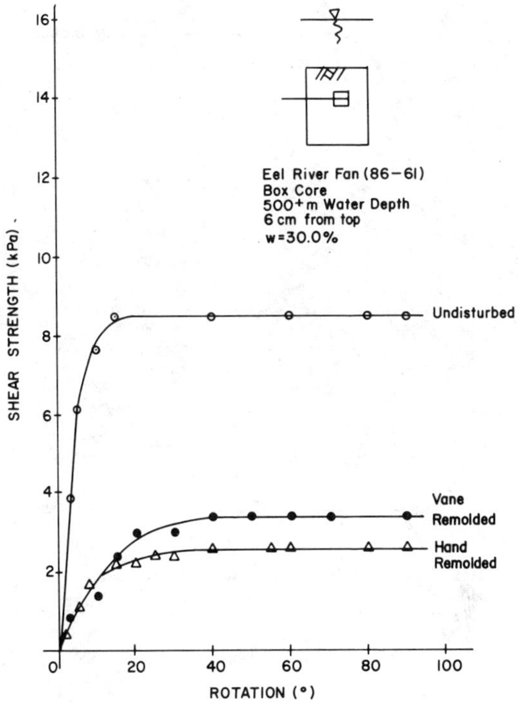

FIG. 5—*Relationship between undisturbed, hand remolded, and vane remolded laboratory miniature vane strengths, Eel River Fan (86-61), 6 cm depth, horizontal vane orientation.*

and unconfined compression tests) on remolded soil. A review of Fig. 14 shows that trends through the data from this study (dashed lines) do support the concept of a log linear relationship for each clay but not a single unique relationship as presented by Sullivan et al. [*13*]. The "uniqueness" only appears to apply for (1) a range of liquidity index, (2) a range of undrained shear strength, and (3) a specified group of clays. Various equations describing the linear relationship between LI and the logarithm of the remolded shear strength have been presented by various authors but are generally limited to the ranges in liquidity index or shear strength over which they are applicable [*13,14*].

Case Studies

Test results from three sites are presented in Figs. 9 to 13. In each case, laboratory miniature vane tests were performed to determine both the undisturbed and vane remolded shear strengths. In addition, the variation of bulk density, water content, plastic limit, and liquid limit are also presented as a function of depth.

A. North Pacific, Deep Sea Drilling Project Site 576A

The site was located near the Shatsky Rise in the Northern Pacific as shown in Fig. 9. The site consisted of two units. The upper unit was subsequently divided into two subunits (1A and 1B). Subunit 1A (0 to 28 m) consisted of a yellowish brown to brown pelagic clay

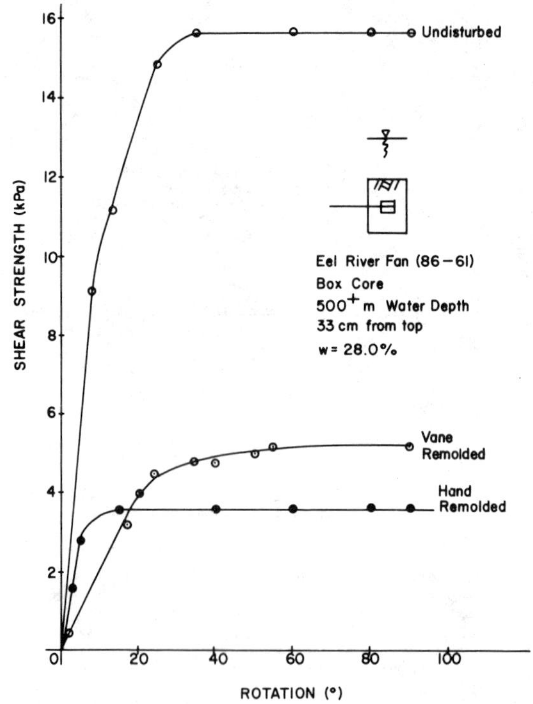

FIG. 6—*Relationship between undisturbed, hand remolded, and vane remolded laboratory miniature vane strengths, Eel River Fan (86-61), 33 cm depth, horizontal vane orientation.*

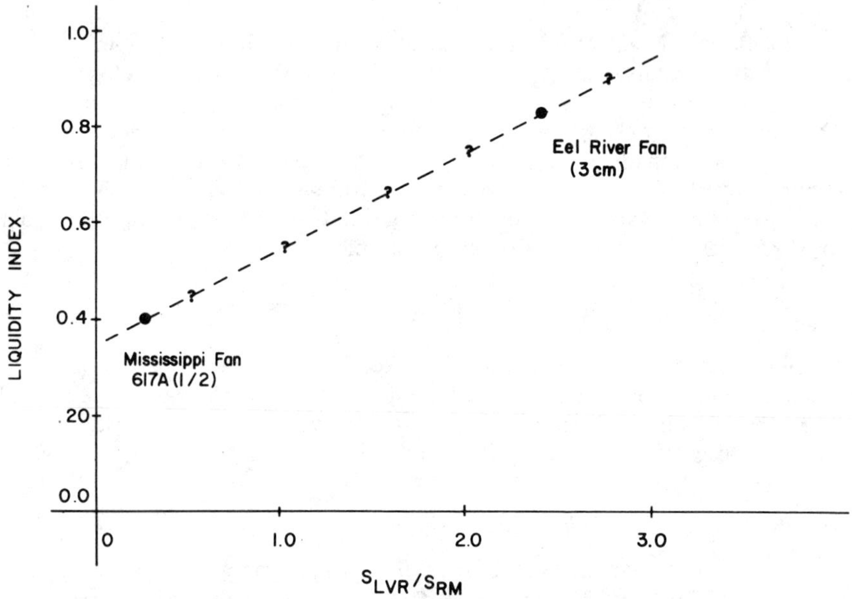

FIG. 7—*Liquidity index versus* S_{LVR}/S_{RM}.

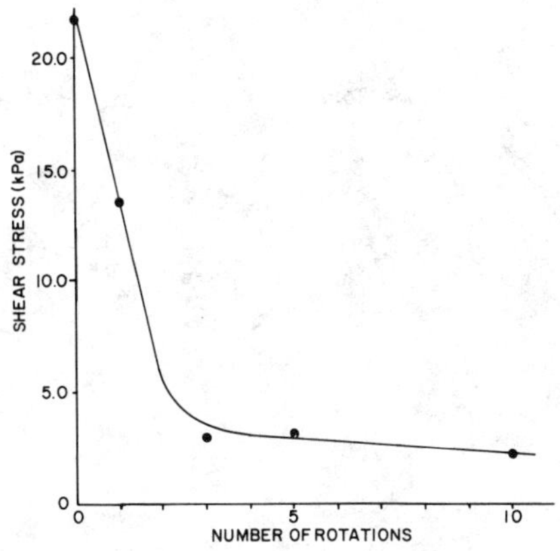

FIG. 8—*Vane remolded laboratory miniature vane strengths versus number of vane rotations, Mississippi Fan, DSDP? Leg 96, Site 617A, Core 1-2, vertical vane orientation.*

of Pleiocene and Quaternary age. Shipboard scientists estimated that this unit is largely of eloian origin. Subunit 1B (28 to 55 m) consisted of a dark brown pelagic clay, which is very fine grained. Unit II extended from a depth of 55 m to the bottom of the hole and consisted of interbedded dark brown pelagic clay similar to Subunit 1B and pale brown nannofossil ooze. A summary of the bulk density, water content, liquid, and plastic limits as well as the average and remolded laboratory vane shear strengths are summarized in Fig. 10. A review of Fig. 10 shows that the undisturbed vane shear strength is greater at all depths than the remolded vane shear strength and that the S_t is variable over depth ranging from approximately 3 at a depth of 10 m to 1.4 at a depth of 42 m.

B. Mississippi Fan, Deep Sea Drilling Project Site 616B

The site was located in the lower part of the Mississippi Fan alongside the fan channel as shown in Fig. 11. The site consisted of two units. The upper unit consisted of a thin (25-cm) layer of yellow brown marly foraminiferal ooze. Unit II extended below this layer to the bottom of the hole (0.24 to 370 m). This unit was subsequently divided into four sequences. Sequence 1 (0.25 to 65 m) consisted of silt laminated muds and fine grained turbidites with highly inclined laminae. Sequence 2 (65 to 146m) consisted of homogeneous very fine grained mud and clay with fewer thin silt laminae. Sequence 3 (146 to 250

TABLE 1—*Summary of selected recommendations for the number of revolutions to determine the vane remolded shear strength.*

Reference	Number of Vane Revolutions	
	Lab	Field
Arman et al. [*6*]	6	10
Skempton [*9*]	. . .	6 or more
Eden and Hamilton [*10*]	4	. . .
Pyles [*11*]	3	3

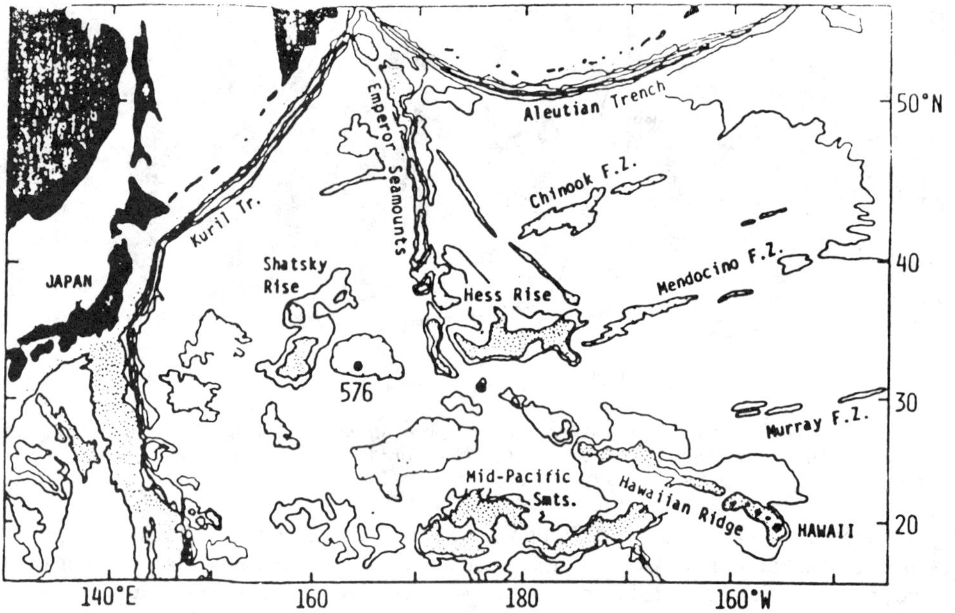

FIG. 9—*Location map of DSDP Site 576A in the Northwest Pacific.*

m) consisted of interbedded medium to fine grained sands, silty sands, lignite bearing muds and fine grained silt mud turbidites. Sequence 4 (250 to 370 m) consisted of homogeneous muds. A summary of the bulk density, water content, liquid and plastic limits, as well as the average and remolded laboratory vane shear strengths are summarized in Fig. 12. A review of Fig. 12 shows that the undisturbed laboratory vane shear strength is greater at all depths than the remolded vane shear strength and that the S_t varies from 3.5 at a depth of 70 m to 6.5 at a depth of 200m.

C. Mississippi Fan, Deep Sea Drilling Project Site 617A

The site was located near the upper portion of the Mississippi Fan in the toe area of the Massingale Slide as shown in Fig. 13. The site consisted of two units. The upper unit is thin (25 cm) consisting of an olive brown foraminifer mud. This unit overlays Unit II, which is made up of terrigenous muds and silts. Unit II is subsequently broken up into three sequences. Sequence 1 (0.25 to 46 m) consists of homogenous muds and muds with silt laminae. Sequence 2 (46 to 84 m) is composed of silt laminated mud. Sequence 3 (84 to 192 m) consists of mud with silt laminations. A summary of the bulk density, water content, liquid and plastic limits as well as the average and remolded laboratory vane shear strengths are summarized in Fig. 9. A review of Fig. 9 shows that the undisturbed laboratory vane shear strength is greater at all depths than the remolded laboratory vane shear strength and that the S_t varies from 10 at 2 m to 14 at a depth of 50 m.

A comparison of these case studies with previously published sites is presented in Table 2. A review of this table shows that for the cases presented the range of sensitivity of marine sediments varies between 2 to 14. In addition, the sensitivity is shown to increase with depth.

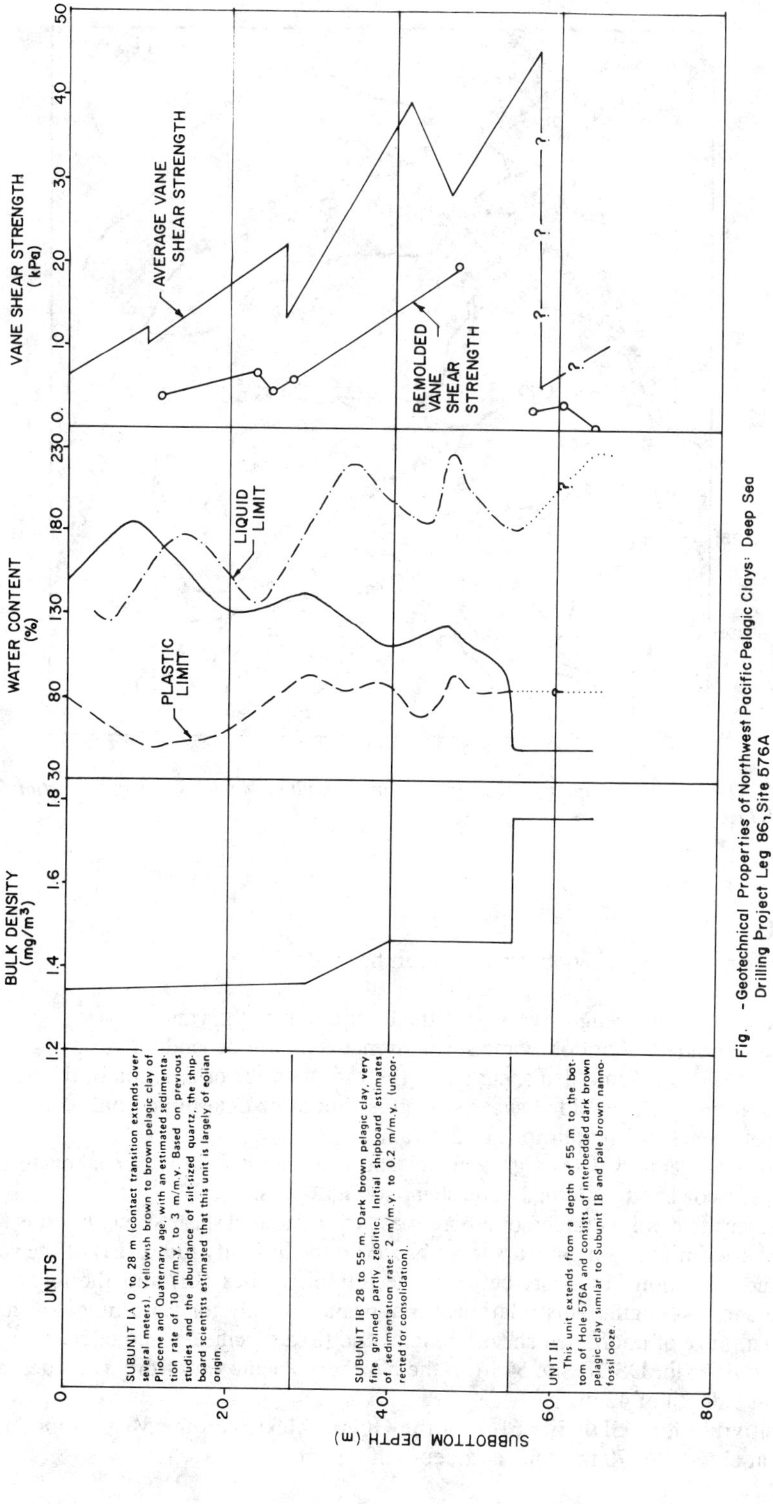

FIG. 10—*Geotechnical properties of Northwest Pacific Pelagic clays: DSDP, Leg 86, Site 576A* [15].

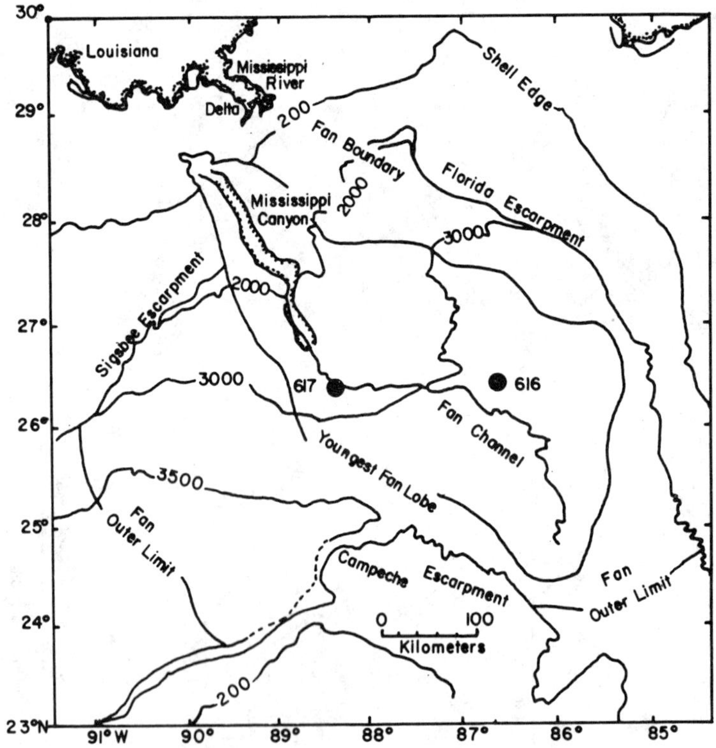

FIG. 11—*Location map of Mississippi Fan in the Gulf of Mexico showing Sites studied, DSDP Leg 96.*

Summary

In summary the following observations can be made.

1. Peak vane shear strengths vary with the orientation of the vane.
2. Hand remolded laboratory vane shear strengths do not depend on vane orientation.
3. Laboratory vane remolded shear strengths vary with the orientation of the vane.
4. Laboratory vane remolded shear strengths should be determined only after a minimum of between 3 to 5 revolutions of the vane.
5. Field vane remolded tests give higher shear strengths than either laboratory vane remolded tests or hand remolded vane shear strength tests.
6. Whether the hand remolded or laboratory vane remolded shear strength test is largest, based on very limited data, appears to be based on the liquidity index (LI) of the soil.
7. A linear relationship exists between the liquidity index (LI) and the logarithm of remolded shear strength. This relationship appears to apply for (1) a range of liquidity index, (2) a range of undrained shear strength, and (3) a specified group of clays.
8. Sensitivities for DSDP Site 576A in the Northern Pacific vary from 3 at a depth of 10 m to 1.4 at a depth of 42 m.
9. Sensitivities for DSDP Site 616B in the Gulf of Mexico on the Mississippi Fan vary from 3.5 at a depth of 70 m to 6.5 at a depth of 200 m.

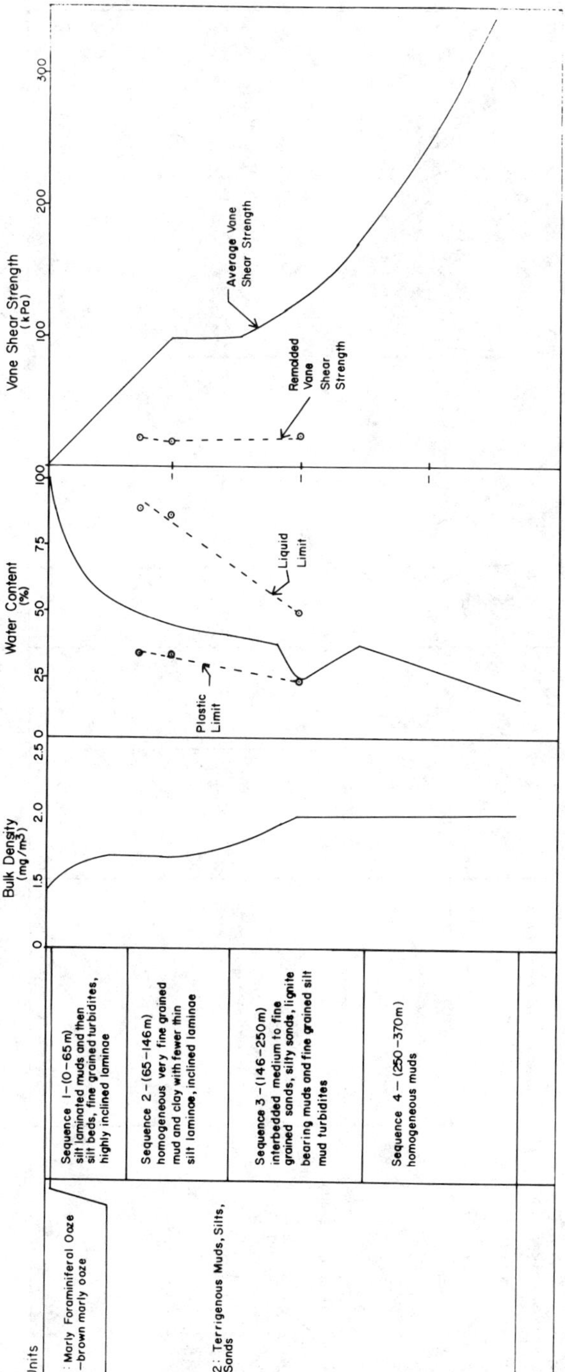

FIG. 12—*Geotechnical properties of Mississippi Fan sediments: deep sea drilling project, Leg 96, Site 616B* [16].

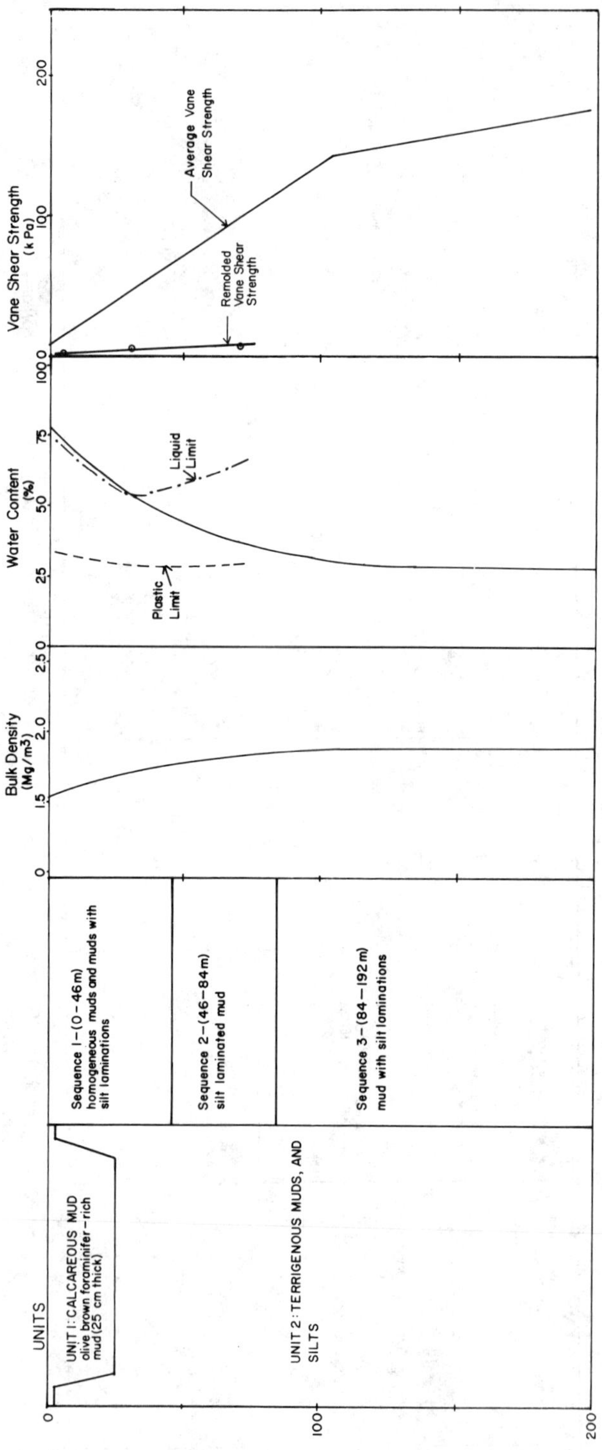

FIG. 13—*Geotechnical properties of Mississippi Fan Sediments, DSDP, Leg 96, Site 617A* [16].

TABLE 2—*Location and undrained strength properties of selected marine sediments* [17].

Sample	Water Depth, m	Location	Water Content, %	Average S_u, kPa		Sensitivity
CRUX 3	unknown	Pacific	232	12		10
CRUX 201			224	8		10
STYX 9-1C	5398	15°40′S, 172°00′W	133	13		2
STYX 9-2	5060	11°55′S, 169°32′W	284	5		4
STYX 9-3	4300	8°01′S, 166°35′W	148	6		3
STYX 9-5A	5069	8°36′N, 154°37′W	247	7		4
STYX 9-5B	5032	8°36′N, 154°18′W	388	9		4
STYX 10-1	5517	23°50′N, 143°58′W	123	4		2
STYX 10-1C	5508	23°25′N, 144°05′W	118	4		3
DSDP 576A	6217	Northern Pacific	98 to 216	upper 10 m: S_u = 0.6		2 to 10
				10 to 25 m: S_u = 0.28	. . .	. . .
		32°0′N, 164°30′W		below 25 m: S_u = 0.22		. . .
		Mississippi Fan-				
DSDP 616B	2983	Gulf of Mexico	50	70 m: 20		3.5
		26°24′N, 86°36′W	25	200 m: 130		6.5
		Mississippi Fan-				
DSDP 617A	2467	Gulf of Mexico	75	2 m: 10		10
			42	50 m: 72		14
		26°24′N, 88°24′W				

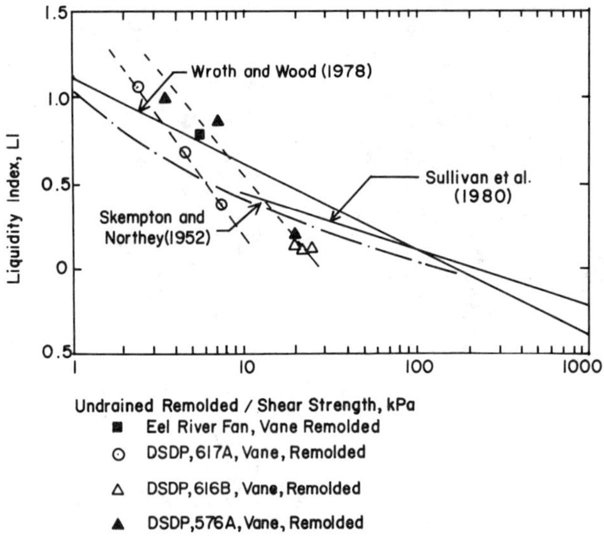

FIG. 14—*Remolded shear strength versus liquidity index.*

10. Sensitivities for DSDP Site 617A in the Gulf of Mexico on the Mississippi Fan vary from 10 at the surface to 14 at a depth of 50 m.

Acknowledgments

The data reported were derived from various studies performed by students in the Department of Environmental Resources Engineering at Humboldt State University. In particular the authors would like to acknowledge the assistance of Gretchen Rau, Brad Shipley, Michael Smith, John Steude, and Jeff Reese. In addition, the authors would also like to thank Dr. Jeff Borgeld, Department of Oceanography at Humboldt State University for box cores from the Eel River Fan. In conclusion, the authors would like to thank Dr. Adrian F. Richards for his insight into marine soil mechanics.

References

[*1*] Terzaghi, K. and Peck, R. B., *Soil Mechanics in Engineering Practice,* 2nd ed., John Wiley and Sons, New York, 1967.

[*2*] Keller, G. H., Lambert, D. N., and Bennett, R. H., Society of Economic Paleontologists and Mineralogists, Special Publication 27, 1979, pp. 131–151.

[*3*] Carius, T. and Richards, A. F., "San Diego Trough Geotechnical Test Area," *Marine Geotechnology,* Vol. 1, No. 4, 1976, pp. 345–370.

[*4*] Richards, A. F., *Deep-Sea Sediments, Physical and Mechanical Properties,* A. L. Inderbitzen, Ed., Plenum Press, New York, 1974, pp. 271–292.

[*5*] Skempton, A. W., "Long-Term Stability of Clay Slopes," *Geotechnique,* Vol. 14, No. 2, 1964, pp. 77–101.

[*6*] Arman, A., Poplin, J. K., and Ahmad, N., "Study of the Vane Shear," *Symposium on In Situ Measurement of Soil Properties,* Vol. 1, American Society of Civil Engineers, 1975, pp. 93–120.

[*7*] Fenske, C. W., "Deep Vane Tests in Gulf of Mexico," *Symposium on Vane Shear Testing of Soils, STP 193,* American Society for Testing and Materials, Philadelphia, 1956, pp. 165–25.

[*8*] Law, K. T., "Triaxial Vane Tests on a Soft Marine Clay," *Canadian Geotechnical Journal,* Vol. 16, Feb. 1979, pp. 11–18.

[9] Skempton, A. W., "Vane Tests in the Alluvial Plain of the River Forth Near Grangemouth," *Geotechnique,* Vol. 1, No. 2, 1948, p. 111.

[10] Eden, W. J. and Hamilton, J. J., "The Use of a Field Vane Apparatus," *Symposium On Vane Shear Testing of Soils, STP 193,* American Society for Testing and Materials, Philadelphia, 1956, pp. 41–53.

[11] Pyles, M. R., "Vane Shear Data on Undrained Residual Strength," *Journal of the Geotechnical Division, American Society of Civil Engineers,* Vol. 110, No. 4, Paper 18706, 1984, pp. 543–547.

[12] Bjerrum, L., "Geotechnical Properties of Norwegian Marine Clays," *Geotechnique,* Vol. 4, No. 2, 1954, pp. 49–69.

[13] Sullivan, R. A., Wright, S. J., and Senner, D. W. F., "Evaluation of Design Parameters From Laboratory Tests," *Offshore Site Investigation,* Graham and Trotman Ltd., 1980, pp. 201–216.

[14] Wroth, C. P. and Wood, D. M., "The Correlation of Index Properties with Some Basic Engineering Properties of Soils, *Canadian Geotechnical Journal,* Vol. 15, No. 2, 1978, pp. 137–145.

[15] Geotechnical Consortium, "Geotechnical Properties of DSDP Sediments from Hole 576A, Leg 86," in Initial Reports of the Deep Sea Drilling Project, Vol. 86, Scripps Institution of Oceanography.

[16] Shipboard Scientists Report, "Mississippi Fan, Sites 616B and 617A," Deep Sea Drilling Project, Leg 96, 1984.

[17] Noorany, I., "Shear Strength of Some Deep-Sea Pelagic Sediments," *Strength Testing of Marine Sediments: Laboratory and In-Situ Measurements, STP 883,* R. C. Chaney and K. R. Demars, Eds., American Society for Testing and Materials, Philadelphia, 1985, pp. 251–257.

Petrus L. M. Veneman[1] *and Tuncer B. Edil*[2]

Micromorphological Aspects of the Vane Shear Test

REFERENCE; Veneman, P. L. M., and Edil, T. B., "**Micromorphological Aspects of the Vane Shear Test,**" *Vane Shear Strength Testing in Soils: Field and Laboratory Studies, ASTM STP 1014,* A. F. Richards, Ed., American Society for Testing and Materials, Philadelphia, 1988, pp. 182–190.

ABSTRACT; Micromorphological studies of the mode of failure using thin sections of soil obtained during laboratory vane shear testing indicate that the actual failure surface is not a sharply defined cylindrical surface, but rather a shear zone. The diameter of the failure surface at large angles of vane rotation in a soft and a very soft silty clay appeared to be equal to the diameter of the vane device. The shear structures that develop as the vane rotates may or may not fully develop to a continuous circular surface when the maximum torque is reached depending on the consistency of the soil. The results indicate that undrained shear strength calculations based on the assumption of a fully developed cylindrical surface may somewhat underestimate the actual soil strength, but that the magnitude of this deviation is soil type dependent.

KEY WORDS: undrained strength, soil failure, laboratory vane shear, micromorphology

The vane shear test (hereafter referred to as VS-test) is widely used to determine the in-situ undrained strength of cohesive soils. The first large scale VS studies were performed in Sweden, and the results of that research were reported by Cadling and Odenstad [*1*]. Somewhat earlier, in 1948, Skempton [*2*] presented the results of some VS-tests in England. Both papers indicated that the VS-test results, especially in soft clays, approximated the calculated in-situ shear strength for actual soil failures more closely than data obtained from compression tests in the laboratory. Leussink and Wenz [*3*] determined that test embankment and loading test results in mineral soil agreed well with the VS strength values. Vane shear test results from an organic-rich deposit however were unreliable due to distortions caused by the fibrous character of the soil. Bjerrum [*4*] cited several extensive studies of actual failures in mineral soils, which demonstrated that the VS-test overestimated the undrained soil shear strength. He concluded that the discrepancy between the vane and the field shear strengths was larger in high plasticity clays and therefore proposed the use of an empirical correction factor, based on the relationship between plasticity index and the undrained shear strength [*4,5*]. Application of this factor, however, is not always satisfactory, as La Rochelle et al. [*6*] reported disagreement with Bjerrum's correction factors based on field observations.

Possible causes for the difference between the various test results are soil anisotropy, differences in strain rate, soil type, and sample disturbance before testing. Most vane units have a height/diameter ratio of two [*7*], which means that most of the shear strength is mobilized along a vertical cylindrical surface. Cadling and Odenstad [*1*] found that a rate

[1] Associate professor of soil science, Department of Plant and Soil Sciences, University of Massachusetts, Amherst, MA 01003.

[2] Professor of civil and environmental engineering, University of Wisconsin, Madison, WI 53706.

of 6°/min resulted in the lowest shear strength values, and this value is now the standard strain rate for the field vane test (ASTM Method for Field Vane Shear Test in Cohesive Soil [D 2573]). Disturbance caused by insertion of the vane was found to be dependent on the thickness of the wings. La Rochelle [6] reported that thicker wings resulted in lower undrained shear strength values because of greater soil disturbance and because of the induced increase in pore-water pressure in the soil surrounding the vane. This pressure has not dissipated when torque is applied shortly after the insertion of the vane. Hence, the time interval between the moment of vane intrusion and the time of failure is also of importance in this regard.

Most researchers assume the existence of a circular failure surface when calculating the undrained shear strength. This assumption is based on observations by Cadling and Odenstad [1] who studied the shape of the surface of rupture by inserting sheets of wet tissue-paper on which a spider-web-like pattern was drawn, in between slabs of soil. By comparing the disturbance in the patterns on the tissue-paper in a series of VS-tests with increasing rotation of the vane, it was concluded that the diameter of the cylinder of rupture closely coincided with the diameter of the vane. Skempton [2] found that shear strength values measured with the unconfined compression test (hereafter called U-test) were lower than those determined with the in-situ VS-test. Reasonable agreement was obtained when the diameter of the failure surface was multiplied by a factor of 1.05. This correction factor, named "effective diameter" by Skempton, was an empirical coefficient based on the assumption that the U-test data represented the true in-situ shear strength values. Later researchers [7] assumed that Skempton actually observed the diameter of the failure surface to be 5% greater than the diameter of the vane unit. Arman et al. found that the failure surface was circular in cross-section with the same diameter as that of the vane unit [7]. Adjacent to the failure surface, they also noticed a very thin, partially sheared zone. They concluded that the actual diameter of the failure surface was slightly larger than the vane diameter, but that the radius of this failure zone was soil type dependent. Wilson [8] noted through a series of photographs of the shearing planes that at the instant of maximum torque the failure surface is not circular in plan, but almost square. Only after considerable deformation takes place does a cylindrical surface form.

In a vane test failure can be expected to start in front of the edge of each wing and to advance gradually across the whole surface of rupture. Cadling and Odenstad [1] in their studies with tissue paper noted that the deformation in front of each wing seemed to be somewhat greater than behind it, but concluded that the effect of progressive failure was only slight and therefore could be ignored. One of the assumptions for the calculation of undrained strength of soils is that the maximum applied torque has to overcome the fully mobilized shear strength along a cylindrical surface. Hence, the occurrence of progressive failure may influence the final soil strength value.

In view of the above, the main objective of this study was to investigate the shape and diameter of the failure surface in the vane shear test and its relationship to degree of vane rotation.

Materials and Methods

Soil samples were prepared in the laboratory under controlled conditions using a slurry consolidometer as described by Sheeran and Krizek [9]. In this device a slurry with a high initial water content is compressed in a stepwise loading manner to allow drainage of excess water.

Two basic types of soil were used in the testing procedures each prepared from Grundite, mixed with 0% and 40% Ottawa sand, respectively. Grundite contains a considerable amount of illite, a nonexpanding aluminosilicate clay mineral. The slurry without any sand (hereafter simply referred to as illite) had an initial moisture content of 250%, while the slurry with the 40% sand (hereafter called illite/sand) was started at a lower water content (100%) to minimize possible segregation of the soil particles. The slurries were stepwise consolidated at loads of 4, 7, 14, 28, 56, 113, 225, and 450 kPa. Drainage was permitted for several days before the next load was applied. This was found to be sufficient to dissipate most of the excess pore pressure for the load increments after the first load increment. The samples were unloaded, without access to water, following the reversed loading schedule. The illite/sand mixture was only consolidated to a maximum load of 113 kPa because of equipment failure. After disassembly of the consolidometers, the resulting blocks of soil were immediately subsampled (about 6 subsamples for each soil) with polished sections of thin-wall Shelby tubing (47-mm in diameter) that were carefully pushed into the soil along the vertical axis. The resulting soil columns with a height of approximately 85 mm encased in Shelby tube sections and wrapped in plastic sheets were kept in sealed plastic bags before testing.

Unconfined compression tests [*10*] at strain rates of 0.8 mm/min^{-1} were performed on a representative soil column of each of the two soil types. Vane shear tests were also conducted with a laboratory device, manufactured by Farnell and Co., Hatfield, England. This apparatus employs a small electric motor in combination with a worm-gear to rotate one end of a calibrated spring at a rate of 10°/min with a vane attached to the opposite side of the spring. This fixed rate of rotation of this equipment is much slower than the rate used in the laboratory vane tests in recent years [*11,12*]. The undrained soil strength s_u values for both soils were obtained with a vane 12.5-mm in diameter and 12.5-mm in height with a wing thickness of 0.8 mm. This type of vane proved too small for subsequent stages of this investigation because it did not form adequately large failure surfaces, therefore, somewhat larger stainless steel vanes with a diameter of 18.5 mm, height of 18.5 mm, and a wing thickness of 0.8 mm, were constructed. These vanes had an area ratio of 22% and a perimeter ratio of 5.5% [*6*]. These values are comparable to those of other commercially available laboratory vanes.

A series of experiments were carried out such that a different degree of vane rotation was induced in each of the subsamples retained in the Shelby tubes. A vane was pushed into the soil, after which the vane was turned a predetermined number of degrees at a rotation rate of 10°/min. The calibrated spring attachment was replaced by a shaft allowing the vane to be directly rotated by the worm-gear and the angle of rotation to be directly measured. The vane then was fixed in place without allowing further rotation for about 15 min to allow dissipation of built-up pore-water pressure in the soil. This was needed to prevent a relaxation of the vane. The sample, including the inserted vane, was then rapidly frozen with dry ice. Freezing was achieved rapidly to enhance the formation of amorphous ice in the pores thus minimizing impact on the fabric. We believe that this procedure for specimen preparation probably has disturbed the fabric somewhat but not to such a degree to obscure the changes induced by the vane. A new vane and subsample then was used for the next test involving a different angle of rotation of the vane. For the illite soil the series of vane rotations consisted of 0°, 10°, 17½°, 25°, 45°, and 70°. For the illite/sand mixture the series included 0°, 10°, 17½°, 25°, 50°, and 75° of vane rotation.

All specimens were freeze-dried under vacuum and impregnated with plastic (Castolite-AP) according to the procedures described by Guertin and Bourbeau [*13*]. The soil specimens, imbedded in plastic, were cut along a plane going through the mid-height of the vane and glued onto a glass slide (50 by 50 mm in size). The specimens then were ground to produce thin sections suitable for microscopic examination. The finished thin sections, with a thickness of 30 to 50 μm, were placed in a slide projector, and the projected patterns of soil cracks were carefully traced on paper. Portions of several thin sections broke during the cutting and grinding process because of the presence of the stainless steel vanes.

Grain size distribution, Atterberg limits, and specific gravity were determined in accordance with laboratory procedures described by Lambe [*10*].

Results and Discussion

Grain size distribution curves for both soils are shown in Fig. 1. The illite curve indicates a weathered uniform soil, and the illite/sand mixture logically shows a gap-graded soil. Other physical properties of the tested soils are presented in Table 1. Both soils had a degree of saturation estimated to be about 90%. The illite soil can be described as soft silty clay, while the illite/sand mixture is a very soft silty clay according to ASTM, Classification of Soils for Engineering Purposes (D 2487). The stress-strain responses of both soils in the unconfined compression test are presented in Fig. 2. The laboratory VS-test normally measures only the maximum torque, which is converted to the undrained shear strength, but by observing the rate of deflection of the calibrated spring a stress-strain curve can be obtained, which is shown in Fig. 3. The angle of apparent rotation is the rotational angle of the drive shaft attached to the electric motor. There was no correction made for the twist of the spring. Figure 3 shows that shear strength values obtained with the laboratory VS-test are slightly higher than those determined with the unconfined compression test (Fig. 2).

The angle of apparent rotation at failure for the illite soil and the illite/sand mixture are 86° and 30°, respectively (Fig. 3). These values were obtained with a 12.5- by 12.5-mm standard miniature vane, while the thin-section studies were performed with 18.5-mm-diameter vanes. Furthermore, the angles of rotation indicated for the thin sections (Figs. 4 and 5) are absolute values, not the apparent ones, because the spring attachment was removed and the angle of rotation of the vane was directly measured. Therefore, it is reasonable to assume that the degree of vane rotation at failure in the thin-sections would be significantly smaller than the 86° and 30° given in Fig. 3, respectively, for the two soils. A

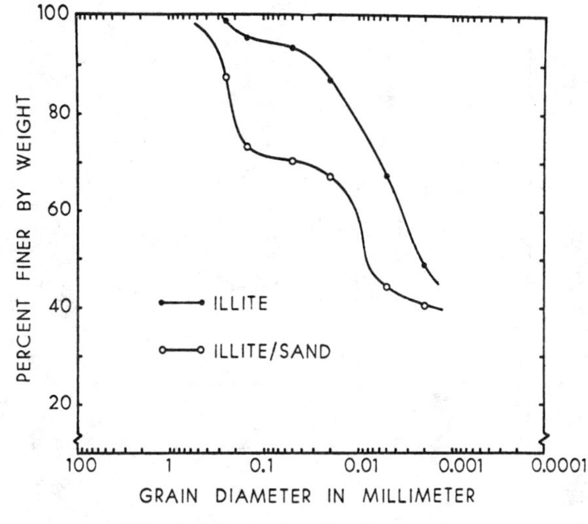

FIG. 1—*Grain size distribution chart.*

TABLE 1—*Properties of the soils tested.*

Property	Illite	Illite/Sand
Water content, %	34	35
Liquid limit, %	47	37
Plastic limit, %	28	20
Plasticity index	19	17
Liquidity index	0.3	0.9
Specific gravity	2.77	2.73
Dry unit weight, kN/m^3	13.0	13.2
Degree of saturation, %	86	93

consideration of the angle of rotation of the larger vane required to cause the same amount of relative circumferential movement as the smaller vane at failure, indicates that the degree of rotation of the larger vane would be about 58° and 17½° for the two soils in Figs. 4 and 5, respectively. These angles should be lowered some more to account for the twist of the spring included in the angle of apparent rotation. Figure 4 shows that before total failure is reached at about 45° rotation, fine cracks develop, originating from the tip of the wings of the vane and tangential to them. A comparison of these cracks at different angles of vane rotation indicates that they cannot be attributed to freezing during the specimen preparation. The configuration of the inside of the final shear surface appears roughly circular, with the same diameter as the vane device. Wilson [*8*] found that the failure surface is initially almost square, and there is some evidence of this observation in Fig. 4. The larger radius of the outer surface is caused by shrinkage of the soil upon drying and should be ignored. Examination of the thin-sections with a petrographic microscope [*14*] revealed the presence of a very thin zone of reoriented clay along the circular shear surface. This agrees with earlier observations by Arman et al. [*7*]. A minor error (about 16%) will be introduced in the strength calculations if it is assumed that failure takes place along a well

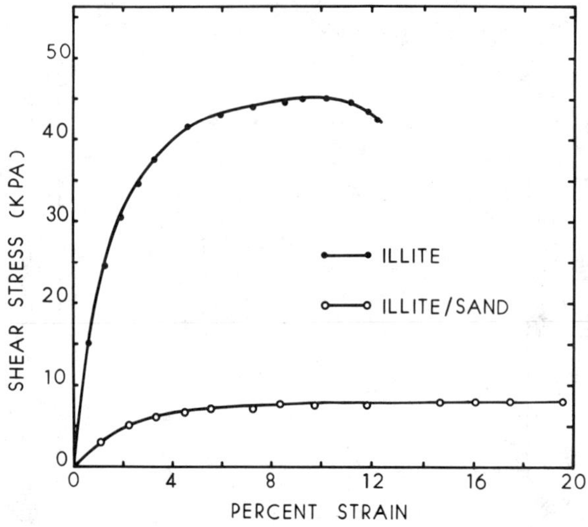

FIG. 2—*Stress-strain relationships in the unconfined compression test.*

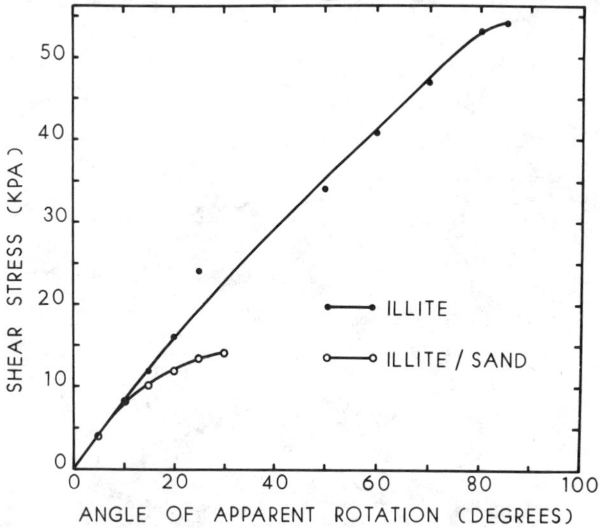

FIG. 3—*Stress versus angle of apparent vane rotation in the laboratory vane shear test. The angle of apparent rotation is the angle of rotation of the drive shaft without any correction for the twist of the spring.*

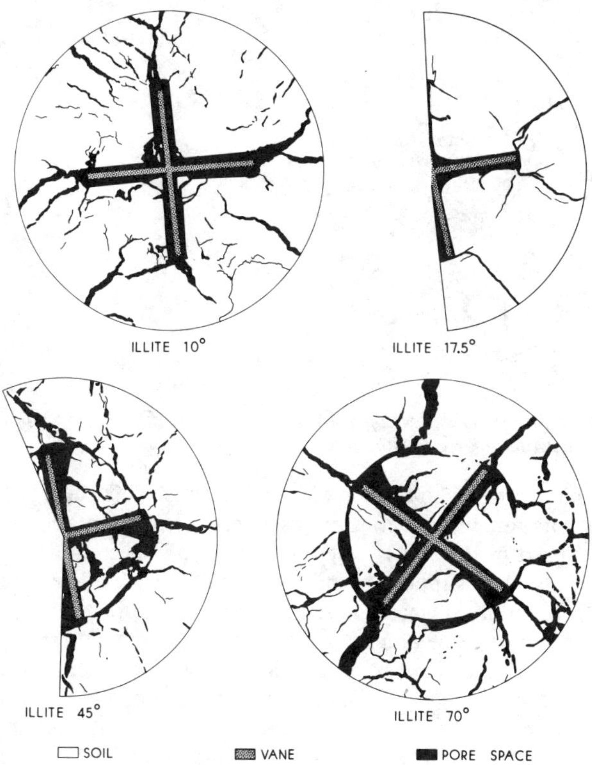

FIG. 4—*Patterns of fissures developed in the illite soil. The number under each figure indicates the actual degree of vane rotation.*

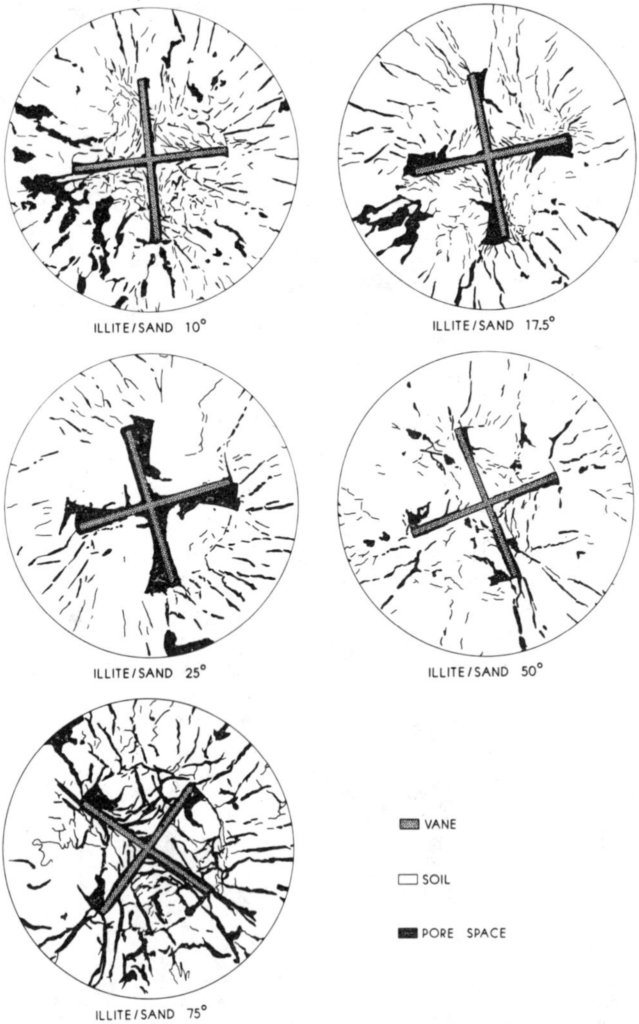

FIG. 5—*Fracture patterns developed in the illite/sand mixture. The number under each circle shows the actual degree of vane rotation.*

defined circular surface with the same radius as the vane. This effect appeared to be local at small angles of rotation without continuous, large-scale reorientation patterns. The rotational stresses induced in the soil by the vane eventually become large enough to cause local soil failures caused by slipping of soil particles over each other. The attractive forces are reduced on these failure surfaces, which results in the formation of fine fissures in the soil. Thin sections appear to be an excellent tool to study the pattern of these failure surfaces, and therefore provide clues about the failure conditions induced in the soil by the vane shear device. Shrinkage during fabric specimen preparation caused by freezing is believed to create some cracks. However, the vane-induced patterns can be delineated by comparing successive thin-sections. Some of the thin-sections, such as the one corresponding to 0° of vane rotation, unfortunately were destroyed during the grinding process. The absence of these thin-sections prevents a clear assessment of the magnitude of shrinkage cracks.

The soft, illite/sand mixture exhibited a quite different behavior. At a rotation angle of 17½° when the maximum torque is probably reached, no failure surface is evident in Fig. 5. Increased rotation shows some evidence of a developing surface of failure, but it is not until the vane has been turned 75° that a definite failure surface can be recognized. Compression testing of soft soils often exhibits the same lack of a clear surface of rupture. The explanation of this effect lies in the consistency of the soil material corresponding to a liquidy index of 0.91 (Table 1). The thin sections in this case indicate that there is only a remote relation between the development of an indentifiable failure surface and the achievement of maximum torque. The diameter of the circular failure surface of the illite/ sand mixture when it eventually fully develops at about 75° also appears to be about equal to the diameter of the vane itself.

Summary and Conclusions

In a laboratory investigation of the vane shear test, two soils, both classified as low plasticity silty clays according to ASTM D 2487, were subjected to different angles of vane rotation. The two soils prepared by slurry consolidation had nearly the same water content but very different consistencies as indicated by the liquidity index. The undrained shear strength values as obtained from both the unconfined compression test and the vane shear test reflected this difference in the consistency of the two soils and indicated slightly higher values for the vane shear test.

The shear structures that develop as the vane rotated were studied by the optical thin-section technique. The thin sections prepared at varying angles of vane rotation revealed different responses for the two soils. The failure surface is not a sharply defined cylinder but is actually a thin shear zone. This surface develops initially as local shear structures originating from the tip of the wings of the vane; at large angles of vane rotation (greater than 45°) it becomes a continuous, roughly circular shear surface. In the stiffer soil, the continuous failure surface is well defined at a vane rotation of about 45°, which corresponds to the maximum torque. In the softer soil, however, the continuous failure surface develops after very large vane rotations (about 75°) and well beyond the angle at which the maximum torque is achieved. The diameter of the failure surface in both soils appears to be equal to the diameter of the vane. However, the undrained shear strength calculations based on the assumption of a fully developed cylindrical shear surface may not be valid in all cases and may underestimate soil strength.

References

[*1*] Cadling, L. and Odenstad, S, "The Vane Borer," *Royal Swedish Geotechnical Institute Proceedings*, No. 2, Stockholm, Sweden, 1959.

[*2*] Skempton, A. W., *Geotechnique*, Vol. 1, 1978, No. 2.

[*3*] Leussink, H. and Wenz, K., "Comparison of Field Vane and Laboratory Shear Strength of Soft Cohesive Soils," *Proceedings of the Geotechnical Conference, Oslo*, Vol. 1, Norwegian Geotechnical Institute, Oslo, Norway, 1967.

[*4*] Bjerrum, L., "Embankments on Soft Ground," *ASCE, Specialty Conference on the Performance of Earth and Earth-Supported Structures, LaFayette, Proceedings*, Vol. II, 1972, pp. 1–54.

[*5*] Bjerrum, L., "Problems of Soil Mechanics and Construction on Soft Clays," *Proceedings of the 8th International Conference on Soil Mechanics and Foundation Engineering*, Moscow, U.S.S.R., Vol. 3, 1973, p. 111.

[*6*] La Rochelle, P., Roy, M., and Tavenas, F., "Field Measurements of Cohesion in Champlain Clays," *Proceedings of the 8th International Conference on Soil Mechanics and Foundation Engineering*, Moscow, Vol. 1.1, 1973, pp. 229–236.

[*7*] Arman, A., Poplin, J. K., and Ahmad, N., "Study of the Vane Shear," *ASCE, Specialty Conference on In-situ Measurement of Soil Properties*, Vol. 1, 1975, pp. 93–120.

[*8*] Wilson, N. E., *Laboratory Shear Testing of Soils, STP 361,* American Society for Testing and Materials, Philadelphia, 1964, pp. 377–389.
[*9*] Sheeran, D. E. and Krizek, R. J., "Preparation of Homogeneous Soil Samples by Slurry Consolidation," *Journal of Materials,* Vol. 6, No. 2, 1971, pp. 356–373.
[*10*] Lambe, T. W., *Soil Testing for Engineers,* John Wiley and Sons, Inc., New York, 1951.
[*11*] Lee, H. J., "State of the Art: Laboratory Determination of the Strength Testing of Marine Soils," *Strength Testing of Marine Sediments: Laboratory and In-Site Measurements, STP 883,* American Society for Testing and Materials, Philadelphia, 1985, pp. 181–250.
[*12*] Richards, A. F., *Deep-Sea Sediments Physical and Mechanical Properties,* A. L. Inderbitzen, Ed., Plenum Press, New York, 1974, pp. 271–292.
[*13*] Guertin, R. K. and Bourbeau, G. A., *Canadian Journal of Soil Science,* Vol. 51, 1971, pp. 243–248.
[*14*] Cady, J. G., "Petrographic Microscope Techniques," *Methods of Soil Analysis,* Part 1, Agron. 9, C. A. Black, Ed., American Society of Agronomy, Madison, WI, pp. 604–631.

Part IV: Laboratory Vane New Test Methods

Sibel Pamukcu[1] *and Joseph Suhayda*[2]

Low-Strain Shear Measurement Using a Triaxial Vane Device

REFERENCE: Pamukcu, S. and Suhayda, J., **"Low-Strain Shear Measurement Using a Triaxial Vane Device,"** *Vane Shear Strength Testing in Soils: Field and Laboratory Studies, ASTM STP 1014,* A. F. Richards, Ed., American Society for Testing and Materials, Philadelphia, 1988, pp. 193–208.

ABSTRACT: Laboratory determination of low strain shear behavior or shear modulus of soft saturated clays can be complicated if high-frequency dynamic testing methods are utilized. Cyclic loading can promote fast degradation of moduli for these soils. A monotonic torsional shear device, namely, a triaxial vane device, was equipped with a computer aided data acquisition system to detect low strain shear deformations under quasi-static loading conditions. The average range of electronically measured strain was 10^{-4} to 1%, which was compatible with that of a high-frequency low-strain dynamic testing method, namely, resonant column. Comparison of the dynamic and static moduli reduction curves of artificially prepared soft kaolinite specimens demonstrated the cyclic degradation effects on strain dependent dynamic moduli. For normally consolidated specimens under confining pressure of 100 kPa, the ratio of maximum static shear modulus to maximum dynamic shear modulus was estimated to be around 0.85. This value and the observed trend of dynamic and static moduli reduction curves at low strains are in agreement with the findings of other investigators. The relatively continuous, high-resolution low-strain static data indicated further gain in understanding of low-strain nonlinearity and yielding behavior of soft marine clays.

KEY WORDS: triaxial vane, static moduli, dynamic moduli, low strain shear, soft clays

Nomenclature

A, A'	Applied rotation rate
C	Constant, function of vane dimensions
G	Shear modulus (secant)
$G_{\max}$	Maximum or initial shear modulus
$G_{\max-s}$	Maximum or initial static shear modulus
$G_{\max-d}$	Maximum or initial dynamic shear modulus
K_0	Coefficient of earth pressure at rest
OCR	Overconsolidation ratio
p_p	Past pressure
p_{total}	Total pressure (confining pressure)
r	Radius of vane
S	Slope of straight line portion of sinusoidal calibration curve
SF	Load spring factor
s_u	Undrained shear strength
t_i	Time

[1] Assistant professor, Department of Civil Engineering, Lehigh University, Bethlehem, PA 18015.

[2] Associate professor, Department of Civil Engineering, Louisiana State University, Baton Rouge, LA 70703.

u_{excess} Excess pore-water pressure
u_r Pore pressure ratio ($= u_{excess}/p_{total}$)
x Width of ring deformation zone next to the edge of vane blade
X_i, X_1 Input value for the position of the reflected light beam
γ_i, γ Shear strain
θ_i, θ Angular rotation of the vane
τ_i, τ Shear stress

Interpretation of low strain (below 10^{-2}%) data obtained through high-frequency cyclic testing methods can be complicated when soft saturated and sensitive clays are of concern. Cyclic loading causes gradual decrease of the effective stress with the increasing number of cycles of loading. This promotes the degradation of the backbone curve, thus the low strain modulus G_{max} [*1,2*]. The degradation effects are usually attributed to stress reversal, progressive pore-pressure build-up, strength deterioration, and deterioration of the clay structure caused by remolding [*3–5*]. Low-strain properties and behavior can be estimated through quasi-static or low-frequency cyclic testing methods using data above 10^{-2}% strain amplitude also. However, this too can be complicated for soft clays. Plastic deformation may start at low-strain ranges with little or no nonlinear elastic region, and the analytical model used to describe the data may not represent such low strain behavior adequately. The highly nonlinear nature of soft saturated deposits and influence of cementation at low-strain amplitudes in some of these deposits emphasize the need to evaluate low stress-strain behavior of such soils.

This paper presents the results of an investigation undertaken to improve the ability to evaluate the low-strain shear stress-strain behavior and properties of soft saturated clays through a quasi-static testing method in an attempt to eliminate the complicating effects of cyclic testing on such clays. A triaxial vane device was modified to accommodate an electronic optical data acquisition system to measure shear strains below 10^{-2}% during monotonic loading. Collection of low strain data and part of the test control was done using a microcomputer. Artificially prepared duplicate soil specimens, normally consolidated at the liquid limit, were tested using the new procedure and a resonant column device. Dynamic and static low strain behavior and properties were compared. The new procedure resulted in high resolution measurement of shear stress-strain at low-strain amplitudes compatible with those measured in high-frequency cyclic testing. These results indicated that, triaxial vane testing, in conjunction with the electronic data acquisition described in here, can be useful in determining low-strain static shear modulus, and low-strain shear behavior of soft saturated clays.

Background

Evaluating Low-Strain Properties of Soft Clays

Investigators have often addressed the difficulties associated with evaluating the low-strain properties and behavior of soft saturated clays, especially those found in an offshore environment. Such clays generally exhibit plastic deformation starting at low-strain levels and yield over a considerable range of strain. According to the concepts of critical state soil mechanics, under undrained conditions a yield condition would be indicated by a sharp increase in pore pressure and decrease in effective stress [*6*]. This behavior has been demonstrated by undrained compression of laboratory prepared kaolinite specimens [*6*]. For instance, in predicting the mechanism of flow deformation of soft marine clays subjected to cyclic loading of waves, the material may be considered to undergo a series of yield states accumulating excess pore pressures to failure condition [*7,8*]. The total shear strain necessary for such clays to accumulate the pore pressure to failure or "remolded" state can be on the order of 10^{-1}% [*7*]. This indicates that as the material undergoes a series of yield states, some of these states may correspond to low-strain ranges. It usually is not clear from typical stress-strain curves at what amplitude of strain the material may undergo yielding if yielding does occur at low strains. Higher resolution measurement of the response at low strain ranges can help to detect such yielding as well as nonlinear behavior at these strains.

Low-strain range is commonly identified as below 10^{-2}% strain amplitude, and the low-strain modulus G_{max} is measured or estimated at 10^{-4}% shear strain [*9*]. In the laboratory, high-frequency equipment, such as the resonant column, have been used to determine shear moduli and damping of soils at strains ranging from 10^{-4}% to 10^{-2}%. Resonant column produces strains compatible to those in geophysical methods. Both resonant column and geophysical methods deliver moduli values that best comply with high-frequency loading conditions.

For strains ranging from 10^{-2} to 1%, low-frequency equipment, such as cyclic triaxial, simple shear, and torsional shear equipment, have been used. The data obtained from low-frequency equipment can also be used to estimate G_{max}. This is best accomplished if the data obtained above the 10^{-2}% strain level have not experienced significant degradation, or the data can be described adequately by an analytical model, or both. Analytical models of a dynamic stress-strain relationship, such as Hardin-Drnevich, Ramberg-Osgood models [*2,9*], are available to use with data and extrapolate G_{max} value at 10^{-4}% strain amplitude. Some torsional shear equipment are also capable of applying quasi-static loading [*10*], which eliminate cyclic degradation effects. Soils with negligible low-strain nonlinearity or soils, which exhibit linear variation of moduli with strain below 10^{-2}%, (referred to as "non-linear elasticity" [*4,6*]) should be expected to deliver the best estimations of G_{max} using this method.

Degradation of shear moduli at constant strain amplitude with increasing number of cycles of loading have been demonstrated in a number of studies [*2,3,11,12*]. Determination of low-strain moduli in the laboratory can be complicated by pore-pressure buildup, moduli degradation, and strength deterioration caused by cyclic loading. G_{max} varies strongly with number of cycles of loading, shear strain and confining pressure. Dyvik et al. [*13*] reported that pore pressure is also a basic indication of the level of degradation of shear modulus of a soil specimen subjected to cyclic loading. Isenhower and Stokoe [*4*] showed that strain rate effects become pronounced at strain amplitudes greater than 10^{-2}% for San Francisco Bay mud. The results of a study conducted by Ishihara and Yasuda [*5*] on alluvial clay show that specimens having zero or small initial shear stress (a fraction of static failure stress), exhibit "strength deterioration," which is attributed to pore-pressure buildup, breakdown of inherent soil structure and stress reversal during cyclic loading. The effect becomes more pronounced when stress reversal is at the origin or at an initial stress level close to the origin forcing the specimen to experience negative stresses at each cycle. Kavazanjian and Hadj-Hamou [*14*] conducted an investigation in which they correlated and compared dynamic and static behavior using results of resonant column tests and empirical methods. The study concluded that most laboratory static shear tests prohibit definition of static stress-strain relation below 10^{-2}%; therefore low-strain properties are to be estimated using empirical relations [*9*].

Incorporation of Vane Shear into Low-Strain Testing

Vane shear testing is used to estimate undrained shear strength of soft saturated clays both in-situ and in the laboratory [*15*]. It is known to perform especially well in soft seabed clays, and results compare well with other in-situ testing methods. A study on San Francisco Bay mud shows substantial agreement in undrained shear strength profiles obtained using laboratory vane, unconsolidated undrained (UU) triaxial, and stress history and normalized soil engineering properties [SHANSEP] (OCR = 1) method [*16*]. With the advent of knowledge on vane shear mechanism [*17,18*], pore-pressure distribution, and disturbance [*19,20*], effective stress state on vane shear plane [*20–23*] and shear rate effects [*24*], the laboratory vane shear device is being upgraded from a practical tool to a better and reliable soil shear testing device. The triaxial vane apparatus makes it possible to simulate some in-situ stress conditions on specimens. Inherent problems of classical lab-vane testing, such as upheaving of soil surface during vane insertion, drying of soil surface during slow-rate long-term tests, and boundary effects caused by size and rigidity of the soil container are eliminated in the triaxial vane device. The triaxial vane was first developed at Norwegian Geotechnical Institute in 1965 by Kenney and Landva [*25*] in which isotropic, anisotropic, and K_0 stress conditions could be simulated, and isotropic consolidation carried as well. Law [*21*] modified a triaxial cell and a vane machine to measure effects of lateral and vertical pressures on torque measurements. The vane was designed to be detachable from the vane rod so that friction pertaining to seals and rod could be measured separately and corrections be made in vane shear measurements accordingly.

The basic design of the triaxial vane unit used in this study is similar to Law's design with some operational differences. One major difference is the use of an optical method to measure rotation of the vane. Although the optical transducers are more complex and therefore harder to use, they provide higher resolution than any other method. It was therefore necessary to use this technology to achieve the required sensitivity of measurements. The optical detector was used to trace the rotation of the vane only. The consecutive positional values from the detector provided the input to calculate the shear strains. Torque values were calculated using the applied constant rate of rotation and the time elapsed at each sampling of positional input. Data were collected and displayed real-time in graphical form via a microcomputer.

Methodology

Artificial specimens of Georgia kaolinite consolidated from slurry to near the liquid limit of the clay were tested. Slurry was prepared using distilled, deaired water. Consolidation time was estimated using a practical method based on finite-strain theory [*26*]. Soil specimens, K_0 consolidated to 100 kPa pressure in large consolidation cells, were used to trim 50- and 35.6-mm diameter duplicate specimens for triaxial vane and resonant column tests, respectively. Some index properties of the specimens were water content (%) = 63, bulk unit weight (kN/m^3) = 15.75, LL (%) = 64, and PI (%) = 30.

Modified Triaxial Vane Device

The apparatus basically consisted of a large triaxial cell of 100-mm nominal diameter, connected to a constant pressure unit and a laboratory vane machine. Several parts of both the triaxial cell and the vane machine were either modified or redesigned in order to couple the separate units. A schematic diagram of the triaxial cell assembly is given in Fig. 1. A stainless steel piston, identical in size to the original loading piston of the cell, was used to house the vane rod within a narrow duct drilled throughout the length of the piston. The specimen top cap was designed to house a detachable vane. Two narrow ducts leading to the housing were used to channel water and measure pore-water pressure. Annular spring loaded Teflon® seals (Ball-Seal) were used both inside the piston and the top cap to ensure

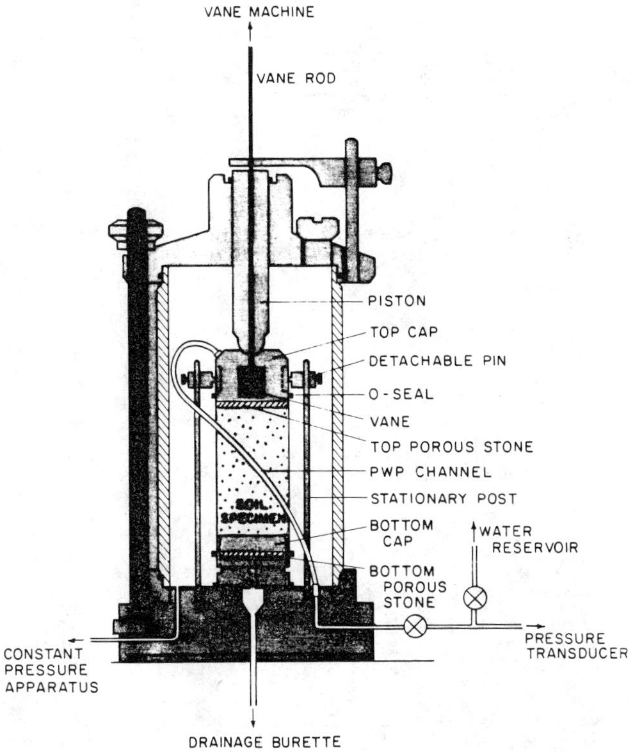

FIG. 1—*Schematic diagram of triaxial cell assembly.*

proper cell pressure and avoid infiltration of water into the specimen while allowing rotation of the vane rod with minimum friction. Details of the piston and the top cap are shown by a schematic diagram in Fig. 2.

The vane would initially rest on recesses inside the housing. Four perpendicular slits situated around a circular hole at the bottom of the cap would allow insertion of the vane into the specimen. The vane would be turned to meet the slits and pushed through at the appropriate time during testing. Two side rods with detachable pins and a bottom adapter that housed the lower end of the specimen were used to prevent the rotation of the specimen when operating the vane. The side rods were fixed to the base of the cell, and they secured the top cap from rotation via detachable pins free to move vertically only. If any rotation would occur because of horizontal clearance between the pins and the slits on the top cap, the graphical display of positional input versus time would indicate this by a sudden jump from one reading to the other between consecutive sampling. That pattern of data would be recognized and eliminated electronically during subsequent analysis.

A conventional laboratory vane machine was modified slightly to join with the triaxial cell. The original torque wheel was replaced by another one larger in diameter, and the original electric motor was replaced by a variable speed, reversible motor with remote control panel. The variable speed motor and the enlarged torque wheel facilitated the low-speed operation of the vane machine thus ensuring low rotation rate for the vane. This condition was observed to prevent excessive pore-pressure buildup in front of the vane blades. As part of data acquisition, the vane machine was equipped with a ¼ wave first

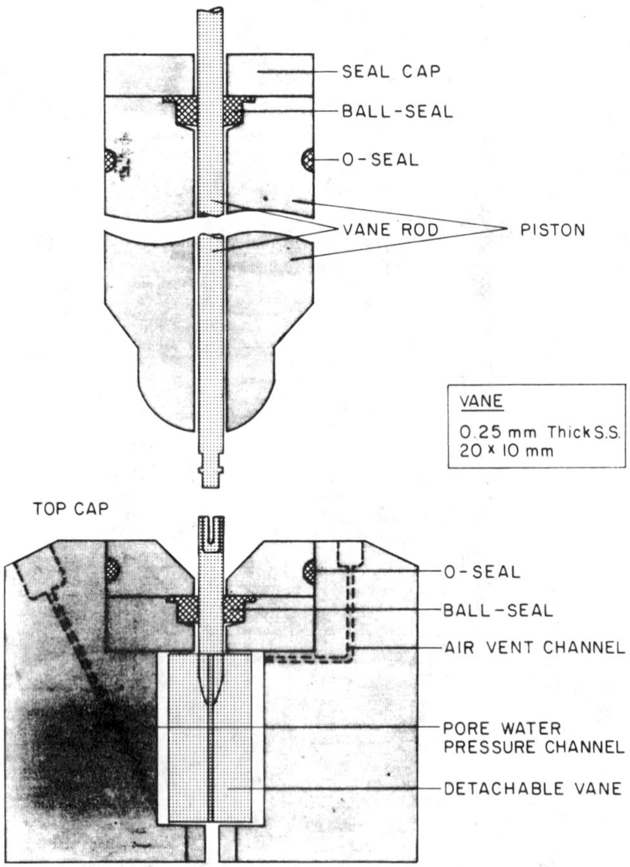

FIG. 2—*Details of piston and the top cap.*

surface plane mirror fixed perpendicularly at the center of the horizontal scale of the vane machine. The mirror, 51 mm in diameter and 13 mm thick, was mounted on a miniature mount that allowed fine position adjustment by rotating about vertical and horizontal axes. Rotation of the vane was directly indicated by the rotation of the mirror by way of a shaft. Triaxial vane setup with parts of the data acquisition system is shown in Fig. 3.

The shear mechanism of the vane has been studied by a number of investigators [*17,18,20,27*]. Soil deformation has been observed to take place in a very narrow ring zone next to the edge of the vane blade [*20,27*]. For small rotations of the vane, the shear strain at the edge of a vane blade γ can be related to the angle of rotation of the blade θ by the following equation

$$\gamma = \theta r/x \tag{1}$$

where θ = angle of rotation in radians, r = radius of the vane, and x = width of the ring deformation zone. The equation would be valid for small rotations of the vane, which is prior to formation of cracks at the edges of the blades and full mobilization of shear around the vane.

FIG. 3—*A view of the vane mirror and mount assembly on the triaxial vane machine.*

The width of the ring deformation zone can be complicated to establish for any given combination of soil, vane size, and boundary conditions. Using 200-mm-diameter normally consolidated kaolinite specimens and a 40-mm-diameter by 80-mm-length laboratory vane with blade thickness of 1.5 mm, one investigation reported the measured thickness of the ring deformation zone to be less than 5 mm [*20*]. This shows that the thickness of the ring deformation zone can be less than ¼ times the radius of the vane. In the study reported here, the diameter of the kaolinite specimens used in the triaxial vane device was 50 mm, and the vane had dimensions of 10 mm diameter and 20 mm length with blade thickness of 0.25 mm. Although the boundary conditions and state of stresses are different in these two tests, the ratio of the size of the specimens to the size of corresponding vanes are similar. The vane blade thickness in the latter one is significantly less, which would indicate less disturbance to the soil specimen. Considering the finding of the previous investigation [*20*] as initial estimate, the thickness of the deformation zone was finally assumed to be equal to the radius of the vane because of the reasons discussed below.

Using Eq 1, with x equal to r, the shear strain developed at the edge of the vane blade on the horizontal plane, was calculated to be equal to the angular deformation of the vane. Since the measured angular rotations were small, maximum being on the order of 0.01 radian, use of Eq 1 was warranted. The secant shear moduli values, ($G = \tau/\gamma$), calculated using the shear strain values from Eq 1, were on the same order of magnitude as the resonant column values. In the resonant column testing, current to the driving coils was attenuated to achieve lower strain amplitudes thus delay degradation effects. Therefore, the resonant column measurement of strain dependent shear moduli were not expected to be degraded significantly to produce as much as two-fold difference between the static and dynamic measurements. Absence of large differences between the two moduli measurements of normally consolidated duplicate specimens (Fig. 4), and the observed compliance of data with results of other investigators [*12,14,28*] strengthened the validity of the assumption made in Eq 1.

Data Acquisition System

A general layout of data acquisition system is shown in Fig. 5. This system was put together to detect and record minute angular displacement of the vane via mirror. Interfacing the detection system to a microcomputer made it possible to collect large amount of data with the required precision of small displacements. As indicated in Fig. 5, the detection and acquisition system comprised of a plane first surface mirror, an autocollimator, a multiplexer and analog/digital (A/D) unit, and a microcomputer. A beam of infrared light emitted from a high power diode inside the autocollimator is reflected through a beam-splitter and then collimated by lens. This parallel beam is reflected off the mirror under test back through the lens, the beam splitter, and an ambient light filter to a detector assembly at the back of the autocollimator. The detector assembly is a two axis position sensor, which has electrode connections (photodiodes) at four 90° positions. Details of detection are shown in Fig. 6. As the incident beam strikes the detector, the amplitude of the analog signals generated by the photodiodes determines the position of the beam. The proximity formulas given in Fig. 6 are used to determine successive X and Y positions of the centroid of the light beam. A sinusoidal response is obtained as the light beam propagates from one sensor to the other along one axis. The horizontal detection signals, X1 and X2, were used to evaluate the angular displacement of the vane. The magnitude of the current flow in each of the photodiodes, which is proportional to spot proximity, is reported to the controlling computer via an A/D converter and multiplexer (optical position indicator, OP-EYE). The digital values are then converted to ASCII code for transmission to the computer. Communication between OP-EYE and the computer is done through use of a few operational commands interpreted by OP-EYE. Two assembly language routines were incorporated to the main Fortran 77 data acquisition/analysis software package to read/

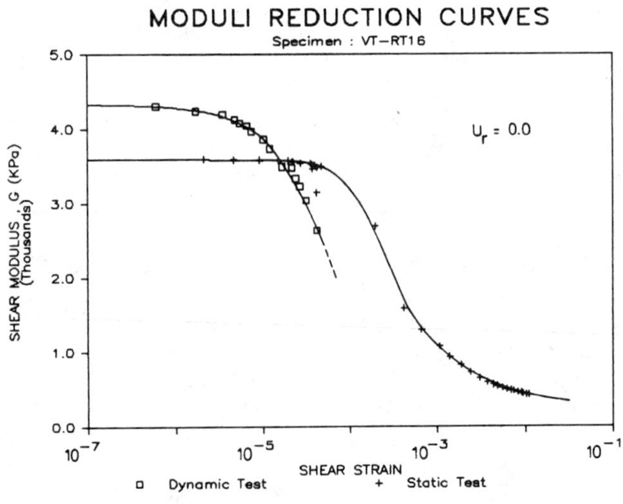

FIG. 4—*Dynamic and static moduli reduction curves compared.*

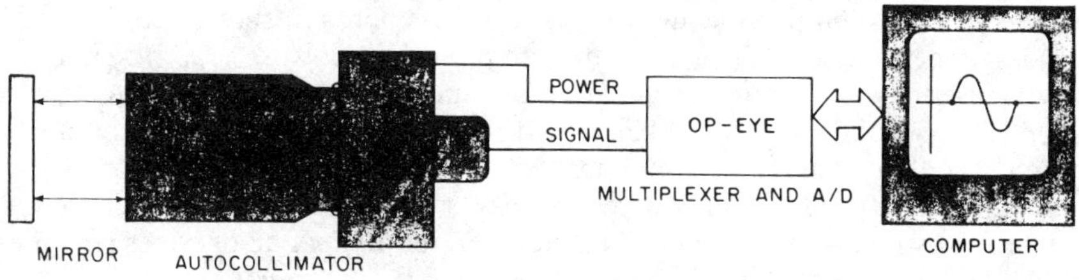

FIG. 5—*Data acquisition assembly.*

write to RS-232C ports. The data acquisition/analysis software package consisted of the following main segments:

(1) calibration and collection,
(2) noise filtering, and
(3) data manipulation and analysis.

During testing, positional inputs from the optical sensors were displayed graphically in real time and stored. Analysis programs were used to manipulate large amounts of data by way of interactive graphics, curve fitting, and regression methods.

Testing Procedure

Following trimming, all the K_0 consolidated specimens were reconsolidated under 100 kPa of triaxial pressure. Testing pressures were 100, 200, and 300 kPa for three different sets of specimens, respectively. Triaxial consolidation was monitored by observing pore-pressure dissipation at the top of the specimen and drainage at the bottom. The value of the B parameter ranged from 0.95 to slightly over 1.0, indicating nearly 100% saturation.

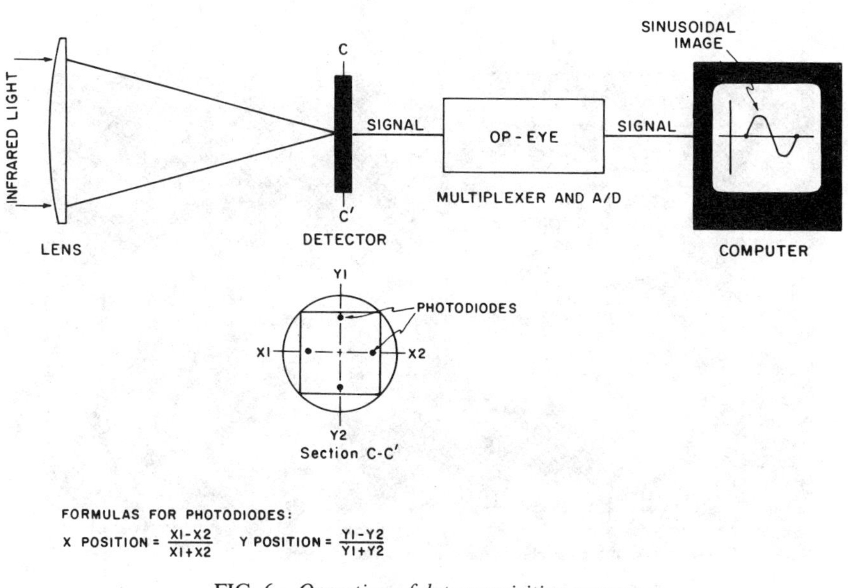

FIG. 6—*Operation of data acquisition system.*

This procedure made it possible to incorporate excess pore-pressure effects into the analysis and data pertaining to three different pore-pressure ratios (u_{excess}/p_{total}) were obtained. Vane was inserted to about mid-height of the specimen shortly after the confining pressure was applied. Consolidation to 100 kPa would progress with the vane already in the soil. Duration of confinement for the excess pore pressure at the top of the specimen to dissipate to approximately 10% of the confining pressure was 24-h on average. This period was well above the 4 h estimated by Kimura and Saitoh [*20*] for dissipation of excess pore pressure generated by insertion of vane in soft saturated clays.

The rate of applied rotation was 0.0125°/s, which is considerably below the threshold value of 1°/s, as estimated by Matsui and Abe [*17*], below which local pore-pressure migration occurs around the vane. This in turn is predicted to result in overestimation of strength values. However, Matsui and Abe have also reported that this effect did not influence low strain parameters. Several factors contributed to the choice of slow rate of rotation application. The frequency of data sampling (2/s) could not be increased because of limitations of the data acquisition hardware. Therefore slower rate of rotation was used to collect a larger number of data points within a small interval of rotation. About 1500 data points could be collected within the shear strain range of 1%. The strain rate increased with increasing angular rotation of the vane. This could also be observed through the real-time graphics as the positional data was plotted with increasing difference between consecutive points for the same time interval. Using Eq 1, the calculated interval of strain below the strain amplitude of 10^{-3}% generally remained constant on the order of 10^{-4}%. With data sampling frequency of 2/s, the strain rate was estimated to be around 0.7%/h (0.003°/s) below the 10^{-3}% strain amplitude.

Figure 7 illustrates the progression of a typical data acquisition sequence using the interactive calibration/acquisition software. The two sinusoidal shaped curves represent the forward and reverse rotation of the mirror uncoupled from the vane rod, thus the vane. This is part of the calibration process in which the slope of the straight line portion of the initial sinusoidal curve is used to calculate strains. The value of this slope is expressed in arc second. The middle 75% of the straight line section of the sinusoidal image is utilized

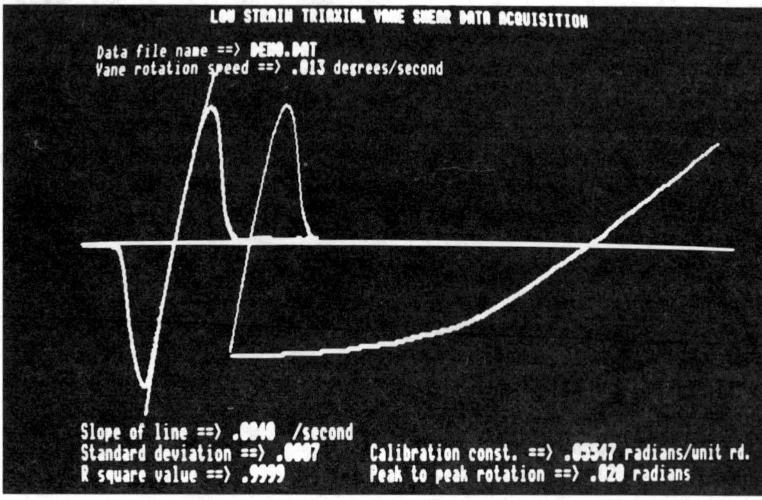

FIG. 7—*OPCALIB—calibration and data collection.*

both for calibration and testing to avoid the nonlinearity in the extremes of the detector's range. The second sinusoidal image is the trace of the reverse rotation of the mirror, which is used to locate a start-up point for data collection within the linear range. Once the desired position of the mirror is set, the vane rod, with the vane already inside the soil, is connected to the machine for testing. The curve on the right is the completed trace of mirror rotation with the vane attached. The coordinates of each point on this curve is (time, positional input value). Time values are acquired along with the positional data through the real-time clock of the data acquisition system. The following relations were used to calculate the successive shear strains γ_i and shear stresses τ_i.

$$\gamma_i = (X_i - X_1) * A/S \tag{2}$$

where

X_i = positional input value (dimensionless),
X_1 = initial positional input value,
A = applied rotation rate (rad/s), and
S = slope of the straight line portion of calibration curve, s^{-1}.

$$\tau_i = (SF/C) * ((A' * t_i) - \theta_i) \tag{3}$$

where

SF = load spring factor, kN/degrees,
C = constant, function of vane dimensions, m^3,
A' = applied rotation rate, degrees/s,
t_i = time s, and
θ_i = angular rotation of the vane, degrees.

For the range of low stresses applied, the spring load factor *SF* was taken to be constant.

At the end of each test, the vane rod is detached from the vane, the mirror repositioned, and data collection is repeated using only the vane rod. This data are utilized to correct vane shear data with respect to seal and vane rod friction. A filtering program is used to eliminate random variations in data that may be caused by noise from the environment. The vane shear and seal-rod friction data curves are manipulated and superimposed by use of graphics cursor as shown in Fig. 8. The latter curve is subtracted from the first and the resultant data are converted into strain and stress values using Eqs 2 and 3, respectively. Figure 9 illustrates the final stress-strain diagram and the bilinear representation.

Figures 7 through 9 are only three of many cathode ray tube (CRT) graphical pictures of a typical interactive testing and analysis sequence. These pictures do not necessarily display all the information but only the necessary parameters and commands for the operator to conduct the test. The following procedural steps are executed during a session of testing:

1. Position the vane assembly over the triaxial cell and run OPINIT program to detect friction interference for calibration.
2. Run OPCALIB program and follow the interactive procedure instructed by the program to calibrate, collect, and store low-strain data (Fig. 7).
3. Maintain testing until failure condition is observed, and manually record the peak and residual strength dial readings.
4. Detach the vane rod from the vane, reposition the mirror, and repeat Step 2 to obtain calibration data.

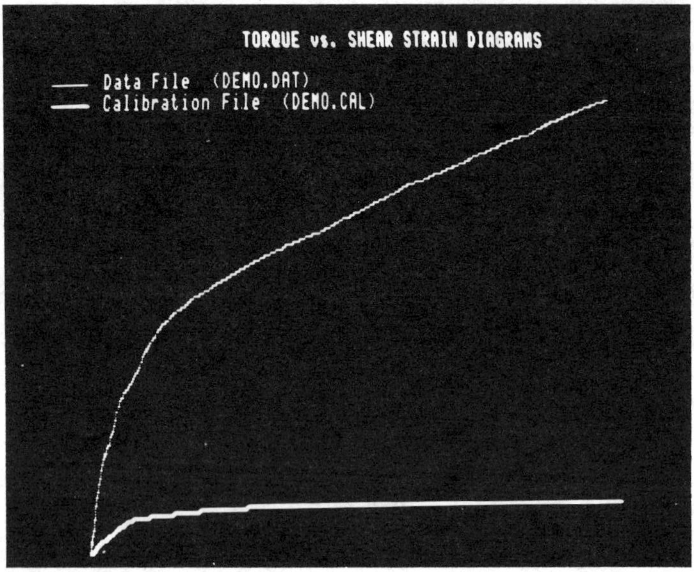

FIG. 8—*OPAUX3—superimposed torque versus shear strain data for vane shear and seal friction.*

5. Run the filtering program (OPAUX1), which employs the running mean method to filter mechanical and electrical noise from data.

6. Manipulate the vane shear and friction data curves by use of graphics curser in two separate programs, namely, OPAUX2 and OPAUX3 (Figs. 8 and 9), to obtain the resultant stress-strain curve. Graphical manipulation and subsequent computation of stress and strain values involve a series of curve fittings, linear regression methods, and application of calibration constants and corrections.

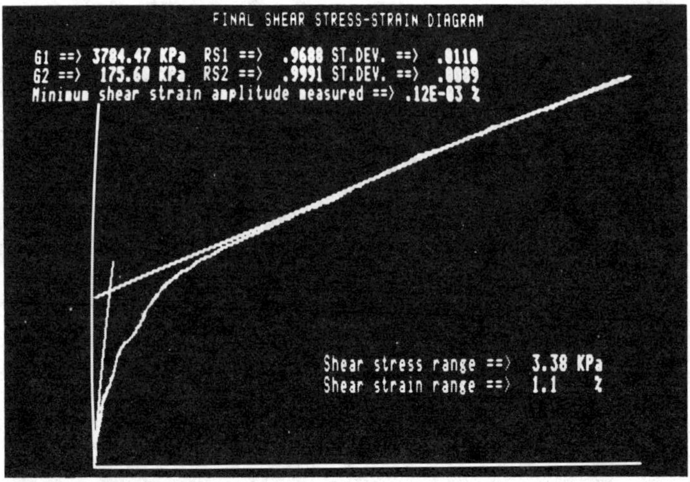

FIG. 9—*OPAUX3—final shear stress-strain curve with bilinear representation.*

Discussion of Results

Figure 4 illustrates typical shear strain versus shear moduli reduction curves for dynamic and static tests on normally consolidated duplicate specimens under 100 kPa confining pressure. The shear modulus values are all secant values except for the estimated maximum value for both curves. The data points on static curves were sampled automatically using a computer program that would match nearest shear strain values of static data to dynamic data. The points on the extension of the static curve are sampled at equal intervals. Hyperbolic curves were fit to the dynamic data.

As observed in Fig. 4, the static moduli reduction curve falls below the dynamic curve. This is similar to other investigators findings [*12,28*] in which empirical relations or low-frequency torsional shear test data were compared with resonant column test data. The difference between high-frequency and low-frequency shear moduli data can be explained by strain rate difference between the two modes of testing [*28*]. The variation between static and dynamic shear behavior of clays is also due to strain rate effects as shown by other investigators [*29,30*]. Influence of cyclic degradation can be observed on the dynamic curve at around 10^{-3}% strain amplitude.

Figure 10 illustrates the variation of normalized static maximum shear modulus, G_{max-s}/G_{max-d}, with pore-pressure ratio u_r. The static modulus values are normalized by corresponding dynamic modulus values from duplicate specimens tested under same conditions. Pore-pressure ratios of 0.5 and 0.67 represent the conditions when the drainage valve is closed at the completion of consolidation under 100-kPa pressure, and the confining pressure is increased to 200 and 300 kPa, respectively. Actual testing would begin within 1 h of application of new confining pressure. In these series of tests the dynamic moduli data seemed to be unaffected by the pore-water pressure increase, while static modulus values decreased. One explanation for this behavior can be possible air migration into the specimen through the latex membrane during the period of increased confining pressure in the resonant column. Such an occurrence would delay full development of excess pore-water pressures and consequently promote results unaffected by the induced state of stress. The increased confinement, on the other hand, can temporarily increase the stiffness of the specimen until the pore water pressure is fully developed. Confirmation of these explanations is not possible because pore-water pressures were not measured during resonant col-

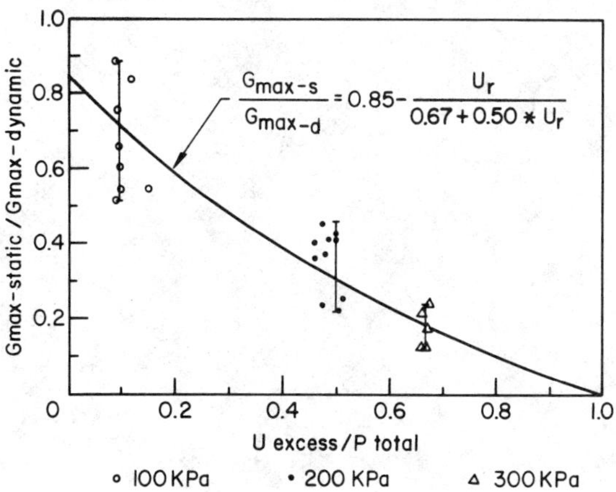

FIG. 10—*Variation of* G_{max-s}/G_{max-d} *with pore-water pressure ratio* u_r.

umn testing. However, the systematic trend of the dynamic data, and other reported complications caused by air migration under similar conditions render support to such an explanation. Data in Fig. 10 are represented by a hyperbolic curve calculated by taking the numerical average of each data cluster and assuming zero stiffness under zero effective stress. Maximum dynamic shear modulus values G_{max-d} were used as reference values in normalization, since they did not show marked variation with pore-pressure ratio. For practical purposes, the data and the fitted curve can be considered to represent the variation of maximum static shear modulus normalized by the maximum dynamic modulus of the clay measured at its normally consolidated state. The variation of the initial shear modulus with pore-water pressure ratio is similar to the results of a study on dynamic response of Gulf of Mexico clay [*13*]. Using the fitted curve, G_{max-s}/G_{max-d} value for the soft kaolinite clay specimens, normally consolidated under 100 kPa pressure, was estimated to be 0.85.

The triaxial vane testing was extended to failure following the collection of low strain data for each specimen. The average value of undrained shear strength s_u measured for normally consolidated specimens under 100 kPa confining pressure was 29 kPa. Because of possible pore-water pressure migration around the vane at larger rotations, this value of s_u was suspected to be an overestimation. Using two approaches (Skempton's s_u/p_p versus PI relation [*31*] and an approach defined by Anderson and Lukas [*32*], an average empirical value of 25 kPa was approximated for s_u. This indicated possible overestimation of 16%. The G_{max-s}/s_u ratio for these specimens was approximated to be 112. This value is lower than the typical values reported in literature all of which are dynamic modulus normalized by s_u.

The fine resolution of stress-strain data made it possible to measure the static shear moduli values at low strains. This resolution is shown in Fig. 11. As observed from the figure the data show a relatively continuous, highly structured, nonlinear stress-strain curve. This curve represents close to 100 data points free from larger random variations. A straight

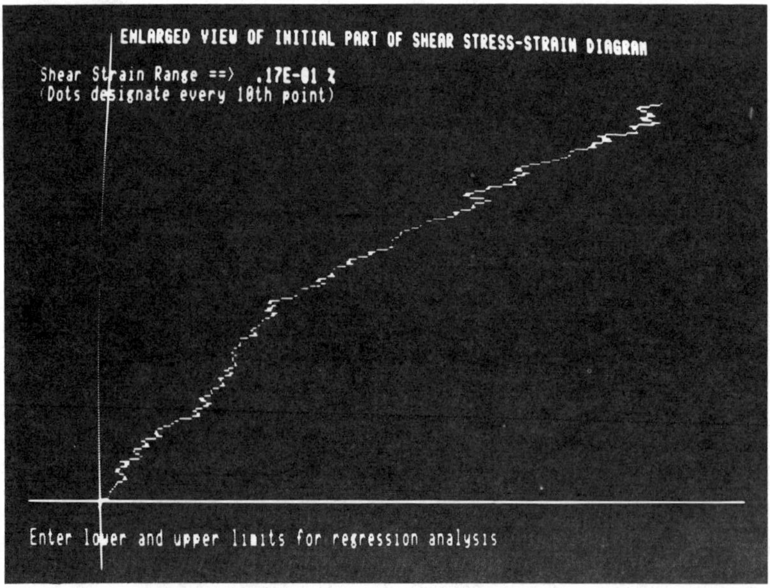

FIG. 11—*Enlarged view of the low-strain range (a CRT picture from OPAUX3).*

line is fit to the initial part of the curve up to 10^{-3}% shear strain to predict the maximum modulus. Data of this nature help recognition of low-strain yielding in soft clays. It can also be useful in understanding some features of microstructure deformations and dislocations during shear in such clays.

Conclusions

A computer aided triaxial laboratory vane equipment was developed to measure low-strain quasi-static shear stress-strain of soft saturated clays. The average range of electronically measured low-strain data were between 10^{-4} to 1% shear strain. Various assumptions were utilized in the calculation of stresses and strains. Moduli reduction curves obtained from these tests when compared with moduli reduction curves of duplicate specimens tested in the resonant column showed agreement with findings of other investigators. The degradation of dynamic moduli occurred at around 10^{-3}% strain amplitude, however, did not influence the calculation of maximum dynamic modulus. The triaxial vane shear test could also be continued to failure to measure the undrained shear strength.

References

[1] Pamukcu, S. and Suhayda, J. N., "Evaluation of Shear Modulus for Soft Marine Clays, Mississippi Delta," *Strength Testing of Marine Sediments: Laboratory and In-Situ Measurements, STP 883,* American Society for Testing and Materials, Philadelphia, pp. 352–362.

[2] Idriss, I. M., Dobry, R., and Singh, R. D., "Nonlinear Behaviour of Soft Clays During Cyclic Loading," *Journal of the Geotechnical Engineering Division, ASCE,* Vol. GT12, Dec. 1978, pp. 1427–1447.

[3] Shibata, T. and Iwazaki, Y., "Seismic Problems of Soft Clay Deposits," *Soft Clay Engineering,* Chapter 11, E. W. Brand and R. P. Brenner, Eds., Elsevier Scientific Publishing Company, 1981.

[4] Isenhower, W. M. and Stokoe, K. H. II, "Strain-Rate Dependent Shear Modulus of San Francisco Bay Mud," *Proceedings, International Conference on Recent Advances in Geotechnical Earthquake Engineering and Soil Dynamics,* University of Missouri, Rolla, MO, 1981, pp. 597–602.

[5] Ishihara, K. and Yasuda, S., "Cyclic Strengths of Undisturbed Cohesive Soils of Western Tokyo," *Proceedings, International Symposium on Soils Under Cyclic and Transient Loading,* Swansea, Jan. 1980, pp. 57–66.

[6] Parry, R. H. G. and Wroth, C. P., "Shear Stress-Strain Properties of Soft Clay," *Soft Clay Engineering,* Chapter 4, E. W. Brand and R. P. Brenner, Eds., Elsevier Scientific Publishing Company, 1981.

[7] Pamukcu, S. and Suhayda, J. N., "Dynamic Properties and Critical State Parameters, Mississippi Delta," *Proceedings of ASCE Annual Convention,* Houston, TX, Session No. 52: Evaluation of Seafloor Soil Properties Under Cyclic Loads, Oct. 1983, p. 21.

[8] Egan, J. A. and Sangrey, D. A., "A Critical State Model for Cyclic Load Pore Pressure," *Proceedings of the Specialty Conference on Earthquake Engineering and Soil Dynamics Conference, ASCE,* Vol 1, 1978, pp. 411–424.

[9] Hardin, B. O. and Drnevich, V. P., "Shear Modulus and Damping in Soils: Measurements and Parameter Effects," *Journal of the Soil Mechanic and Foundation Division, ASCE,* Vol. 98, No. SM6, June 1972, pp. 603–624.

[10] Alarcon, A., Chameau, J. L., and Leonards, G. A., "A New Apparatus for Investigating the Stress-Strain Characteristics of Sand," *Geotechnical Testing Journal,* Vol. 9, No. 4, Dec. 1986, pp. 204–212.

[11] Pamukcu, S., Poplin, J. K., Suhayda, J. N., and Tumay, M. T., "Dynamic Sediment Properties, Mississippi Delta," *Proceedings, Conference on Geotechnical Practice in Offshore Engineering, ASCE,* University of Texas, Austin, TX, 27–29 April 1983, pp. 111–132.

[12] Isenhower, W. M., "Torsional Simple Shear/Resonant Column Properties of San Francisco Bay Mud," M.S. thesis, University of Texas at Austin, TX, 1979.

[13] Dyvik, R., Zimmie, T. F., and Schimelfenyg, P., "Cyclic Simple Shear Behaviour of Fine Grained Soils," Norwegian Geotechnical Institute, Publication 149, 1983, 6 pp.

[14] Kavazanjian, E. Jr., and Hadj-Hamou, T. A., "Estimating Dynamic Properties From Static

Tests," *Proceedings, International Conference on Recent Advances in Geotechnical Earthquake Engineering and Soil Dynamics,* University of Missouri, Rolla, MO, 1981, pp. 1–6.

[*15*] Bjerrum, L. "Problems of Soil Mechanics and Construction on Soft Clays," *Proceedings of the Eight ICSMFE,* Moscow, Vol. 3, 1973, pp. 111–159.

[*16*] Clough, G. W. and Denby, G. M., "Self-Boring Pressuremeter Study on San Francisco Bay Mud," *Journal of the Geotechnical Engineering Division, ASCE,* Vol. 106, No. GT1, Jan. 1980, pp. 45–63.

[*17*] Matsui, T. and Abe, N., "Shear Mechanism of Vane Test in Soft Clays," *Soils and Foundations,* Vol. 21, No. 4, Dec. 1981, pp. 69–80.

[*18*] Menzies, B. K. and Merrifield, C. M., "Measurements of Shear Stress Distribution on the Edges of a Shear Vane Blade," *Geotechnique,* Vol. 30, No. 3, 1980, pp. 314–318.

[*19*] Azzouz, A. S. and Baligh, M. M., "Corrected Field Vane Strength For Embankment Design," *Journal of the Geotechnical Engineering Division, ASCE,* Vol. 109, No. 5, May 1983, pp. 730–734.

[*20*] Kimura, T. and Saitoh, K., "Effect of Disturbance Due to Insertion on Vane Shear Strength of Normally Consolidated Cohesive Soils," *Soils and Foundations,* Vol. 23, No. 2, June 1983, pp. 113–124.

[*21*] Law, K. T., "Triaxial Vane Tests on Soft Marine Clay," *Canadian Geotechnical Journal,* Vol. 16, No. 1, Feb. 1979, pp. 11–18.

[*22*] Schmertmann, J. H., "Measurement of In-Situ Shear Strength," *Proceedings, Specialty Conference on In-Situ Measurement of Soil Properties, ASCE,* Raleigh, Vol. 2, 1975, pp. 57–138.

[*23*] Wilson, N. E., "Laboratory Vane Shear Tests and the Influence of Pore-Water Stresses," *Laboratory Shear Testing of Soils, STP 361,* American Society for Testing and Materials, Philadelphia, pp. 377–385.

[*24*] Sharifounnasab, M. and Ullrich, C. R., "Rate of Shear Effects on Vane Shear Strength," *Journal of the Geotechnical Engineering Division, ASCE,* Vol. 111, No. 1, Jan. 1985, pp. 135–139.

[*25*] Kenney, T. C. and Landva, A., "Vane Triaxial Apparatus," *Proceedings, Sixth ICSMFE,* Montreal, Vol. 1, 1965, pp. 269–272.

[*26*] Cargill, W. K., "Prediction of Consolidation of Very Soft Soil," *Journal of the Geotechnical Engineering Division, ASCE,* Vol. 110, No. 6, June 1984, pp. 775–795.

[*27*] Arman, A., Poplin, J. K., and Ahmad, N., "Study of the Vane Shear," *Proceedings of the Specialty Conference on In-Situ Measurement of Soil Properties, ASCE,* Vol. 1, 1975, pp. 93–120.

[*28*] Stokoe, K. H. II, Isenhower, W. M., and Hsu, J. R., "Dynamic Properties of Offshore Silty Samples," *Proceedings of Offshore Technology Conference,* OTC 3771, Houston, TX, 1980.

[*29*] Cheng, R. Y. K., "Effect of Shearing Strain-Rate on the Undrained Strength of Clay," *Laboratory Shear Strength of Soil, STP 740,* American Society for Testing and Materials, Philadelphia, 1981, pp. 243–253.

[*30*] Whitman, R. V., "The Response of Soils to Dynamic Loadings," Defense Atomic Support Agency, U.S. Army Waterways Experiment Station, Report 26, 1970.

[*31*] Skempton, A. W. "Discussion of the Structure of the Inorganic Soil," *ASCE, Proceedings,* Separate No. 478, Vol. 80, 1975, pp. 479.19–479.22.

[*32*] Anderson, T. C. and Lukas, R. G., "Preconsolidation Pressure Predicted Using S_u/P' Ratio," *Laboratory Shear Strength of Soil, STP 740,* American Society for Testing and Materials, Philadelphia, 1981, pp. 502–515.

Marcio S. S. Almeida[1] *and Richard H. G. Parry*[2]

Miniature Vane and Cone Penetration Tests During Centrifuge Flight

REFERENCE: Almeida, M. S. S. and Parry, R. H. G., **"Miniature Vane and Cone Penetration Tests During Centrifuge Flight,"** *Vane Shear Strength Testing in Soils: Field and Laboratory Studies, ASTM STP 1014,* A. F. Richards, Ed., American Society for Testing and Materials, Philadelphia, 1988, pp. 209–219.

ABSTRACT: Correct assessment of soil properties in centrifuge tests require these to be measured during flight, and small vane and penetrometer devices have been developed with this purpose. The vane is fixed in position, but the penetrometer can traverse the length of the model, allowing tests to be made at different positions. A program of tests has been conducted in the centrifuge at 100*g* investigating the use of these methods in consolidated beds of clay. Clay beds consist of an overconsolidated layer with a top stiffer layer overlaying a normally consolidated layer. Curves of point resistance with depth are similar in shape to curves of vane strength with depth, and the stiffer layer obtained as a result of the stress history is indicated in both tests. However measured point resistances are lower than expected, which is attributed to water pressure effects around the cone tip.

KEY WORDS: vane tests, cone penetrometer tests, shear strength, soft clays, centrifuge tests, physical model

An important feature of centrifuge tests of clay models is the correct assessment of clay strength, and with this purpose vane and penetrometer equipment have been used [*1–3*]. However, the correct determination of clay strength requires these to be measured during centrifuge operation, as the strength measured after stopping the centrifuge is lower than in flight because of swelling of the clay model associated with possible cavitation [*4*].

Prototype and penetrometer apparatuses have been developed by Davies [*1*] in connection with centrifuge modelling of embankments on clay foundations. However in-flight strengths measured by Davies were higher than expected, which suggested that strengths mobilized by vane and penetrometer shafts were responsible for a significant part of the measured strengths. There were also uncertainties about vane rotation rates and the drawback that only one strength profile could be measured.

A modified vane and a new penetrometer have been developed by Almeida [*3*], which enables vane and tip resistances to be separated from shaft friction and vane rates of rotation to be directly measured. Results of tests at 1*g* using these devices, and also a miniature piezocone, have been reported [*5*]. These studies included investigations of the effects of testing rates, water pressure in penetrometer tips, and correlations between vane and cone measurements. Pore-water pressure dissipation tests with the piezocone were also performed at 1*g*.

[1] Associate professor, Post Graduation School, COPPE/UFRJ, Federal University of Rio de Janeiro, Cx. Postal 68506, 21945, Rio de Janeiro, RJ, Brazil.

[2] Lecturer, Engineering Department, Cambridge University, Trumpington St., CB21PZ, Cambridge, England.

A new penetrometer apparatus to measure cone resistances at different positions in the centrifuge package was developed as well [3] in connection with centrifuge modelling of stage constructed embankments on soft clays [6]. A detailed description of this apparatus and studies of penetration rates and water pressure effects in clay beds tested at 100*g* have already been reported [7]. Penetrometer tests in sand beds at different relative densities have also been performed [8] at three acceleration fields to study the variation of the angle of shearing resistance with the relative density and stress level.

This paper presents previously unpublished results of vane and penetrometer tests carried out during centrifuge modelling of embankments on soft clays. Vane strengths are compared with theoretical strengths, and empirical cone factors are obtained by correlating vane strength and cone resistance.

Apparatus

The package used for the present study (Fig. 1) consisted of a rectangular strong box in which a clay cake, 0.675 m long by 0.2 m wide by 0.16 m deep, was contained; a vane apparatus to measure the clay strength in flight at the completion of the consolidation run; and a penetrometer that can traverse the length of the strong box.

The vane apparatus (Fig. 2) is an upgraded version of the apparatus used by Davies and Parry [4]. The improvements over that device are (1) blades of smaller heights (18 mm diameter by 14 mm height), which make possible a better definition of the strength profile

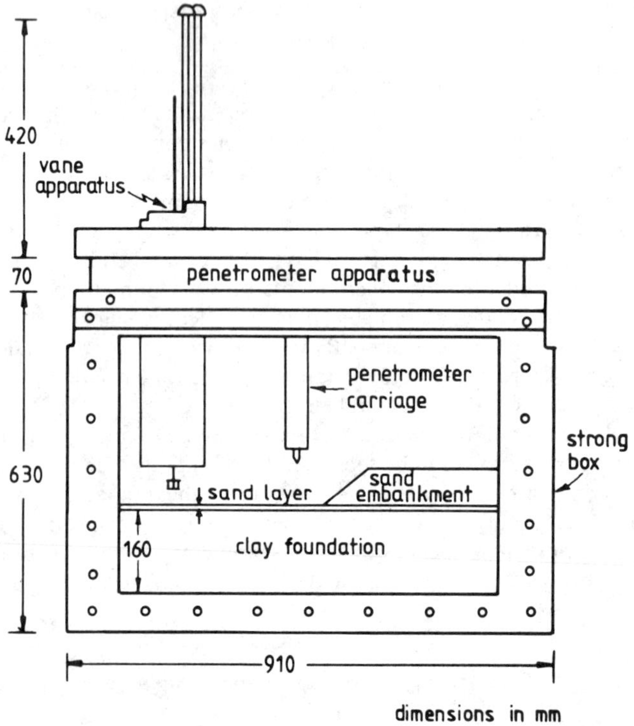

FIG. 1—*Package for centrifuge tests.*

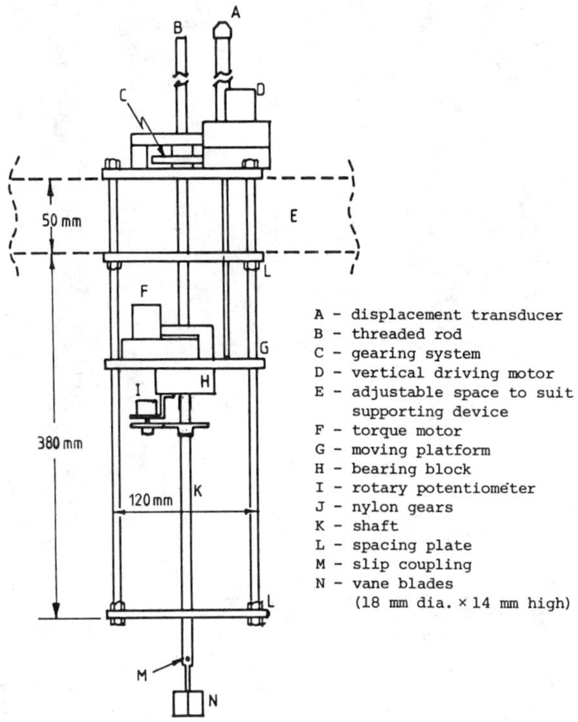

FIG. 2—*Vane apparatus Mark II.*

by providing more data points; (2) direct measurement of the rate of rotation using a rotary potentiometer and nylon gears connected to the vane shaft and potentiometer; and (3) separation of the torques mobilized by blades and shaft using a slip coupling. The vane shaft is hollow at its top and over a length of 10 mm near the top a thinner wall is machined and four strain ganges bonded to measure torque. Early tests [*9*], also reported in a companion paper at this publication [*10*] using blades of 19 mm diameter by 27 mm height produced results very close to tests using the 18 mm diameter, 14 mm height blades adopted here. Because the use of a 27-mm high vane allowed a very limited number of vane strength determinations along the 160-mm-deep clay model, the 14-mm-high vane was used in the tests reported here.

The cone penetrometer apparatus consisted of (1) a horizontal drive system to move the penetrometer across the box, (2) a curved track supporting the carriage, and (3) a carriage supporting the penetrometer, which is driven in the radial direction of the gravity field by a motor mounted on the carriage. Further details about this apparatus are presented elsewhere [*3,7*].

Cone penetrometers, Mark II and Mark III, were used in the tests reported here. Cone penetrometer Mark II (Fig. 3*a*) has two load cells mounted at the extremities of the cone penetrometer. Hence, independent measurements of side friction and total load, including side friction are possible. Cone penetrometer Mark III (Fig. 3*b*), more sensitive than cone penetrometer Mark II, was provided with a rosette load cell mounted at the top of the penetrometer. An internal rod transmits the tip load to the rosette cell. The four webs of the load cell subjected to bending produced signal outputs seven times higher than the load

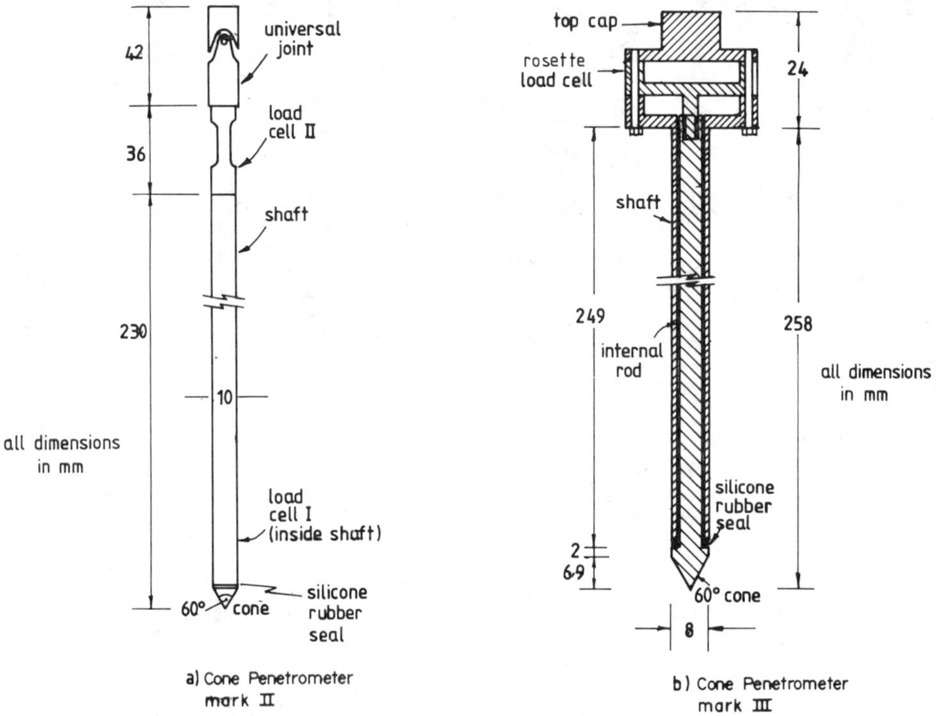

FIG. 3—*Penetrometer probes used during flight.*

cell of penetrometer Mark II. However penetrometer Mark III could not accommodate a second load cell to measure total load. All load cells were provided with fully active bridges arranged for temperature compensation.

Model Preparation

Clay cakes prepared for the present program consisted of a layered foundation of Gault clay overlying kaolin clay. Some properties of Gault clay and kaolin clay are presented in Table 1. Gault clay was used on the top of the model to produce a stiff crust. Both clays were consolidated together from a slurry condition at a water content of twice the liquid limit to a pressure of 54 kPa (Fig. 4*a*). After consolidation was completed miniature pore-pressure transducers were inserted in the model.

TABLE 1—*Mechanical properties of speswhite kaolin and Gault clay.*

Property	Symbol	Speswhite Kaolin	Gault Clay
Liquid limit	LL	69	60
Plastic limit	PL	38	25
Plasticity index	PI	31	35
Specific gravity of solids	G_s	2.61	2.72

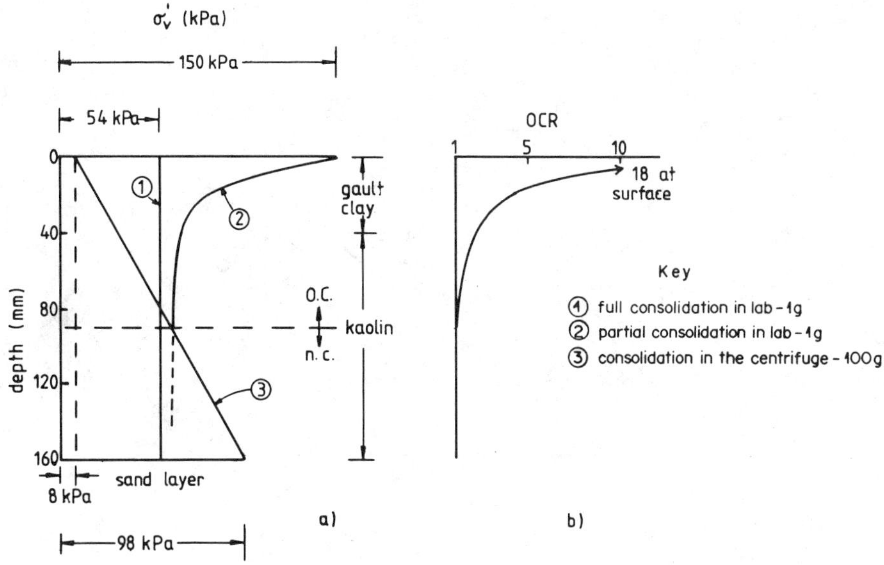

FIG. 4—*Stress history for clay models.*

In order to produce a stiff crust the clay model was subjected to a partial consolidation in the laboratory, with drainage permitted at the top only, and under an applied pressure of 150 kPa. Monitoring of pore pressures inside the clay model allowed the σ_v' profile to be determined, as shown in Fig. 4*a*. The thickness of the clay cake at the end of the consolidation in the laboratory press was about 160 mm, of which the top 40 mm was Gault clay and the bottom 120 mm was kaolin.

Following the consolidation in the laboratory press the clay model was transferred to the centrifuge container where a 9-mm sand layer was placed on its top and was then subjected to an acceleration of 100*g*. Equilibrium pore pressures were achieved after about 9 h of continuous centrifuging, and values of σ_v' at equilibrium are shown in Fig. 4*a*. The resulting overconsolidation ratio varies from 18 at the clay surface to 1 at depth of 90 mm, as shown in Fig. 4*b*. At prototype scale it consists of a 16-m clay foundation, of which the top 9-m layer is overconsolidated and the bottom 7-m layer is normally consolidated.

Vane and Penetrometer Tests

Vane tests were performed at a rate of rotation of 72°/min, which was found to be an appropriate one for tests carried out in kaolin and Gault clay. This rate of rotation practically assures an undrained condition, according to the criterion suggested by Blight [*13*]. The above rotation rate corresponds to an angular velocity of 0.18 mm/s, which is in accordance with the 0.15 mm/s value recommended by Perlow and Richards [*14*]. The average angle of rotation for peak during the tests was 10°.

Undrained strengths of Gault clay and kaolin measured with the vane have been obtained from a laboratory vane shear testing program [*6*], as shown in Fig. 5. It has also been found that kaolin strengths measured with the vane are very close to results of c_u triaxial tests reported by Davidson [*15*], which are plotted in Fig. 6*a*. together with the results of the vane tests. Equations used here to predict undrained strengths c_u of clay cakes

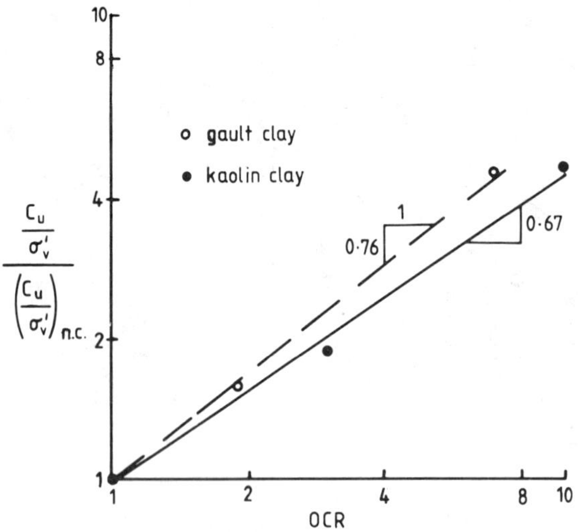

FIG. 5—*Vane strength measured at 1g.*

prepared for centrifuge tests are based on the results of Figs. 5 and 6 and are expressed by

$$c_u/\sigma'_v = 0.22\ (\mathrm{OCR})^{0.67} \tag{1}$$

for kaolin, and

$$c_u/\sigma'_v = 0.22\ (\mathrm{OCR})^{0.76} \tag{2}$$

for Gault clay, where OCR is the overconsolidation ratio. The powers in Eqs 1 and 2 are the slopes in the diagram of Fig. 5, and 0.22 is the normalized undrained strength of the

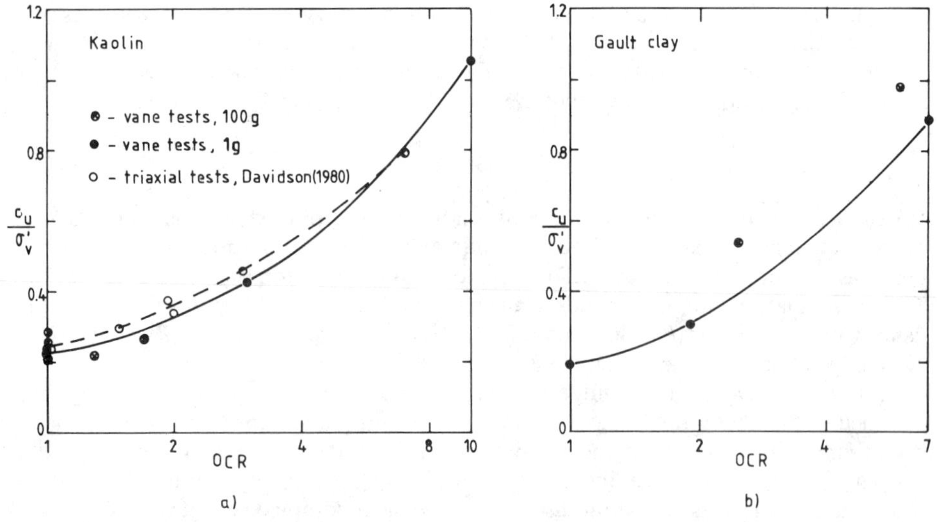

FIG. 6—*Comparison between tests performed at 1g and 100g.*

normally consolidated reconstituted clays (OCR = 1) as shown in the plots of Fig. 6. The value 0.22 has also been adopted by other research workers at Cambridge [*16,17*] dealing with the present reconstituted clays.

Figure 7 presents results of a vane strength profile, which is compared with the strength profile predicted using Eqs 1 and 2 plus the data of σ_v' and OCR given in Fig. 4. The agreement between the two curves is generally good, which suggested that the stress history shown in Fig. 4 is achieved during the tests. Therefore the resulting clay foundation (Figs. 4 and 7) consists of a more stiff 5-m top overconsolidated layer (OCR > 2), an intermediate lightly overconsolidated (1 < OCR < 2) layer from 5 to 9 m, and a normally consolidated clay below 9 m. Stress histories similar to this are commonly observed in soft clay sites [*11,12*].

An alternative way of checking values of the in-flight measured vane strengths c_u is to plot the normalized 100*g* values of c_u/σ_v' against OCR in the diagram of the 1*g* tests. This is shown in Fig. 6*a* for tests in kaolin and in Fig. 6*b* for tests in Gault clay. It is seen that kaolin data compare well, but Gault clay data show greater values of c_u/σ_v' for tests performed at 100*g*. Nevertheless all sets of tests have good general agreement and show the same trend.

Penetrometer tests were performed at a rate of 5 mm/s, which was a convenient rate for recording of data during tests in very shallow models. Results of vane and penetrometer tests plotted in the same diagrams are presented in Fig. 8 for three different clay models, used in centrifuge tests of embankments on clay foundations, and values are tabulated in Table 2. Profiles with depth of point resistance q_c and undrained strength c_u are similar in shape, which is an expected feature. However values of q_c are considerably lower than expected, as discussed below.

Results of vane and penetrometer tests are usually correlated using the expressions

$$N_c = \frac{q_c - \sigma_v}{c_u} \tag{3}$$

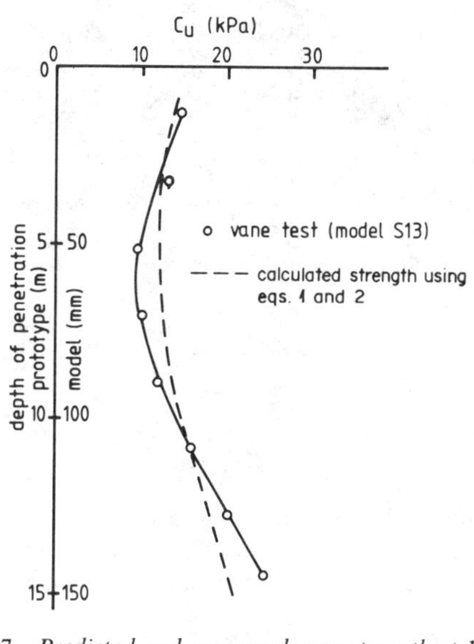

FIG. 7—*Predicted and measured vane strength at 100*g.

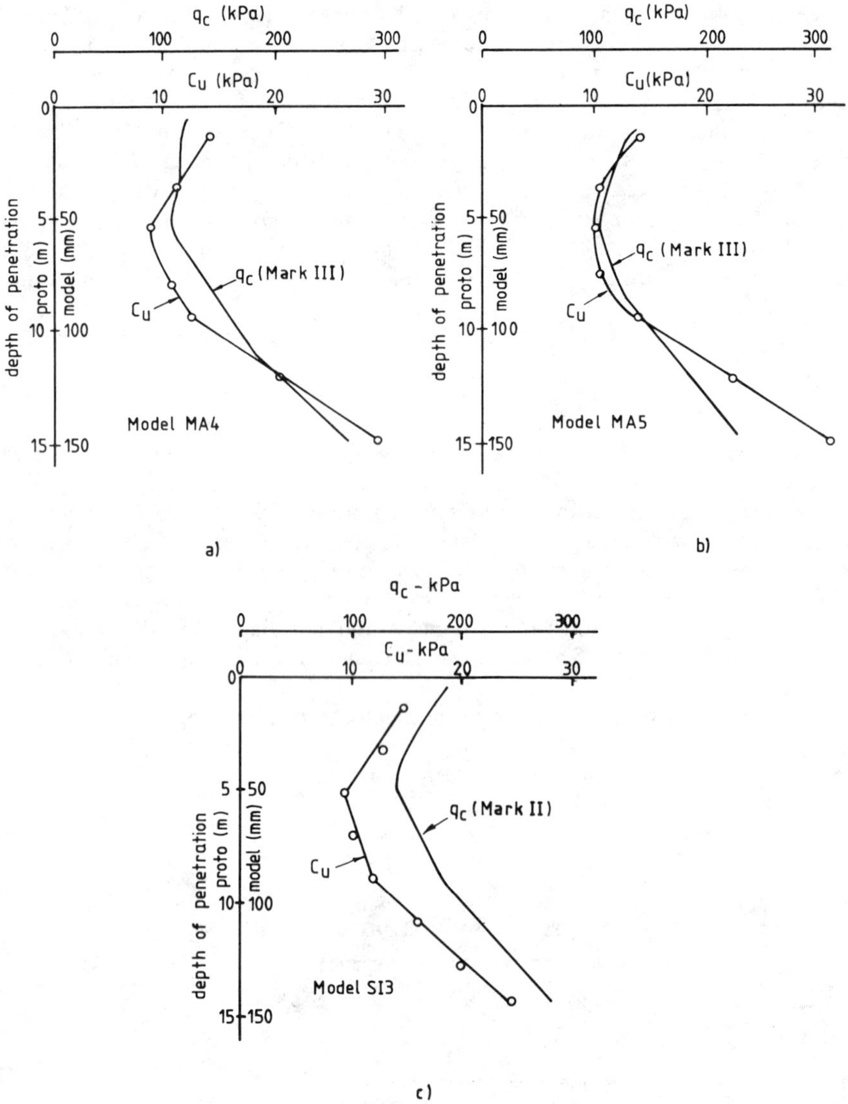

FIG. 8—*Penetrometer and vane tests.*

or

$$N_k = \frac{q_c}{c_u} \tag{4}$$

Computations of the empirical cone factors, N_c and N_k, for the three tests are shown in Table 2. It should be noted that

1. Values of N_k vary between 8 and 13 for tests in the centrifuge and between 9 and 15 for 1g laboratory tests [7].

TABLE 2—*Correlation between vane and penetrometer tests.*

Clay Model	Depth, mm	σ_v, kPa	c_u, kPa	q_c, kPa	N_k	N_c	q_T,[a] kPa	N_{kT}	N_{cT}
	13	38	14.7	174	11.8	9.2	236	16.0	13.5
	32	66	12.9	150	11.5	6.4	214	16.6	11.4
	51	96	9.4	142	14.5	4.3	212	22.6	12.3
S13[b]	70	126	10.1	160	15.8	3.4	244	24.2	11.7
	89	155	11.9	182	15.8	2.8	281	23.7	10.7
	108	184	16.0	218	13.8	2.3	337	21.1	9.6
	127	214	20.3	253	12.6	2.0	391	19.3	8.8
	144	240	24.5	282	11.8	2.0	438	17.9	8.1
	13	38	14.2	118	8.3	5.7	215	15.1	12.5
	33	69	11.6	115	9.9	4.0	230	19.8	13.9
MA4[c]	53	98	9.0	108	12.0	1.1	238	26.5	15.6
	79	138	10.9	140	12.8	0.2	315	28.9	16.3
	93	160	12.7	162	12.8	0.8	365	33.5	18.8
	120	204	20.9	205	9.9	0.1	461	22.0	12.3
	147	245	29.7	265	8.9	0.7	584	19.7	11.4
	13	38	14.0	130	9.3	6.6	232	16.6	13.9
	33	69	10.4	115	11.0	4.4	228	21.9	15.3
	53	98	10.0	105	10.5	0.7	231	23.1	13.3
MA5[b]	73	130	10.5	120	11.4	<0	275	26.2	13.8
	93	160	14.0	145	10.4	<0	335	23.9	12.5
	120	204	23.0	190	8.3	<0	434	18.9	10.0
	147	245	31.5	232	7.4	<0	528	16.8	9.0

[a] q_T assuming $\Delta u_b = 0.63\ q_T$ in Eq 5.3.
[b] Test using penetrometer Mark II.
[c] Test using penetrometer Mark III.

2. Values of N_c decrease with depth (or with decreasing OCR), which is consistent with patterns for $1g$ tests, including those performed by Francescon [*16*], as seen in Fig. 9*a*.

The slightly higher values of N_c and N_k measured in the laboratory compared with the centrifuge may be explained by the different boundary conditions existing in both types of tests, but mainly caused by water pressure effects existing around the cone, as discussed below.

Corrected Point Resistances

When penetrometer tests are performed, pore pressures acting at the recessed top of the cone decrease measured point resistances, as pointed out previously [*18*]. The equation used to compute corrected point resistances q_T based on measured point resistances q_c and on measured pore pressures u_b acting at the recessed base of the cone tip is

$$q_T = q_c + (1 - a)u_b \tag{5}$$

where a is the net ratio as defined by Campanella et al. [*19*], which had values of 0.39 and 0.63, respectively, for penetrometers, Mark II and Mark III. The pore pressure u_b consists of the hydrostatic pore pressure u_o plus the excess of pore pressure Δu_b. However, piezocone measurements were not performed during centrifuge tests, thus actual values of q_T could not be computed.

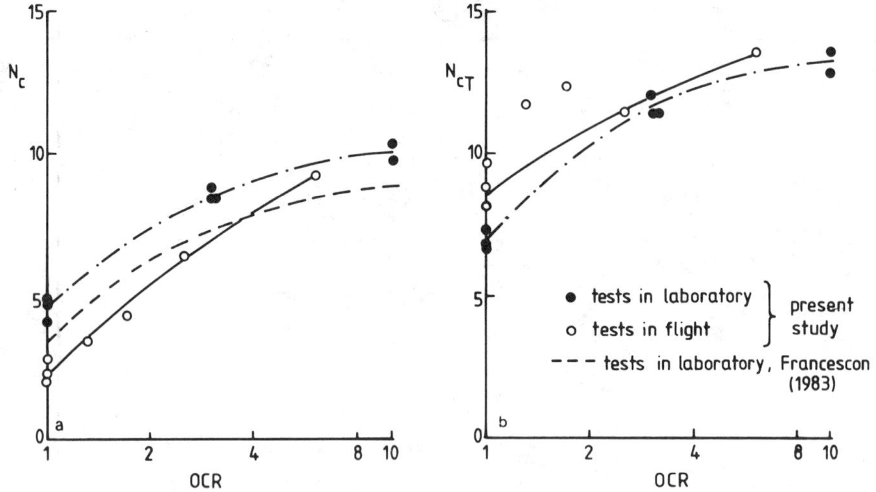

FIG. 9—*Empirical cone factors.*

Just to give an idea of likely values of cone factors (N_{ct} and N_{kT} computed from "corrected" point resistances q_T), use was made of results of piezocone tests carried out at the normal gravity field [*5*]. For these tests the pore pressure was measured at the cone tip and the typical value of the pore pressure ratio obtained was $\Delta u/q_T = 0.72$. Making the arbitrary assumption of base pore pressures to be 10% less than tip pore pressures as obtained in Ref *18* and substituting these into Eq 5, values of q_T were computed and are shown in Table 2. Corresponding values of N_{cT} and N_{kT} computed from q_T are given in Table 2, and it is seen that all corrected values have increased considerably. New corrected values of N_{cT} for centrifuge tests, SI3, are shown in Fig. 9*b*. In both these tests the same probe mark II was used. As observed in Fig. 9*b*, the agreement between values of N_{cT} obtained at 1*g* and at 100*g* is fairly good.

Summary and Conclusions

To measure clay strength during centrifuge tests, miniature vane and penetrometer apparatuses have been developed. A bed consisting of Gault clay overlying kaolin was consolidated from slurry. After consolidation in the laboratory and during centrifuge operation, a clay bed was formed with a 90-mm overconsolidated layer on the top of a 70 mm normally consolidated layer. Vane strengths showed that a stiff 50-mm crust was obtained from the induced stress history. Vane strengths also compared well with theoretical strengths.

Curves of point resistances with depth are similar in shape to curves of vane strength with depth. However, point resistances are low and consequently empirical cone factors are lower than expected, which have been attributed to water pressure effects around the cone tip. Corrected point resistances assuming pore pressures measured in 1*g* tests were computed, and new cone factors increased considerably.

Acknowledgments

The authors are grateful to Mr. P. W. Turner and to Dr. R. G. James from Cambridge University for assistance for the developments of the equipment and also during the centrifuge tests and to Dr. G. C. Sills, Oxford University, for comments to the paper.

The first author was supported by the Brazilian Research Council (CNPq) during his studies at Cambridge.

References

[*1*] Davies, M. C. R., "Centrifugal Modelling of Embankments on Clay Foundations," PhD thesis, Cambridge University Engineering Department, Cambridge, England, 1981.
[*2*] Pincent, B. and Tchocothe, F., *Proceedings of the International Conference on Soil Mechanics and Foundation Engineering,* Vol. 1, Stockholm, Sweden, 1981, pp. 741–744.
[*3*] Almeida, M. S. S., "Stage Constructed Embankments on Soft Clays," Ph.D. thesis, Cambridge University Engineering Department, Cambridge, England, 1984.
[*4*] Davies, M. C. R. and Parry, R. H. G., "Determining the Shear Strength of Clay Cakes in the Centrifuge Using a Vane," *Geotechnique,* Vol. 32, No. 1, pp. 59–62.
[*5*] Almeida, M. S. S. and Parry, R. H. G., *Geotechnical Testing Journal,* Vol. 8, No. 1, March 1985, pp. 14–24.
[*6*] Almeida, M. S. S., Davies, M. C. R., and Parry, R. H. G., *Geotechnique,* Vol. 35, No. 4, Dec. 1985, pp. 425–441.
[*7*] Almeida, M. S. S. and Parry, R. H. G., *Symposium on Application of Centrifuge Modelling to Geotechnical Design,* Manchester, England, April 1984, pp. 47–65.
[*8*] Almeida, M. S. S., *Symposium from Theory to Practice on Deep Foundations,* Porto Alegre, Vol. 1, Nov. 1985, pp. 175–186.
[*9*] Cheah, H., "Site Investigation Techniques for Laboratory Soil Models," Masters of Philosophy thesis, Cambridge University Engineering Department, Cambridge, England, 1981.
[*10*] Davies, M. C. R., Almeida, M. S. S., Cheah, H. C., and Parry, R. H. G., "Studies with Centrifuge Vane and Penetrometer Apparatus in a Normal Gravity Field," *International Symposium on Laboratory and Field Vane Shear Strength Testing,* American Society for Testing and Materials, Philadelphia, 1988.
[*11*] Leroueil, S., Tavenas, F., Trak, B., La Rochelle, P., and Roy, M., *Canadian Geotechnical Journal,* Vol. 15, 1978, pp. 54–65.
[*12*] Ortigão, J. A. R., Werneck, M. L. G., and Lacerda, W. A., *Journal of the Geotechnical Engineering Division, Transactions of ASCE,* Vol. 109, 1983, pp. 1460–1479.
[*13*] Blight, G. E., *Canadian Geotechnical Journal,* Vol. 5, No. 3, pp. 142–149.
[*14*] Perlow, M. and Richards, A., *Journal of Geotechnical Engineering Division, Transactions of the ASCE,* Vol. 1, No. 1, Jan. 1977, pp. 19–32.
[*15*] Davidson, C. S., "The Shear Modulus of Clays," Part II Research Report, Cambridge University Engineering Department, Cambridge, England, 1980.
[*16*] Francescon, M., "Model Pile Tests in Clay," Ph.D. thesis, Cambridge University Engineering Department, England, 1983.
[*17*] Clegg, D. P., "Model Piles in Stiff Clay," Ph.D. thesis, Cambridge University Engineering Department, Cambridge, England, 1981.
[*18*] Baligh, M. M., Assouz, A. M., Wissa, A. Z. E., Martin, R. T., and Morrisson, M. J., *Proceedings of the ASCE Conference on Cone Penetration Testing and Experience,* St. Louis, MO, Oct. 1981.
[*19*] Campanella, R. G., Robertson, P. K., and Gillespie, D., *Canadian Geotechnical Journal,* Vol. 20, 1983, pp. 23–35.

Tetsuro Tsutsumi,[1] *Yoshio Tanaka,*[1] *and Toshiaki Tanaka*[2]

Initial Stage Hardening Characteristics of Marine Clay Improved Cement

REFERENCE: Tsutsumi, T., Tanaka, Y., and Tanaka, T., **"Initial Stage Hardening Characteristics of Marine Clay Improved Cement,"** *Vane Shear Strength Testing in Soils: Field and Laboratory Studies, ASTM STP 1014,* A. F. Richards, Ed., American Society for Testing and Materials, Philadelphia, 1988, pp. 220–229.

ABSTRACT: Laboratory vane shear strength tests were conducted in order to study the various characteristics of the initial hardening stages of treated soils, such as the influence of moderate heat portland cement (MP) and blast-furnace slag (slag) mixture ratios, regional differences in marine clays, the influence of the quantity of additive to the slow hardening cement, and the influence of the water/cement ratio, and so forth, with respect to slow hardening cement, which is considered suitable for operations by the deep mixing method of soil stabilization in Japan. The results were that when the quantity of MP in the mixture is decreased, the shear strength of the treated soil decreases. The progress of hardening is closely related to the reactivity of the clay. The influence of the quantity of additives appears after hardening has progressed to a certain extent. When the water/cement ratio is increased, the shear strength decreases. Such characteristics were made clear, and it also became clear that the vane shear strength test is the most suitable method for studying these initial stage hardening characteristics.

KEY WORDS: deep mixing method, soil stabilization, discharge, slow hardening cement, blast-furnace slag, portland cement, vane shear strength tests, marine alluvial clay, clay minerals, lime reaction capacity, shear strength

The method referred to as the deep mixing method of soil stabilization has been established and is practiced in Japan [*1,2*]. With this method, 150–200 kg/m^3 of cement in milk form is supplied to soft clay on the seabed, the columns of stable treated soil are formed forcible in their original positions using mixing blades, and these are used as foundations for revetments, wharfs, and other structures (Fig. 1). There are two versions of this method, one in which the cement milk is discharged upon penetration of the mixing blades (discharge at pushing down), the other upon removal of the blades (discharge at pulling out), but the discharge at pushing down method is considered preferable as it improves the mixture precision of the improved substance and firmly unites the lap surfaces. With discharge at pushing down, the hardening of the treated soil must be suppressed for 2 to 4 h after the cement milk is injected into the ground or it is difficult to remove the mixing blades. When regular cement milk such as that of ordinary portland cement or blast-furnace slag cement is added to the soft clay on the seabed, hardening progresses by a rather great extent 1 to 2 h after mixture, so operations by discharge at pushing down have been considered impossible.

Various cements that harden more slowly than ordinary portland cement have been

[1] Soil Cement Research Department, Mitsubishi Mining and Cement Company, Ltd., 297 1-chome, Kitabukuro-machi, Omiya-shi, Saitama T330, Japan.

[2] Penta Ocean Construction Company, Ltd., Tokyo, Japan.

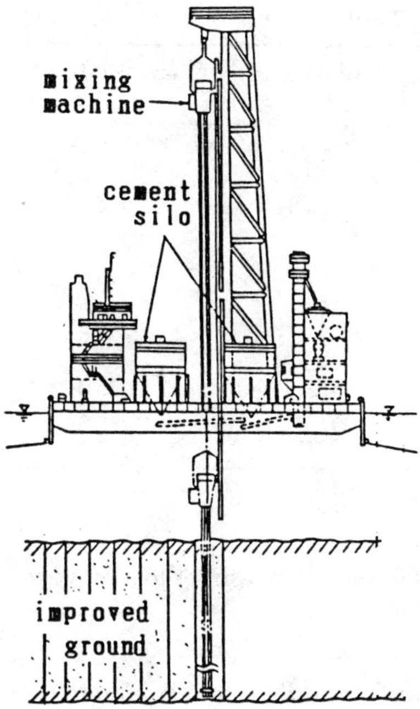

FIG. 1—*Illustration of deep mixing method.*

developed (herein referred to as slow hardening cements) in order to improve the limitations posed by this type of operation [*4*]. The slow hardening cements described in this report are suitable mixtures of blast-furnace slag powder (slag) and moderate heat portland cement (MP) [*5,6*]. This report includes an elucidation through laboratory vane shear strength tests of the initial stage hardening characteristics of soils treated using slow hardening cement, and as the vane shear strength test was determined to be an extremely effective means of showing the early stage hardening characteristics of treated soils, the results of these tests are also described.

Outline of Experiment

Soil Samples Used for Tests

Four types of marine alluvial clays representative of clays found in Japan were used as soil samples for the tests. All are extremely soft clays having natural water contents equal to or greater than their liquid limits.

(1) Hiroshima Bay, Kure Port marine alluvial clay (HIROSHIMA clay),
(2) Tokyo Bay, Yokohama offshore marine alluvial clay (YOKOHAMA clay),
(3) Tokyo Bay, Chiba offshore marine alluvial clay (CHIBA clay), and
(4) Osaka Bay, Izumi offshore marine alluvial clay (OSAKA clay).

Table 1 shows the index test results for the above soil samples.

Of the four soil samples, Osaka clay had the highest natural water content, followed in order by Chiba clay, Yokohama clay, and Hiroshima clay. Also, Yokohama clay had the lowest clay content.

TABLE 1—*Index tests on soils tested.*

Measuring Items	HIROSHIMA Clay	YOKOHAMA Clay	CHIBA Clay	OSAKA Clay
Natural water content, %	75.2	79.1	97.7	115.2
Wet density, g/cm^3	1.544	1.523	1.465	1.383
Specific gravity of soil particle	2.66	2.63	2.62	2.66
Liquid limit, %	64.6	80.5	86.2	98.5
Plastic limit, %	29.0	36.1	39.2	43.0
Plasticity index	35.6	44.4	47.0	55.5
Sand particle, %	3.7	7.6	0.8	2.2
Silt fraction, %	44.3	57.4	32.2	32.3
Clay fraction, %	52.0	35.0	67.0	65.5
pH	7.7	8.7	8.2	8.4
Organic matter content, %	3.9	4.1	3.9	2.8

The samples were passed through 2 mm sieves to remove coarse particles before the tests described below were conducted.

Chemical Composition and Physical Properties of Materials

As shown in Table 2, ordinary portland cement (NP), moderate heat portland cement (MP), and blast-furnace slag powder (slag) were used for these experiments.

Four types of slow hardening cements (MSC) were used, these being different mixtures of MP and slag at the following ratios: 30:70, 25:75, 20:80, and 15:85.

Specimen Preparation for Vane Shear Tests

Specific quantities of cement milk (mainly consisting of mixtures with a water/cement ratio of 60%) were added to the soil samples, after which they were mixed in a Hobart mixer for 10 min. Next, the treated soils were put into steel molds, which had a diameter of 7.5 cm and height of 10 cm. They were pounded and packed on a concrete table to remove as many air bubbles as possible. Next, the tops of the test pieces were covered with polyethylene films and left standing for a specific number of hours, after which the vane shear strength tests were performed. The preparation, curing, and measurements on the specimen were conducted in thermostatic ovens at 20 ± 3°C. Seawater was used as the mixing water for the cement milk.

TABLE 2—*Physico-chemical properties of cement.*

Kinds	Chemical Composition, %						Specific Gravity	Specific Surface, cm^2/g
	SiO_2	Al_2O_3	Fe_2O_3	CaO	MgO	SO_3		
NP	22.7	5.0	2.9	63.8	1.4	2.0	3.14	3360
MP	23.7	3.9	4.1	63.8	1.0	2.0	3.21	3020
Slag	34.1	13.1	0.4	42.1	8.0	. . .	2.91	4050

Laboratory Vane Shear Strength Test Method [7]

A vane with a width of 2 cm and a height of 4 cm was used. The vane was pressed into the specimen gently just before measurements were performed, to a depth of 7 cm so that it reached the center of the specimen. Next, the specimen was rotated at an angular velocity of 0.1°/s (1.7×10^{-3} rad/s), and the angle of rotation and rotational moment were measured.

Test Results

Some factors that are thought to govern the strength of treated soils using slow hardening cement in the initial stages are the water content, the gradation, the consistency limit, the type and quantity of clay minerals included in the soil, the quantity of additives in the cement, the MP and slag mixture ratio, and the time elapsed since mixing. The influence of these factors on the vane shear strength of the treated soils was investigated, and the results are described below.

Slow Hardening Cement Reaction Mechanism

Before discussing the results of the vane shear strength tests, we will explain in brief the mechanism of why cements consisting of a mixture of MP and slag display slow hardening characteristics [6].

Figure 2 is a type of diagram of the slow hardening cement reaction mechanism. When slow hardening cement is mixed with soil, hydration of the MP and the water in the soil begins, and calcium hydroxide ($Ca(OH)_2$) is formed. The $Ca(OH)_2$ causes an immediate ion exchange reaction with the clay minerals in the soil, resulting in an increase in the cohesion immediately after mixing. Hydration of the cement progresses, hardening also progresses. As time elapses, the surplus $Ca(OH)_2$ works as an alkali stimulant for the slag, hydration of the slag progresses, and the strength increases.

The main factors for the initial hardening discussed in this report can be thought to be

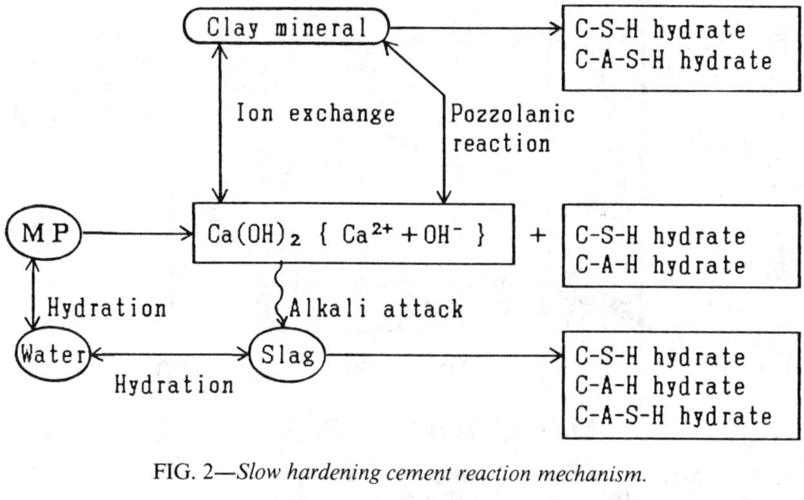

FIG. 2—*Slow hardening cement reaction mechanism.*

the ion exchange reaction between the $Ca(OH)_2$ and the clay minerals and the initial hydration of the cement. However, one can surmise that the hardening of the treated soil to which slow hardening cement was added is quite slow because the quantity of MP in the mixture was small and MP in which hydration is slow was used. On the other hand, when only MP was added, large quantities of $Ca(OH)_2$ were formed directly after the MP was added to the soil, the reaction with the clay minerals was remarkably rapid, and the hydration of the cement itself was also fast, so hardening began to progress from the early stages.

Thus, as the reaction conditions of slow hardening cement and portland cement are naturally different, the manifestations on the strength are also different.

Relationship Between Angle of Rotation and Shear Stress

Figure 3 shows one example of the relationship between the angle of rotation and shear stress illustrated by the results of vane tests, in this case for HIROSHIMA clay by itself (stirred soil) and when 180 kg/m^3 of slow hardening cement consisting of MP and slag at a ratio of 25:75 was added to the HIROSHIMA soil sample.

The treated soil contains no coarse impurities or large particles, and so the stress strain curve is smooth. The angle of rotation at the peak shear strength is practically constant, being betwen 20 and 30° (0.35 to 0.53 rad). Furthermore, for samples having a high shear strength the angle of rotation at the peak shear strength is small and the peak is clear, while with samples having a low peak shear strength the peak is not clear. The peak strength is used below to evaluate the characteristics of the treated soil.

Figure 3 also shows the peak shear strength of the soil sample to which no slow hardening cement was added (stirred soil), and for samples with cement added the peak shear strength of the treated soil increases with increasing time after mixing.

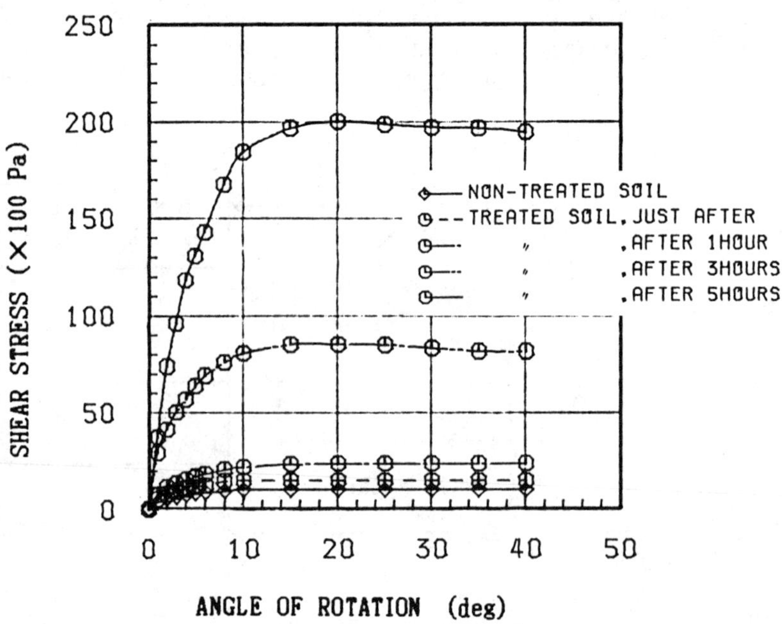

FIG. 3—*Relationship between angle of rotation and shear stresses (HIROSHIMA clay, MP: slag = 25:75, 180 kg/m^3, and* W/C = *60%).*

Influence of MP and Slag Mixing Ratio

Figure 4 shows variations in time of the shear strength of treated soils in which ordinary portland cement itself and the MP and slag mixing ratios were varied.

The shear strength of treated soils to which NP was added increases more rapidly than soils treated with slow hardening cement. The strengths increase markedly with time. Thus, if soil improvement is attempted with discharge at pushing down, hardening will have progressed when it is time to remove the mixing blades and removal may be impaired.

On the other hand, it is clear that with slow hardening cement, the shear strength does not increase substantially for several hours after mixing. The shear strength varies greatly depending on the MP and slag mixing ratio; the lower the quantity of MP, the more increases in shear strength will be suppressed. As shown by the reaction mechanism in Fig. 2, this can be thought to be a result of the ion exchange reaction between the $Ca(OH)_2$ and the initial hydration of the cement itself in the early stages of hardening of the treated soil. As decreasing the quantity of MP acts to decrease the effects of the two factors mentioned above, it is clear that the slag and MP mixing ratio greatly influences decreases in the shear strength.

Differences Caused by Soils

Figures 5*a* and 5*b* show the shear strengths of treated soils to which 180 kg/m^3 of slow hardening cement with MP:slag ratios of 30:70 and 25:75 was mixed for the various marine alluvial clays typical in Japan.

As can be seen, the shear strength of HIROSHIMA clay directly after mixing is low but increases greatly with time. The shear strength of YOKOHAMA clay is high directly after mixing but increases relatively little afterward. The shear strength of OSAKA clay is low

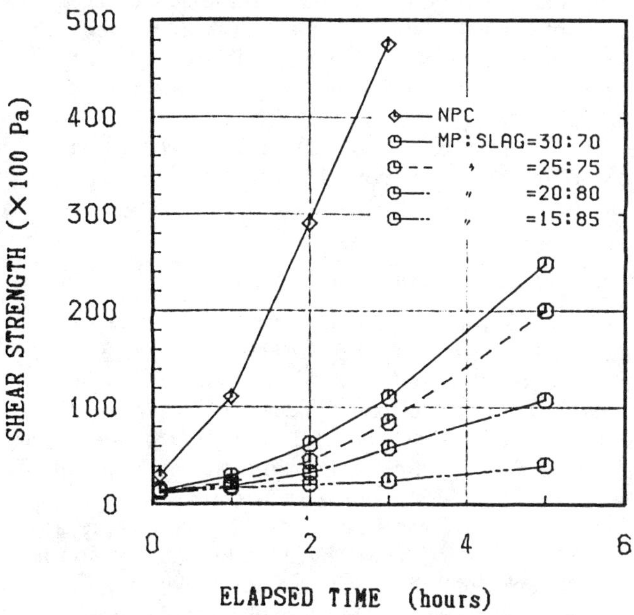

FIG. 4—*Influence of MP: slag ratio (HIROSHIMA clay, 180 kg/m^3 and* W/C= 60%).

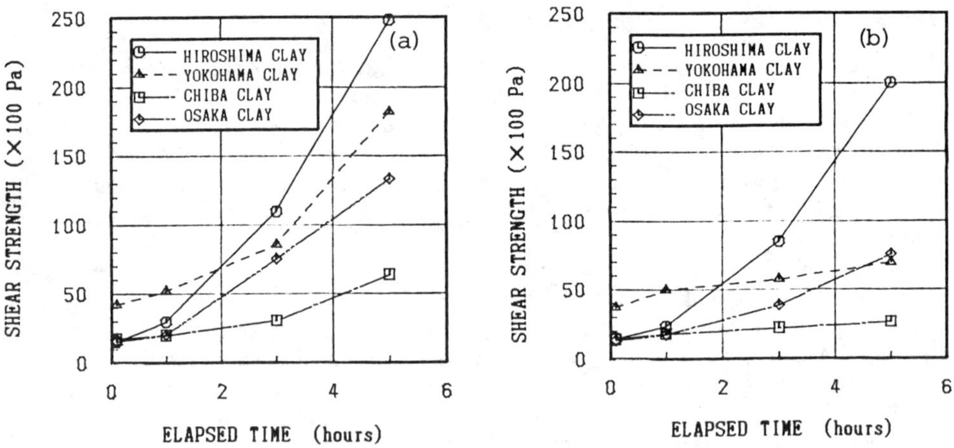

FIG. 5—(a)*Differences caused by soils (MP: slag = 30:70, 180 kg/m^3, and* W/C = *60%) and* (b)*Differences caused by soils (MP: slag = 30:70, 180 kg/m^3, and* W/C = *60%).*

directly after mixing and increases slowly thereafter. The shear strength of CHIBA clay is also low directly after mixing and practically does not increase thereafter.

Thus, it can be seen that even when the same cement is mixed with the soil, the trends in the shear strength differ according to the properties of the soil. One can expect that the types of clay minerals in the soil and their amounts, that is, the reactivity of the clay minerals and $Ca(OH)_2$, as well as the water content of the soil affect the strength increase with time.

Table 3 shows the reaction capacity in terms of the lime reaction capacity of the $Ca(OH)_2$ and the clay when an amount of $Ca(OH)_2$ equivalent to 20% of the dried soil is added to the various soil samples. Of the four, YOKOHAMA clay has the highest lime reaction capacity, demonstrating that it includes clay minerals with high reactivity. The lime reaction capacity is respectively lower for CHIBA, HIROSHIMA, and OSAKA clays, so it can be assumed that their reactivity is also lower.

It can be seen that the lime reaction capacity, that is, the reactivity with lime and the shear strength of the treated soil, are related and that the lime reactivity could be shown as an index of the strength increasing with time.

Influence of Quantity of Slow Hardening Cement Added

Figure 6 shows the relationship between the shear strength and the quantity of slow hardening cement added (MP:slag = 25:75) for HIROSHIMA clay.

The influence of the quantity added is not clear for the first hour after mixing, but becomes clear after 3 to 5 h have elapsed. The larger the quantity added, the higher the shear strength. The same tests were conducted for YOKOHAMA clay and CHIBA clay, but the influence of the quantity of cement additive was not clear after 3 h and becomes more difficult to assess as the MP:slag ratio nears 20:80.

It can be thought that the differences caused by the quantities are largely a result of the degree of progress of hydration of the MP or slag, and that with slow hardening cements containing low proportions of MP or with samples displaying low increases in shear strength, the influence of the quantity of added cement is not noticeable because the hydration of the MP and slag do not progress.

TABLE 3—*Lime reaction capacity of soil samples.*

Kinds of Clay	Lime Reaction Capacity, % 2 h	1 Day	3 Days	7 Days
HIROSHIMA	22.4	34.2	39.4	56.3
YOKOHAMA	38.5	45.6	68.1	82.7
CHIBA	28.4	38.5	45.9	86.0
OSAKA	15.9	23.9	30.6	54.6

Influence of Water/Cement Ratio

Figure 7 shows the change in shear strength when the water/cement ratio is varied between 60 and 100% for HIROSHIMA clay to which 180 kg/m^3 of slow hardening cement with MP:slag ratios of 30:70 and 25:75 was added.

As can be seen in the figure, the obvious result is that when the water/cement ratio added to the soil sample is increased, the shear strength of the treated soil also decreases. However, the influence of the water/cement ratio is little directly after mixing but becomes marked with the passing of time. To compare the values 2 to 4 h after mixing, the shear strength with an MP:slag ratio of 30:70 and a water/cement ratio of 100% is approximately the same as that with an MP:slag ratio of 25:75 and water/cement ratio of 60%, and practically the same results are obtained by increasing the water/cement ratio by 40% and decreasing the proportion of MP by 5%.

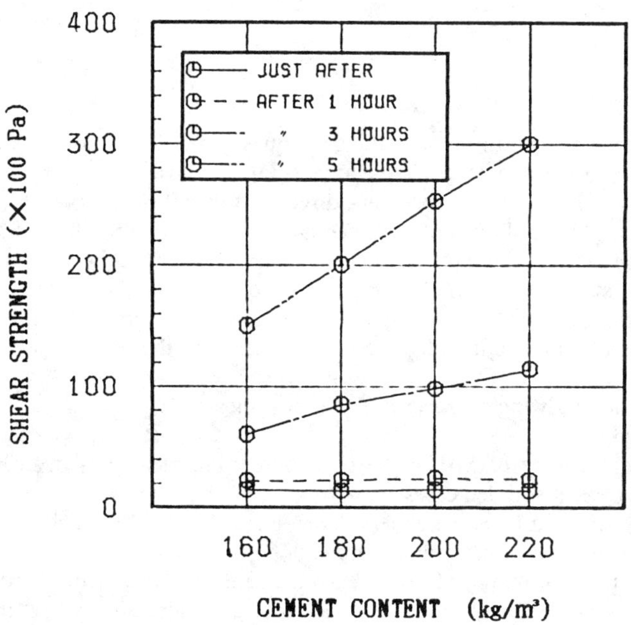

FIG. 6—*Influence of quantity of slow hardening cement added (HIROSHIMA clay, MP: slag = 25:75, and W/C = 60%).*

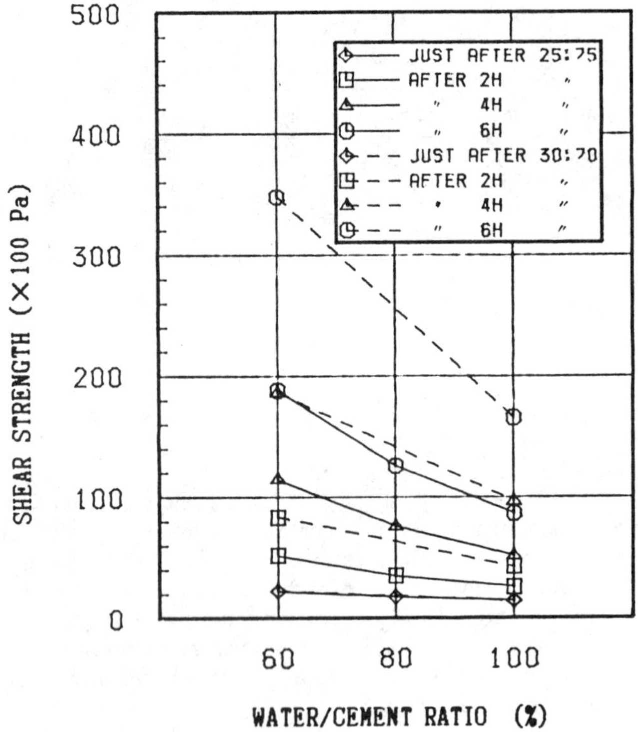

FIG. 7—*Influence of water/cement ratio (HIROSHIMA clay, MP: slag = 30:70, 25:75, 180 kg/m³).*

Conclusions

In order to execute operations with the deep mixing method of soil stabilization and in particular with the method of discharge at pushing down, it is necessary to suppress the hardening of the soil treated with cement directly after mixing so that the mixing blades can be pulled out. Using slow hardening cement consisting of a mixture of MP and slag as a cement, which satisfies these requirements, we have employed the laboratory vane shear strength test to study the various characteristics of the initial stages of hardening, with respect to the mixing proportions of MP and slag, the differences caused by the soil samples, the quantities of cement added, and the influence of the water/cement ratio, among others.

As a result, the following characteristics became clear:

1. When the quantity of MP in the slow hardening cement is decreased, the shear strength of the treated soil decreases.
2. The shear strength of a treated soil depends on the clay but is closely related to the reactivity between the clay mineral and lime.
3. The influence of quantity of slow hardening cement appears after hardening has progressed to a certain extent, with the shear strength increasing when the quantity is increased.
4. When the water/cement ratio increases, the shear strength decreases.

The proctor penetration test (ASTM Test Method for Time of Setting of Concrete Mixtures by Penetration Resistance [C 403]) is also widely used as a method of investigating the initial hardening characteristics of soil treated by added cement milk in Japan, but though this method is suitable for studying the characteristics after hardening has progressed to a certain extent, it is not considered suitable for assessing the hardness characteristics in extremely soft conditions. This experiment has shown that if the laboratory vane shear strength test used in this case is employed, it is possible to show clearly slight differences in the early age hardening characteristics of treated soils.

References

[1] Terashi, M., *Deep Mixing Method of Soil Stabilization, The Foundation Engineering and Equipment,* Vol. 13, No. 2, 1985, pp. 2–9.

[2] Terashi, M., "Practice and Problems of the Deep Mixing Method of Soil Stabilization," *Soils and Foundations,* Vol. 31-8, 1983, pp. 75–83.

[3] *Soft Ground Handbook,* Construction Industry Board of Inquiry Co., Ltd. 1981, pp. 821–837.

[4] Suzuki, S., Moriwaki, T., Yamaguchi, M., and Fujisaki, H., "On the Deep Mixing Method of Soil Stabilization Using Slow Hardening Stabilizers," *Reclamation and Dredging,* No. 129, 1986, pp. 63–75.

[5] Kira, K., Tsutsumi, T., and Tanaka, Y., "Characteristics of Soils Treated with Deep Mixing Method of Stabilization Hardening Stabilizers," *Cement Association of Japan Review of the 39th General Meeting,* 1985, pp. 494–497.

[6] Kira, K., Tsutsumi, T., and Tanaka, Y., "Consideration on Hardening Characteristics of Stabilizer for Deep Mixing Method Using Moderate Heat Cement and Slag," *20th Conference on Soil Engineering Research,* 1985, pp. 1717–1718.

[7] *Technic of Soil Exploration,* Japanese Society of Soil Mechanics and Foundation Engineering, 1980, pp. 240–245.

Part V: Field Vane Comparisons to Laboratory and In-Situ Test Methods

Guy Lefebvre,[1] *Charles C. Ladd,*[2] *and Jean-Jacques Paré*[3]

Comparison of Field Vane and Laboratory Undrained Shear Strength in Soft Sensitive Clays

REFERENCE: Lefebvre, G., Ladd, C. C., and Paré, J.-J., **"Comparison of Field Vane and Laboratory Undrained Shear Strength in Soft Sensitive Clays,"** *Vane Shear Strength Testing in Soils: Field and Laboratory Studies, ASTM STP 1014,* A. F. Richards, Ed., American Society for Testing and Materials, Philadelphia, 1988, pp. 233–246.

ABSTRACT: The purpose of the paper is to compare, at two different sites, the field vane strength with the undrained shear strength measured in laboratory on intact clay specimens cut from block samples. The laboratory tests included triaxial compression, triaxial extension, and direct simple shear tests on specimens anisotropically reconsolidated to the in-situ stresses. The laboratory undrained shear strength was determined on averaged stress strain curves built from the three types of laboratory tests in order to account for strain compatibility. At both sites, the undrained shear strengths obtained by the field vane and by the laboratory tests are nearly identical. The comparison is discussed in terms of field vane correction.

KEY WORDS: shear strength, field vane, laboratory testing, triaxial tests, simple shear tests, sensitive clays

The field vane is a widely used in-situ test for evaluating the undrained shear strength of soft clay deposits. It has been known however for some years that the measured field vane strengths may need to be corrected for use in stability analyses. Empirical corrections have been derived from the back analyses of embankment failures [*1–4*] and as such are influenced by other factors like the end effects of the failure surface [*5*] and the shear strength mobilized in the embankment and in the weathered crust [*6*]. Hence, even though these empirical corrections are based on case histories from around the world, one should question their reliability with unusual soil types and with clay deposits in territories where past experience does not exist [*7*].

This paper compares, for two soft, sensitive marine clay deposits in northern Quebec, the undrained shear strength profiles measured by the field vane with those obtained from consolidated undrained triaxial compression, triaxial extension, and direct simple shear tests run on block samples. The laboratory testing program was conducted to calibrate the vane for design of embankment dams.

[1] Professor of civil engineering, Université de Sherbrooke, Department of Civil Engineering, 2500 Univesité Blvd., Sherbrooke, Quebec, Canada.

[2] Professor of civil engineering, Massachusetts Institute of Technology, Cambridge, MA 02139.

[3] Head, Soil Mechanics Division, Sociéte d'énergie de la baie James, Montreal, Quebec, Canada.

Site Description

The investigation was carried out at two sites for the Nottawoy, Broadback and Rupert rivers (NBR) hydroelectric development scheme in the James Bay territory. The sites referred to as B-2 and B-6 are 50 km apart along the Broadback River and about 700 km northwest of Montreal. The sites are located in the Tyrrel sea basin where the clay was deposited some 8000 years ago. The stratigraphy includes at both sites, from surface to bottom, a layer of peat, the marine clay deposit of interest, a transition zone, a deposit of varved and layered lacustrine clay, and a till layer lying on a granitic bedrock. The pore-water salinity of the marine clay has been reduced since deposition to a concentration less than 1 g/L.

The index properties and stress history for the B-6 and B-2 marine clays are presented respectively in Figs. 1 and 2. At B-6, the marine clay is covered by 0.5 m of peat and is weathered to a depth of about 4 m. Below the crust, the natural water content is fairly constant at 45 to 50% between 4 and 12 m. The plasticity index I_P of 17% at the bottom of the crust decreases with depth, resulting in an increasing liquidity index I_L. The sensitivity S_t evaluated with the laboratory Swedish fall cone is about 30 at 4 m and near 400 at 10 m. The clay size fraction (minus 2 μm) is constant throughout the marine clay at about 75%. A transition from marine to lacustrine clay starts at 12.3 m.

Figures 1 and 2 present the profiles of in-situ vertical effective stress σ'_{vo} and of the preconsolidation pressure σ'_p determined from incremental oedometer tests on specimens cut

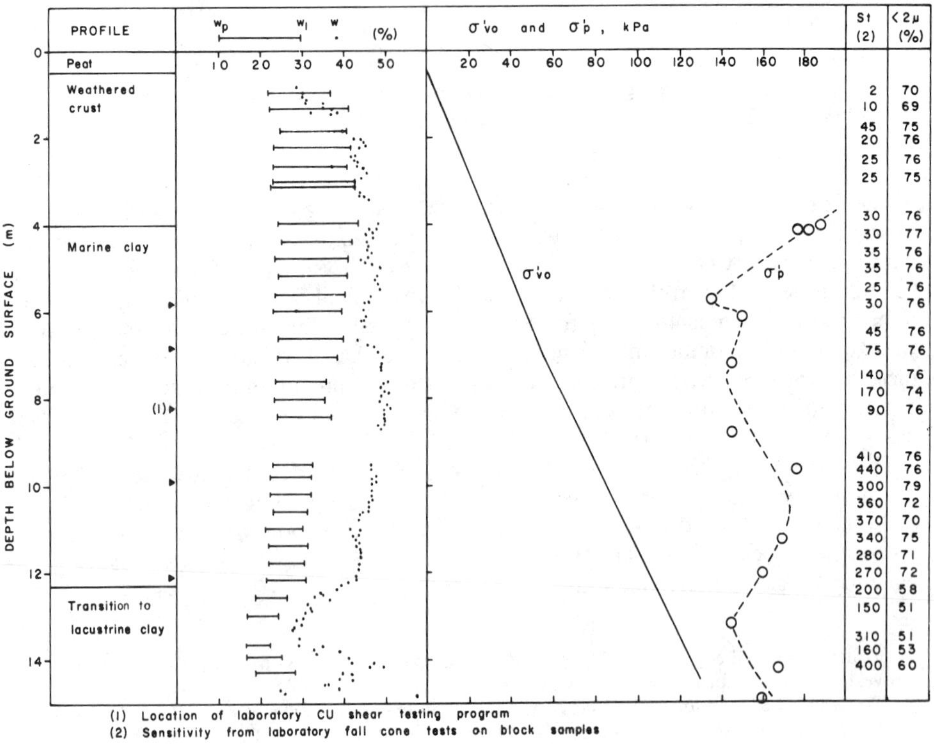

FIG. 1—*Index properties and stress history at Site B-6.*

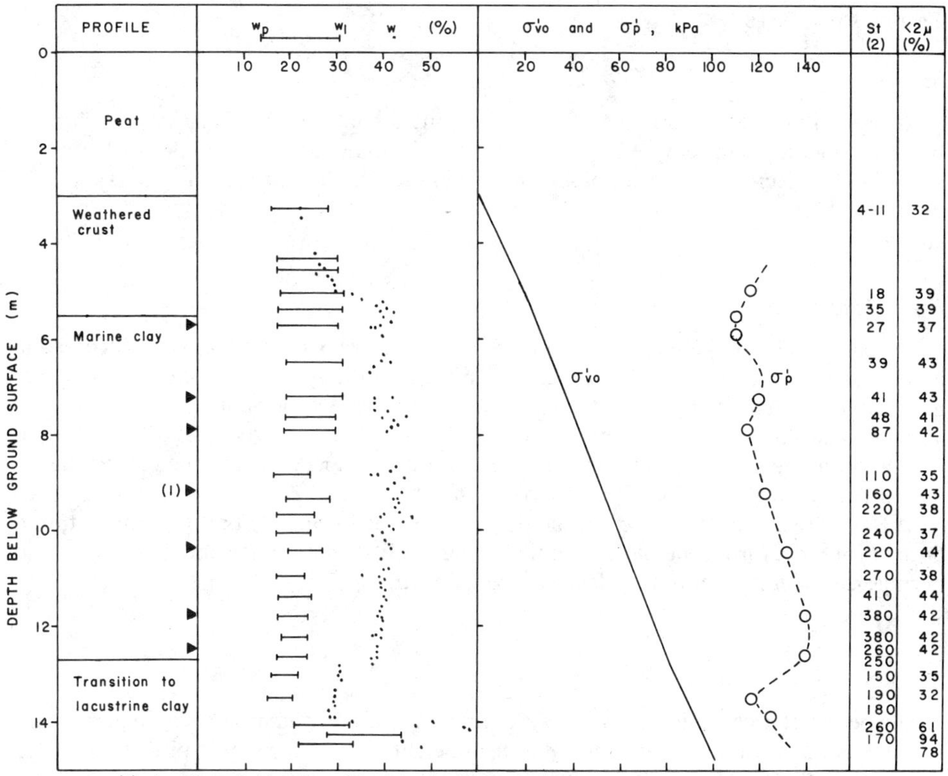

(1) Location of laboratory CU shear testing program
(2) Sensitivity from laboratory fall cone tests on block samples

FIG. 2—*Index properties and stress history at Site B-2.*

from block samples. At B-6, the marine clay is overconsolidated by about 90 kPa ($\sigma'_p - \sigma'_{vo}$), while the lower lacustrine clay is nearly normally consolidated. This is interpreted as an indication that the measured σ'_p of the marine clay does not reflect geological preloading but rather some kind of structure effects. The σ'_{vo} profile of Fig. 1 takes into account a downward gradient of about 0.20 with the water table at ground surface.

At B-2 (Fig. 2), the site is covered by 3 m of peat. Below a 2.5-m-thick weathered crust, the marine clay has a water content near 40% throughout. The I_P decreases with depth from 14% at the bottom of the crust to 6% at 12 m. Because of the increase in liquidity index, the sensitivity increases with depth from about 40 to 400. The B-2 clay is much coarser than at B-6, the clay size fraction being only 40 ± 5%. The profile of preconsolidation pressure in Fig. 2 shows an overconsolidation ($\sigma'_p - \sigma'_{vo}$) of about 70 kPa in the marine deposit. It decreases to 30 or less within the underlying lacustrine clay, pointing again to the existence of some structure effect instead of geological preloading. The ground-water table was at the surface with a downward gradient of only 0.07.

Except for the clay fraction, the marine clay deposits at B-2 and B-6 are fairly similar and exhibit the same trends with depth. For the purpose of comparing field vane and laboratory tests results, a depth interval of 7.0 m going from 5.5 to 12.5 m at both sites was chosen to have a homogeneous soil layer and to exclude the weathered crust.

Testing Procedures

Sampling

At both sites, undisturbed samples were obtained continuously using the Sherbrooke block sampler [*8*]. This device carves and retrieves cylindrical block samples with a diameter of about 25 cm and a height of 40 cm. The block sampling was conducted to about 15 m at both sites.

Field Vane Testing

The field vane tests were performed at 0.5 or 1 m interval with the standard Nilcon vane borer using a pointed vane 13 cm high and 6.5 cm in diameter. The Nilcon vane borer needs no protective casing around the vane and the rods. A slip coupling between the rod and the vane allows a free 15° rotation of the rod to measure soil-rod friction. The torque in relation to the angular rotation is recorded on waxed paper and is applied at the rate of 0.37%/s at the top of the rod (2 handle rotations/s).

At B-6, the vane data used in this study were obtained from eight profiles run within a 5- to 10-m radius from the block sample boring. At B-2, the results come from five vane borings run within 70 m of the block sampling location.

Laboratory Consolidation and Strength Testing

The oedometer consolidation tests were run with the Wykeham Farrance apparatus on specimens 63 mm in diameter and 13 mm in height. A load increment ratio $\Delta p/p$ of 0.5 was applied every 24 h. The 24-h compressive curve was used to estimate σ'_p with the Casagrande technique.

The consolidated-undrained triaxial tests were run strain controlled, with Wykeham Farrance cells modified to reduce piston friction and tied to a data acquisition system. The specimens had a diameter of 36 mm and a height of 71 mm. Porous stones and filter paper were used at each end of the specimens. Five filter paper strips 2 mm wide and inclined at 60° provided lateral drainage. One prophylactic served as a membrane. The pore pressure was measured at the base of the specimen after back pressuring to 100 kPa.

The specimens were anisotropically consolidated to a vertical stress about equal to σ'_{vo} at the depth of the samples in order to reach an in-situ state of stress and measure the initial undrained shear strength, which can be mobilized before any consolidation of the deposit. No data were available on the in-situ K_0; a σ'_3/σ'_1 of 0.55, typical of normally consolidated clay deposits, was adopted for the triaxial tests on the basis that the B-2 and B-6 deposits were geologically normally consolidated. The compression (TC) tests were performed by increasing the axial load and the extension (TE) tests by decreasing the axial load at a fixed rate of 0.006 mm/min.

The CK_0U direct simple shear (DSS) tests were run with the Geonor apparatus [*9*] using a cylindrical specimen 20 mm high and 80 mm in diameter confined in a rubber membrane reinforced by steel wires. The apparatus maintains constant volume during shear by adjusting the normal stress σ'_v. The specimens, confined between two porous stones, were consolidated with a vertical stress equal to the in-situ stress and then sheared by applying a horizontal displacement at a rate of 0.006 mm/min. To avoid slippage, the porous stones were equipped with thin needles 3.6 mm long on a 5 mm grid.

Test Results

Field Vane Results

Figure 3 presents the field vane strengths and the average profile for each site. The scatter is small with a coefficient of variation of about 15% at both sides. The c_u(FV) is nearly constant with depth below the crust at about 34 kPa for B-6 and 23 kPa for B-2.

Laboratory CU Shear Test Results

The consolidated-undrained triaxial compression and extension tests and the direct simple shear tests were performed at every 1 or 1.5 m depth intervals as shown in Figs. 1 and 2. The laboratory testing program was aimed at measuring the effect of anisotropy on the undrained shear strength. Specifically, the compression, extension, and direct shear tests attempted to simulate typical stress systems occurring along a failure plane in an embankment foundation. Figures 4 and 5 present typical normalized shear stress-shear strain and normalized effective stress plot data from the triaxial and direct simple shear tests run on specimens consolidated at in-situ stresses. The shear stress $q = 0.5\ (\sigma_v - \sigma_h)$ in the triaxial or the horizontal stress τ_h in the direct shear and the average effective stress $p' = 0.5\ (\sigma_v' + \sigma_h')$ in the triaxial test or σ_v' in the direct shear test have been normalized by the measured preconsolidation pressure σ_p'. For triaxial tests, the shear strain has been made equal to 1.5 times the axial strain. In the direct simple shear tests, especially those performed at relatively low σ_{vo}', the stress plots often indicate a larger pore-pressure generation at the beginning of the test. This appears to be related to the straining, which is necessary to fully mobilize the porous stone needles embedded in the specimens.

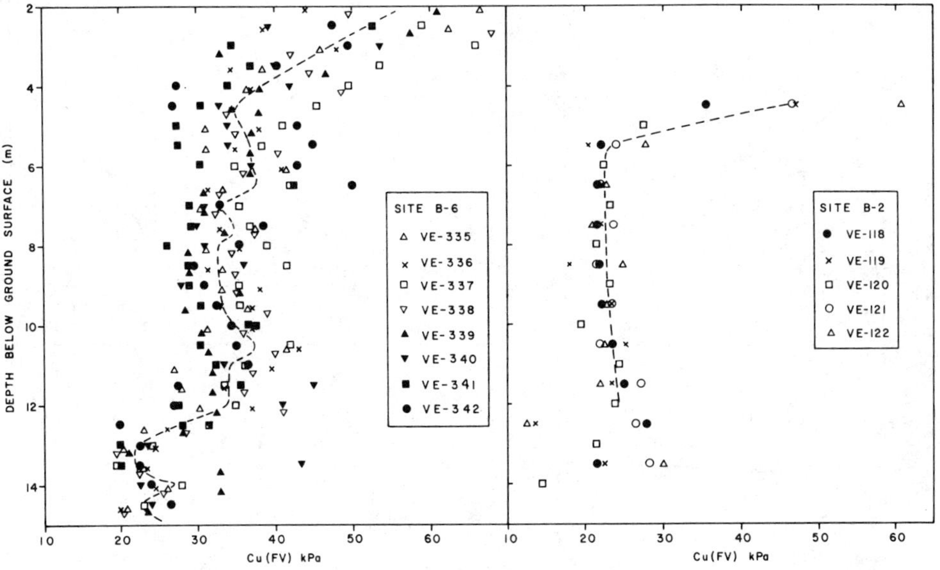

FIG. 3—*Measured and mean field vane strength* c_u*(FV) at B-6 and B-2.*

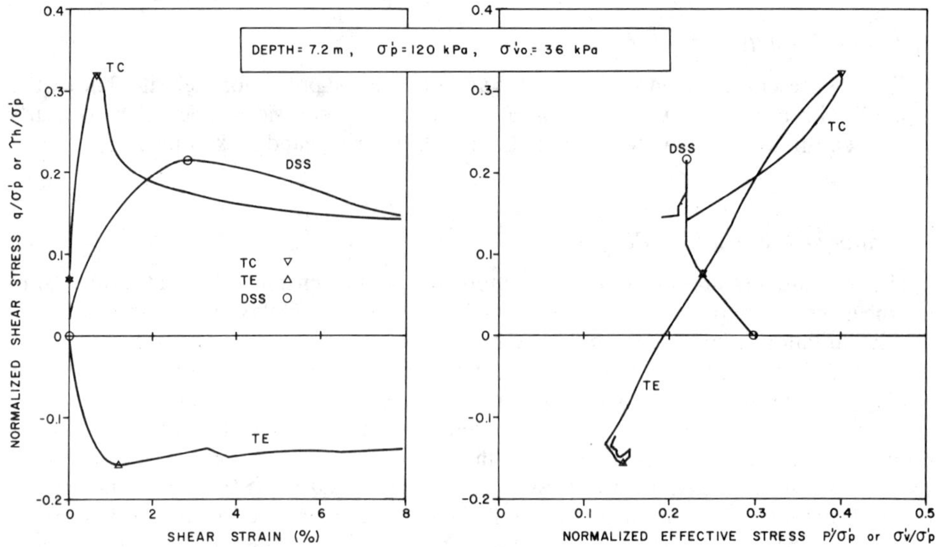

FIG. 4—*Typical normalized shear stress versus shear strain and effective stress plots* CK_0U *tests, B-2.*

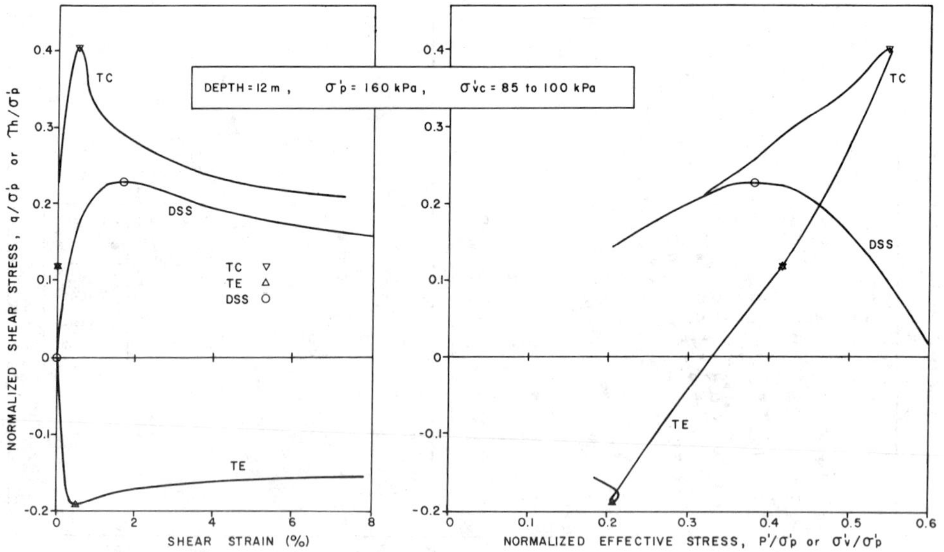

FIG. 5—*Typical normalized shear stress versus shear strain and effective stress plots* CK_0U *test, B-6.*

All specimens in triaxial compression developed a well defined failure plane oriented at about 60° with the horizontal. Specimens in triaxial extension failed by localized constriction. For the B-2 specimens, the extension failure was also characterized by fissuring at about 30° with the horizontal. Shear in triaxial compression produces a high peak resistance at very low strain ($\gamma \simeq 0.5\%$) followed by a substantial strength loss, that is, large strain softening. In contrast, shear in triaxial extension or in direct simple shear produces relatively small strain softening, and the peak strength was reached at larger shear strains than in triaxial compression.

Profiles of peak undrained shear strength measured in triaxial compression, triaxial extension, and direct simple shear tests are presented in Fig. 6 for B-2 and B-6. Since failure planes developed at about 30° with the direction of σ_1 in the triaxial tests, the peak strength has been multiplied by cos 30° in order to obtain the shear strength mobilized on the failure plane, that is, $\tau = q \cos 30°$. In the direct simple shear test, the undrained shear strength has been taken as equal to the maximum applied horizontal stress, that is, $\tau_d = \tau_h$. As shown in Fig. 6, the undrained shear strength differs significantly depending on the applied stress system. Failure does not occur at the same strain in the three types of shear tests, and in addition the stress strain curves in triaxial compression showed a very brittle failure and substantial strain softening. Following Koutsoflas and Ladd's [*10*] strain compatibility procedure as an approximate methodology to account for anisotropy and progressive failure in circular arc and wedge type failure surfaces, it has been assumed that all elements will have the same shear strain at the moment when a rupture surface forms and leads to

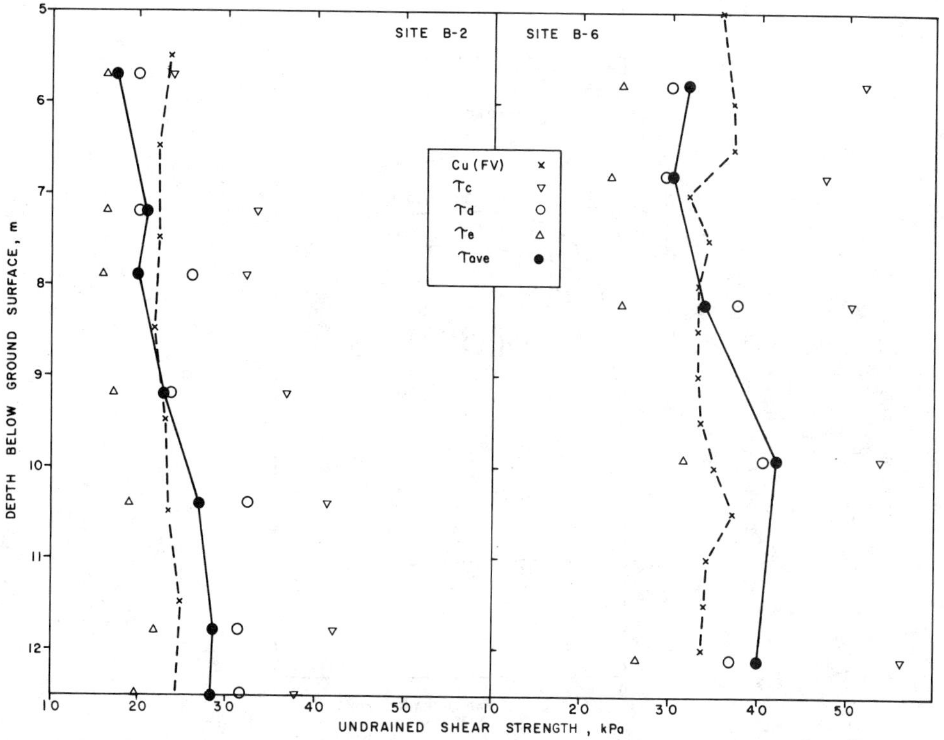

FIG. 6—*Comparison of field vane and laboratory* c_u *strength.*

a gross failure through the foundation, that is, when the maximum average resistance is mobilized. For the purpose of comparison with the vane strength, it was also assumed that the rupture surface is composed of equal segments of compression, direct shear, and extension. At any given strain, the average resistance therefore equals one third of the sum of compression, direct shear, and extension strength.

The strain compatibility procedure is illustrated in Fig. 7, which presents plots of normalized shear stress versus shear strain for a series of tests run on B-6 specimens from depth 6.8 m. For triaxial and extension tests $\tau = q \cos 30°$ while for the direct simple shear $\tau_d = \tau_h$. The upper portion of Fig. 7 plots individual stress-strain curves, and the bottom presents the average mobilized strength. This average resistance increases rapidly below $\gamma = 0.5\%$, reaches a maximum value at $\gamma = 1\%$, and then decreases slightly at larger strains. The average curve defines a maximum $\tau_{ave}/\sigma_p' = 0.207$. For comparison, the average of the peak strengths, that is, $\frac{1}{3}(\tau_c + \tau_d + \tau_e)$, comes to 0.228. Therefore, the treatment for strain compatibility reduced the strength by about 10%. The maximum τ_{ave} generally occurred between $\gamma = 0.5$ and 1.0% for B-6 and at about 0.6% for B-2. It is worthwhile to note that by using this procedure, the sharp peak obtained in triaxial compression has very limited influence on the average shear strength τ_{ave}. The profile of τ_{ave} versus depth is presented for B-2 and B-6 on Fig. 6.

Comparison of Laboratory and Field Undrained Shear Strengths

Figure 6 compares the mean field vane profiles c_u(FV), (from Fig. 3) to the peak laboratory τ_c, τ_d, τ_e and to the shear strength treated for strain compatibility τ_{ave}. On Fig. 8, the

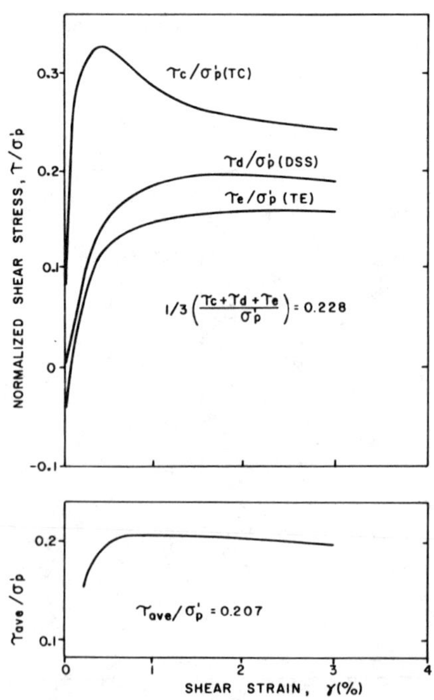

FIG. 7—*Application of the strain compatibility technique to stress-strain data (B-6 depth = 6.8 m).*

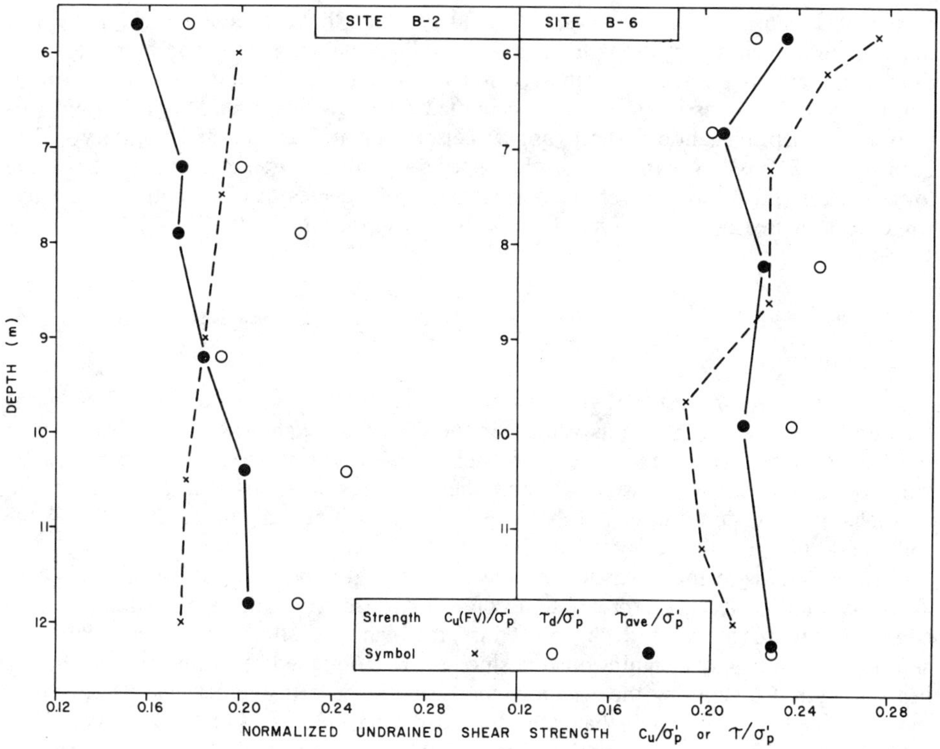

FIG. 8—*Comparison of normalized undrained shear strength.*

comparison is made in terms of profiles of normalized shear strength, c_u/σ_p' or τ/σ_p'. Mean values of the normalized shear strength have been calculated for the portion of the deposit considered in this study (5.5 to 12.5 m) and are presented in Table 1 with the standard deviation for each value.

In general (Fig. 6), the peak strength in triaxial compression τ_c is much higher than the field vane strength while the peak triaxial extension τ_e is much lower. The direct simple shear strength τ_d is intermediate betewen compression and extension and relatively close to the field vane. When compared in terms of mean normalized shear strength for the two deposits (Table 1), the laboratory shear strengths treated for strain compatibility τ_{ave} are identical to the field vane strengths at both sites. The mean DSS strength is also relatively

TABLE 1—*Mean values of normalized undrained shear strengths at B-2 and B-6.*

Site	$c_u(FV)/\sigma_p'$	τ_c/σ_p'	τ_d/σ_p'	τ_e/σ_p'	τ_{ave}/σ_p'
B-2	0.185 ±0.01	0.27 ±0.04	0.215 ±0.025	0.14 ±0.01	0.185 ±0.02
B-6	0.225 ±0.03	0.345 ±0.025	0.23 ±0.02	0.17 ±0.01	0.225 ±0.01

close to c_u(FV) or τ_{ave}. On this basis, one could conclude that for these two clays, the normalized undrained shear strength measured by the field vane and laboratory tests are exactly the same if CU tests are run at in-situ stress on undisturbed block samples, and if anisotropy and progressive failure are accounted for. However, when compared individually versus depth, slight differences appear between the c_u(FV) and the laboratory τ_{ave}. At both sites c_u(FV) is larger than τ_{ave} in the upper portion of the deposit and smaller in the lower portion (Fig. 6). Moreover, the normalized c_u(FV) decreases with depth at both sites, while τ_{ave}/σ'_p increases with depth at B-2 and is nearly constant at B-6 (Fig. 8).

Discussion

Progressive Failure

The only laboratory test that yielded a shear strength comparable to the field vane was the direct simple shear. Specimens tested in the DSS did not show the brittle behavior so pronounced in triaxial compression, probably because of progressive failure associated with nonuniform stress and strain. Because they yielded similar strengths and exhibit similar behavior, one could infer that both tests impose a progressive type of failure to the soil, in addition to stress axis rotation.

Good general agreement was found between the field vane strength and the results of TC, DSS, and TE tests interpreted to account for progressive failure by using the strain compatibility procedure. With this procedure, the high peak strength in triaxial compression tests did not play a significant role since τ_{ave} was obtained at strains beyond the TC peak strength. Considering the generally good experience with the field vane to evaluate the undrained shear strength that is mobilized during failures of embankments on low plasticity soft clays, one can conclude that (1) the high peak that is observed in triaxial compression tests does not contribute to the shear resistance which is mobilized during failure of an embankment and (2) a progressive type of failure develops in a field vane test.

A high peak is also generally observed in drained triaxial compression of undisturbed specimens consolidated to stresses less than σ'_p. Back analysis of many failures of natural slopes in Canada shows, however, that the shear strength mobilized at failure is much lower than the peak strength and is, in fact, equal to the large deformation strength [*11,12*]. It is therefore interesting to note that the very sharp peak observed in drained or undrained triaxial compression is irrelevant to the analysis of stability problems for sensitive clays.

Field Vane Correction

Field vane corrections are normally based on the comparison of field vane strength and of the strength back calculated from an actual failure. If one accepts that the laboratory τ_{ave} computed from τ_c, τ_d, and τ_e treated for strain compatibility correctly models the undrained shear strength, which is mobilized on a circular failure surface in an embankment foundation, the laboratory τ_{ave} can be used to develop a field vane correction for a given clay. The ratio τ_{ave}/c_u(FV) will then be the μ factor, which should be applied to correct the field vane for embankment design. When considering mean normalized shear strengths for the B-2 and B-6 marine clay deposits (Table 1), the field vane yields exactly the same strength as the laboratory $CK_0U\tau_{ave}$, and it could be concluded that the field vane correctly measures the undrained shear strength for those deposits without the need of any correction (μ = 1.0). When considering the strength profiles (Fig. 6) or the normalized strength profiles (Fig. 8) in the two deposits, the vane at both deposits slightly underestimates the undrained shear strength in the lower portion of the deposit ($\mu \simeq 1.15$) and slightly overestimates the undrained shear strength in the upper portion of the deposit ($\mu \simeq 0.85$) if one neglects the low τ_{ave} value obtained at B-2 at depth 5.7 m. This value of $\tau_{ave}/\sigma'_p = 0.155$ should rather be of the order of 0.175 based on the other tests results for the upper portion of the B-2 deposit. The μ factor in the upper and the lower portion of the deposits is still near unity and will in any event average out to a value close to unity for deep failures, as would be expected for embankments with flat slopes or large stabilizing berms, especially given the almost constant strength with depth.

The two deposits are uniform in terms of grain size. The plasticity index and the OCR, however, decrease with depth. The effect of OCR on c_u(FV) and on the in-situ stress ratio K_0 is difficult to assess for structured clays where the measured σ'_p is not thought to be related to geological preloading. If K_0 and the horizontal stress increased with OCR as for clay deposits overconsolidated by preloading, the decreasing OCR with depth at B-2 and B-6 could partially explain the decreasing $c_u(\text{FV})/\sigma'_p$ in Fig. 8. And for the triaxial testing, the selected value of $K_0 = 0.55$ would be too low in the upper portion of the deposits, where the OCR is higher, and hence lead to a slight underestimation of τ_{ave}. This might also explain the low B-2 value at depth 5.7 m where the OCR was the highest at 4.3. On the other hand, the plasticity index decreases from the top to the bottom of the deposits (from 17 to 10% at B-6 and from 14 to 6% at B-2), resulting in a large increase in sensitivity. Therefore, the decrease in $c_u(\text{FV})/\sigma'_p$ (Fig. 8) is also probably related to the decreasing plasticity index and increasing sensitivity with depth.

Comparison with Existing Vane Corrections

Bjerrum [*1,13*] analyzed a number of embankment and excavation failures for which field vane data were available, and plotted the computed factors of safety versus the plasticity index of the clay. A straight line, fitted through the data points, constituted the basis for his now widely accepted field vane correction factor, $\mu = 1/\text{FS}$, shown on Fig. 9*a*

$$c_u \text{ corrected} = \mu c_u(\text{FV})$$

For a plasticity index of 20%, the Bjerrum μ factor equals to 1.0. Geotechnical engineers have been reluctant to apply a μ factor larger than unity for clays with I_P lower than 20% and in practice a μ factor of unity is often used in such cases. The μ factors derived for the B-2 and B-6 deposits from CK_0U tests reconsolidated at in-situ stress are in good agreement with Bjerrum's correction as shown on Fig. 9*a*.

Aas et al. [*4*] after reanalyzing the rationale behind the Bjerrum correction have concluded that the μ factor should be related to $c_u(\text{FV})/\sigma'_{vo}$ rather than to the plasticity index. Considering well documented case records as well as laboratory CK_0U tests results (triaxial compression, triaxial extension, and direct simple shear tests reconsolidated at in-situ stress), they have proposed the correction shown in Fig. 9*b*. One correction curve is presented for "normally consolidated" (NC) clay and another for overconsolidated (OC) clay. The "normally consolidated" curve applies to young, aged, or cemented clay (that is, to all clays that have not been geologically preloaded) and is thus the one which should be used for the B-2 and B-6 deposits. The $c_u(\text{FV})/\sigma'_{vo}$ values versus depth are (by chance) identical at both sites and equal 0.62 at 7 m and 0.36 at 11 m. According to Fig. 9*b*, the μ factor to correct the field vane, using the NC curve, should be 0.6 and 0.75, respectively at these depths. Such correction factors reduce the field vane strengths to values much less than the laboratory τ_{ave} evaluated from CK_0U tests (Fig. 9*b*) and do not appear applicable for the clays considered in this study.

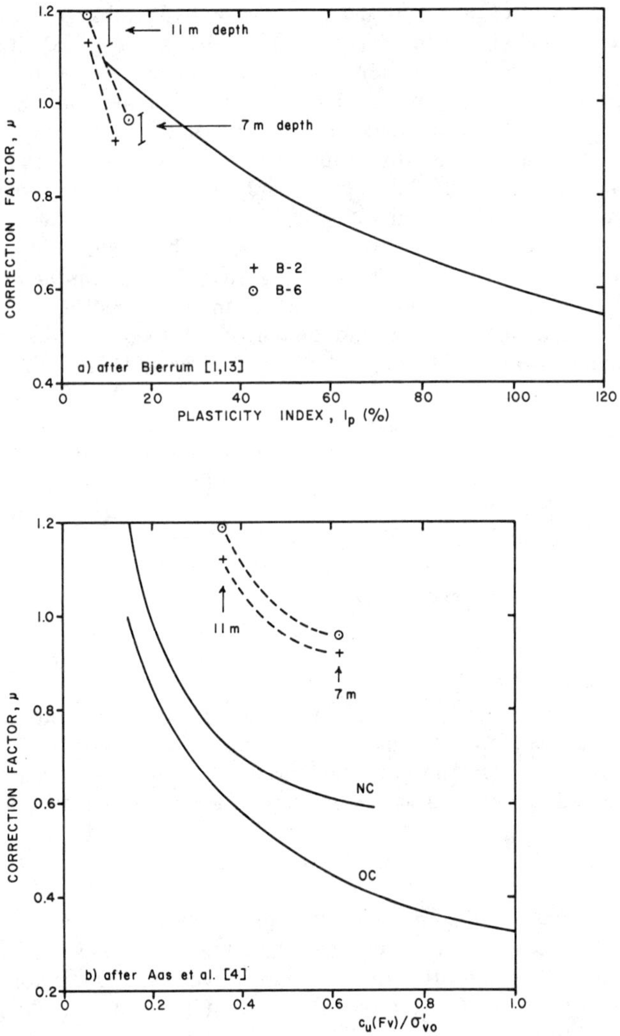

FIG. 9—*Field vane correction factor.*

Summary and Conclusions

The paper compares the shear strength measured by the field vane c_u(FV) with those measured in the laboratory by consolidated-undrained triaxial compression (TC), direct simple shear (DSS) and triaxial extension (TE) tests for two deposits of marine sensitive clays at the sites B-2 and B-6 in northern Quebec. The two sites have similar variations of OCR and I_P, the OCR decreasing with depth from about 4 to 1.5 and the I_P from about 18 to 8% (Figs. 1 and 2). The sites differ in terms of grain size, the clay size fraction (<2 μ) being of the order of 75% at B-6 and 40% at B-2.

The laboratory specimens were cut from block samples and reconsolidated to in-situ stresses using a K_0 of 0.55. The TC, DSS, and TE results were interpreted in terms of shear strength mobilized on the failure plane and averaged using a technique that satisfies strain compatibility. This average strength τ_{ave} is assumed to be appropriate for circular arc embankment stability analyses and is used to determine field vane corrections suitable for the clays tested.

The undrained shear strength determined in DSS is about intermediate between the peak TC and TE strengths. The field vane strength c_u(FV) is much lower than measured in triaxial compression, but is close (slightly lower at B-2) to the DSS strength (see Fig. 6). When compared in terms of mean normalized undrained shear strength for each deposit, the field $c_u(FV)/\sigma'_p$ and the laboratory τ_{ave}/σ'_p are identical at 0.185 for B-2 and 0.225 for B-6. When considering the variation with depth, c_u(FV) is larger than τ_{ave} in the upper portion of the two deposits and lower in the lower portion (see Fig. 8). The normalized c_u(FV) decreases slightly with depth at both sites, probably because of decreasing plasticity and increasing sensitivity. The normalized laboratory τ_{ave} is constant at B-6 and increases with depth at B-2. The effect of OCR is presently difficult to assess in structured clay but may also contribute to the variation in the normalized c_u(FV) and τ_{ave} with depth.

For these two clay deposits of low plasticity and high sensitivity with a medium to low OCR, very good agreement is found between the field vane strength and the laboratory test results on block sample specimens reconsolidated at in-situ stress when anisotropy and strain compatibility are accounted for. If one accepts that the laboratory τ_{ave} represents the undrained shear strength mobilized in an embankment failure, a field vane correction factor can be defined as $\tau_{ave}/c_u(FV)$; it varies at both sites from about 0.85 in the upper portion of the deposit to about 1.15 in the lower portion, with an average of 1.0 for both deposits. These μ correction factors determined as $\tau_{ave}/c_u(FV)$ are in good agreement with Bjerrum's correction [*1,13*] based on plasticity index. The corrections recently proposed by Aas et al. [*4*] based on the ratio $c_u(FV)/\sigma'_{vo}$ lead to μ factors between about 0.6 and 0.75 and are much lower than those determined in this study.

For low plasticity and sensitive clays, such as B-2 and B-6, the field vane appears as a reliable tool for profiling the undrained shear strength for embankment stability. The Bjerrum correction also appears appropriate for these clays even if the medium to low OCR is believed to be due to structuration rather than to geological preloading.

Acknowledgments

The study was supported by the Société d'Énergie de la Baie James (SEBJ), Montreal, Canada. The vane testing was carried out by SEBJ and the sampling and laboratory testing by Université de Sherbrooke personnel. The study was performed as part of the activities of an SEBJ "expert committee," which had the mandate to recommend techniques for designing and building dikes on soft sensitive clays for the NBR hydroelectric project in northern Québec. This committee was composed of J.-J. Paré and J.-G. Lavallée, SEBJ, M. Bozozuk, and K. T. Law, National Research Council of Canada, L. S. Brzezinski, Geocon Inc., O. Dascal, Hydro-Québec, O. Eide, Norwegian Geotechnical Institute, C. C. Ladd, Massachusetts Institute of Technology, G. Lefebvre, Université de Sherbrooke, G. Mesri, University of Illinois, P. Rosenberg, Lupien, Rosenberg, Journeaux et associés Inc. and F. Tavenas, Université Laval.

References

[*1*] Bjerrum, L., "Embankments on Soft Ground: State-of-the Art Report," *Proceedings of the ASCE Specialty Conference on the Performance of Earth and Earth-Supported Structures,* Vol. II, Purdue University, West Lafayette, IN, June, 1972, pp. 1–54.

[*2*] Pilot, G., "Study of Five Embankments on Soft Soils," *Proceedings of the ASCE Specialty Con-*

ference on the Performance of Earth and Earth-Supported Structures, Vol. I, Purdue University, West Lafayette, IN, June 1972, pp. 81–99.

[*3*] Ladd, C. C., Foot, R., Ishihara, K., Schlosser, F., and Poulos, H. G., "Stress-Deformation and Strength Characteristics: State-of-the-Art Report," *Proceedings of the 9th ICSMFE,* Vol. 2, Tokyo, Japan, pp. 421–494.

[*4*] Aas, G., Lacasse, S., Lunne, T., and Hoeg, K., "Use of In Situ Tests for Foundation Design on Clay," *Proceedings of the ASCE Specialty Conference In Situ 86,* Virginia Polytechnic Institute, Blacksburg, VA, June 1986, pp. 1–30.

[*5*] Azzouz, A. S., Baligh, M. M., and Ladd, C. C., "Corrected Field Vane Strength for Embankment Design," *Journal of Geotechnical Engineering, Proceedings of the ASCE,* Vol. 109, No. 5, May, 1983, pp. 730–734.

[*6*] Lefebvre, G., Paré, J.-J., and Dascal, O., "The Undrained Shear Strength in the Surficial Weathered Crust," *Canadian Geotechnical Journal,* Vol. 24, No. 1, 1987, in print.

[*7*] Ladd, C. C., Weaver, J. S., Germaine, J. T., and Sauls, D. P., "Strength-Deformation Properties of Arctic Silt," *Proceedings of the ASCE Conference Arctic 85,* San Francisco, CA, March 1985, pp. 820–829.

[*8*] Lefebvre, G. and Poulin, C., "A New Method of Sampling in Sensitive Clay," *Canadian Geotechnical Journal,* Vol. 16, No. 1, Feb. 1979, pp. 226–233.

[*9*] Bjerrum, L. and Landva, A., "Direct Simple Shear Tests on a Norwegian Quick Clay," *Géotechnique,* Vol. 16, No. 1, 1966, pp. 1–20.

[*10*] Koutsoftas, D. C. and Ladd, C. C., "Design Strength for an Offshore Clay," *Journal of the Geotechnical Engineering Division, Proceedings of the ASCE,* Vol. 111, No. 3, March 1985, pp. 337–355.

[*11*] Lefebvre, G. and LaRochelle, P., "The Analysis of Two Slope Failures in Cemented Champlain Clays," *Canadian Geotechnical Journal,* Vol. 11, No. 1, Feb. 1974, pp. 89–108.

[*12*] Lefebvre, G., "Strength and Slope Stability in Canadian Soft Clay Deposits," *Canadian Geotechnical Journal,* Vol. 18, No. 3, Aug. 1981, pp. 420–442.

[*13*] Bjerrum, L., "Problems of Soil Mechanics and Construction on Soft Clays: State-of-the-Art Report," *Proceedings of the 8th ICSMFE,* Vol. 3, Moscow, U.S.S.R., 1973, pp. 111–159.

James W. Greig,[1] *Richard G. Campanella,*[1] *and Peter K. Robertson*[2]

Comparison of Field Vane Results with Other In-Situ Test Results

REFERENCE: Greig, J. W., Campanella, R. G., and Robertson, P. K., **"Comparison of Field Vane Results with Other In-Situ Test Results,"** *Vane Shear Strength Testing in Soils: Field and Laboratory Studies, ASTM STP 1014,* A. F. Richards, Ed., American Society for Testing and Materials, Philadelphia, 1988, pp. 247–263.

ABSTRACT: Undrained shear strength results from field vane shear tests are compared with those obtained from flat plate dilatometer, screw plate, pressuremeter, and piezocone tests at several sites in the lower mainland near Vancouver, British Columbia, Canada. The sites include deltaic deposits of soft organic soils, clay silts, and moderately to highly sensitive clays. The sites consist primarily of normally consolidated soils, but results in overconsolidated soils are also included. The test procedures and methods of interpretation are briefly described for each in-situ test type in addition to a discussion of the results.

KEY WORDS: vane, in-situ, field, strength, comparison, dilatometers, screw-plate, pressuremeters, cone penetration tests (CPT), piezocone tests (CPTU)

In recent years there has been a growing tendency towards the use of in-situ testing techniques for evaluating engineering soil parameters. Wroth [*1*] attributes this growth to the rapid increase in the variety and quality of in-situ testing instruments in addition to our better understanding of the real behavior of soils and the subsequent realization of some of the limitations and inadequacies of conventional laboratory testing. The high cost of offshore geotechnical investigations and the difficulties associated with the recovery of undisturbed samples have made the use of in-situ testing techniques particularly attractive if not essential.

The soil property most often measured in the field in clay soils is undrained shear strength S_u [*1,2*]. Unfortunately, S_u is not a unique parameter as it depends significantly on the type of test used, the rate of strain, and the orientation of the failure planes [*3*]. There are several methods available for measuring the undrained shear strength of clay in-situ. Campanella and Robertson [*4*] presented a table listing various in-situ test methods and their perceived applicability in determining soil parameters. A list of the methods relevant to the measurement of S_u is reproduced in Table 1. Of the 15 in-situ test methods only 2 methods have a rating of high applicability: the field vane shear test (FVST) and the self-boring pressuremeter test (SBPM). Their high rating is a result of their ability to provide a direct evaluation of S_u. Eight entries have a rating of moderate applicability, all of which estimate the undrained shear strength by empirical or semi-empirical methds. Among these are the flat plate dilatometer test (DMT), the screw plate test (SPLT), the cone penetration test (CPT), and the piezocone test (CPTU).

[1] Research engineer and professor, respectively, Civil Engineering Department, University of British Columbia, 2324 Main Mall, Vancouver, British Columbia V6T 1W5, Canada.

[2] Professor, Civil Engineering Department, University of Alberta, Edmonton, Alberta, Canada.

TABLE 1—*Perceived applicability of in-situ test methods to determine undrained shear strength of clays* [4].

In-Situ Test Method	Rating
Dynamic cone	C
Static cone	
mechanical cone	B
electrical friction cone	B
electrical piezo cone	B
electrical piezo/friction cone	B
Acoustic probe	C
Dilatometer	B
Field vane shear	A
Standard penetration test	C
Resistivity probe	C
Screw plate	B
Impact cone	C
Borehole shear	B
Menard pressuremeter	B
Self-boring pressuremeter	A

NOTE: A = high applicability. B = moderate applicability. C = limited applicability.

Because it has been proven to be a reliable and highly repeatable test method, the FVST is currently the most common method of measuring S_u in-situ. One of its main advantages is the great deal of experience that has been developed over its long history. However, it does suffer some serious disadvantages. The FVST is incremental with tests usually being conducted at 1-m intervals. The soil type in which the test has been performed must be uniform or homogeneous, and the type is estimated from the test resuts or confirmed by an adjacent borehole. Verticality of the instrument and profile are not ensured or measured. To prevent damage to the vane blades, preboring is often required through coarse grained material. For these reasons it is often desired to estimate S_u from other in-situ testing methods.

This paper presents a comparison of S_u values determined from various in-situ test methods with field vane shear test results at several of the University of British Columbia (UBC) research sites.

Field Tests

Test Sites

Field tests were conducted at five sites in the lower mainland region of southwestern British Columbia near Vancouver, Canada. In this paper the sites are referred to as McDonald Farm, Cloverdale, Langley Railway, and Upper and Lower 232nd St. sites. They were selected because of the different material properties such as sensitivity and stress history found at each site. Their locations are shown in Fig. 1. A summary of the material properties of the five sites is presented in Table 2.

McDonald Farm

McDonald Farm is a relatively flat lying area located at the northern edge of Sea Island on Ministry of Transport Land adjacent to Vancouver International Airport, several 100

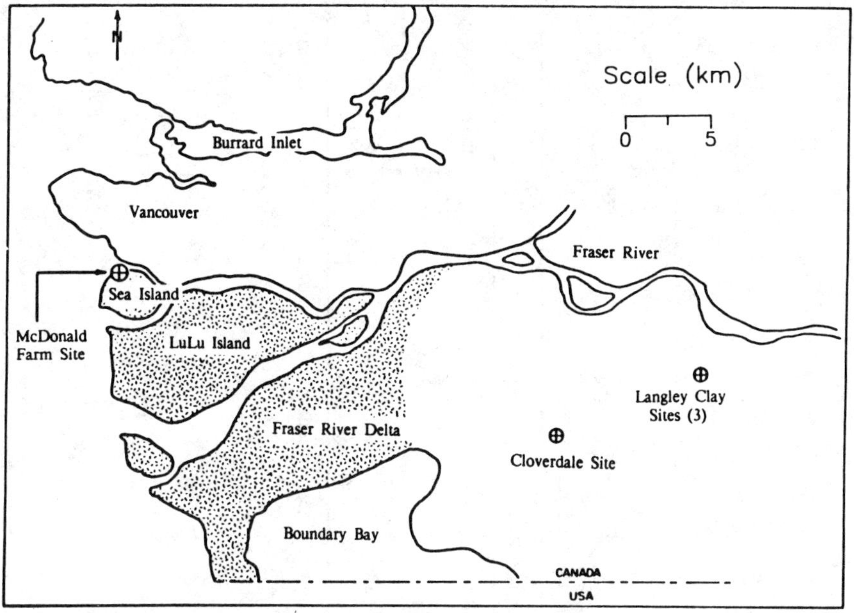

FIG. 1—*General location of the UBC research sites.*

m south of the north arm of the Fraser River. The island is one of several that make up the Fraser River Delta. The general geology consists of deltaic distributary channel fill and overbank deposits, which overlie post glacial estuarine and marine sediments [*5*]. A typical soil profile, shown in Fig. 2, indicates that the stratigraphy consists of a 2-m surface layer of soft organic silty clay overlying 11 m of loose to dense medium to coarse sand with some layers of fine sand. These deposits are underlain by a 2-m transition zone of fine sand and silt followed by a thick (up to 300 m) unit of soft normally consolidated clayey silt. This paper will be concerned only with the clayey silt below 15 m.

Cloverdale Site

The Cloverdale site is located adjacent to the Pacific Highway overpass in Cloverdale, British Columbia, and consists of the Cloverdale sediments. These deposits were laid down in a marine proglacial environment when the land was depressed because of the advancement of the Sumas ice [*5*]. The site is located on level ground approximately 2.3-m above sea level with a stratigraphy consisting of a 2-m surficial fill of wood chips and gravel over 3 m of sensitive organic clay and silt. Below this is approximately 22 m of medium soft sensitive clays and silty clays interbedded with occasional sand lenses. The material between 5 and 16 m is lightly overconsolidated. The high sensitivity of the Cloverdale clay is probably due to leaching after isostatic rebound of the area. A typical soil profile is shown in Fig. 3.

Langley Railway Site

The Langley railway site is located at the base of a 5-m cut adjacent to the Trans Canada Highway. It is approximately 100 m west of the British Columbia Hydro railway overpass near the 232nd St. exit in Langley, British Columbia. The site is located at the eastern

TABLE 2—*Summary of material properties at the UBC research sites.*

		w_l		w_p		w_n		P_I		S_t	
Site	SG	Range	Avg	Range	Avg	Range	Avg	Range	Avg	Range	Avg
McDonald farm	2.8	25 to 42	35	22 to 25	24	23 to 40	34	3 to 20	15	2 to 7	5
Cloverdale site	2.8	. . .	51	. . .	24	. . .	51	. . .	27	8 to 29	17
Langley railway site	2.8	32 to 59	42	16 to 27	21	27 to 53	45	16 to 34	24	7 to 10	9
232nd St. sites	2.8	. . .	40	. . .	20	. . .	45	. . .	19	2 to 19	11

NOTE: SG = specific gravity. w_l = liquid limit, %. P_I = plasticity index, %. w_n = natural water content, %. w_p = plastic limit, %. S_t = sensitivity (field vane).

extent of the Capilano sediments, which consist of raised deltas, intertidal and beach deposits, and glaciomarine sediments [5]. The site profile in Fig. 4 shows that the stratigraphy consists of a 2.5-m surface layer of mixed gravel and sand fill overlying a 7.5-m-thick layer of lightly overconsolidated silty clay with occasional silty sand layers. This in turn is underlain by a deposit of normally consolidated silty clay with occasional silty sand layers. A continuous sample (to 15 m) obtained at the site [6] indicates that the sand content tends to increase with depth.

232nd St. Site

This site is located at the 232nd St. exit of the Trans Canada Highway in Langley, British Columbia, approximately 1 km east of the Langley railway site. The site lies at the western extent of the Fort Langley Formation. This formation has recorded at least three advances and retreats of a valley glacier and consists of interbedded marine, glaciomarine, and glacial sediments [5].

Upper Site—The upper site is situated on a compacted clay fill that forms the approach for the 232nd St. overpass. A profile of the upper site is shown in Fig. 5. The stratigraphy consists of 2.5 m of compacted organic clay fill over a 5-m layer of overconsolidated silty clay, which is underlain by a thick layer of normally consolidated silty clay with occasional sand lenses. Sand content tends to increase with depth.

Lower Site—The lower site is situated slightly above highway level and about 5 m below the elevation of the upper site. The near surface material is overconsolidated because of dessication. A typical profile is shown in Fig. 6.

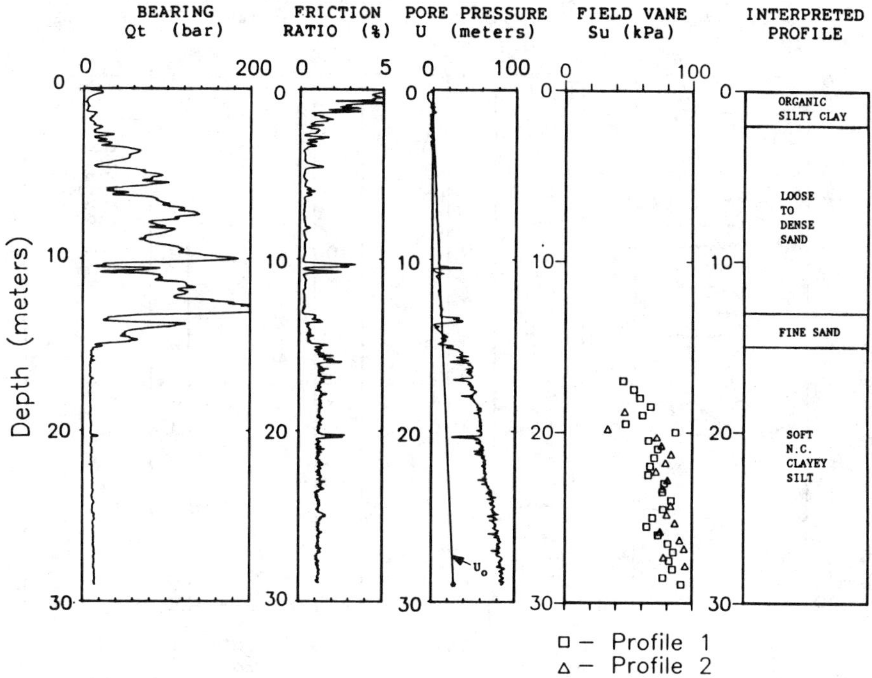

FIG. 2—*Soil profile at the McDonald Farm site.*

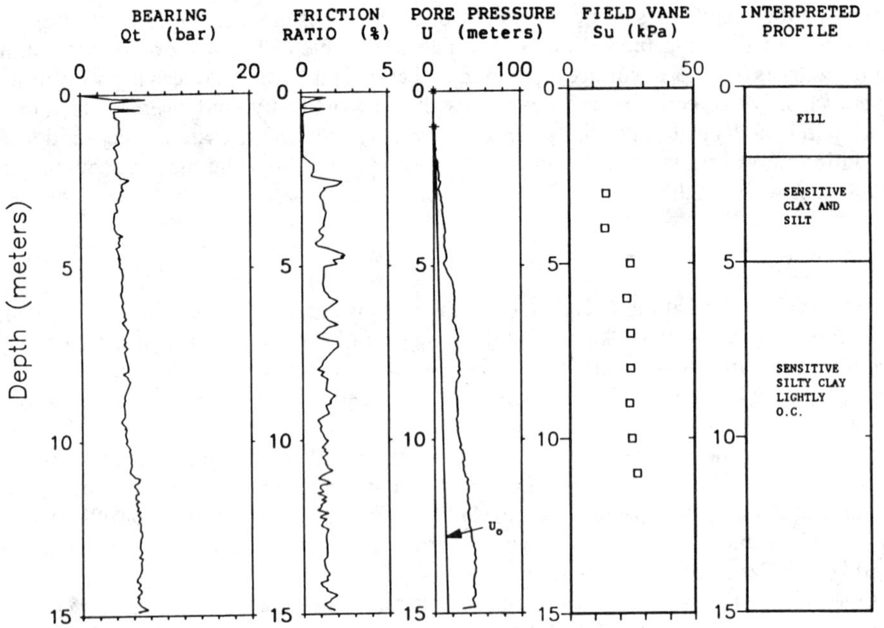

FIG. 3—*Soil profile at the Cloverdale site.*

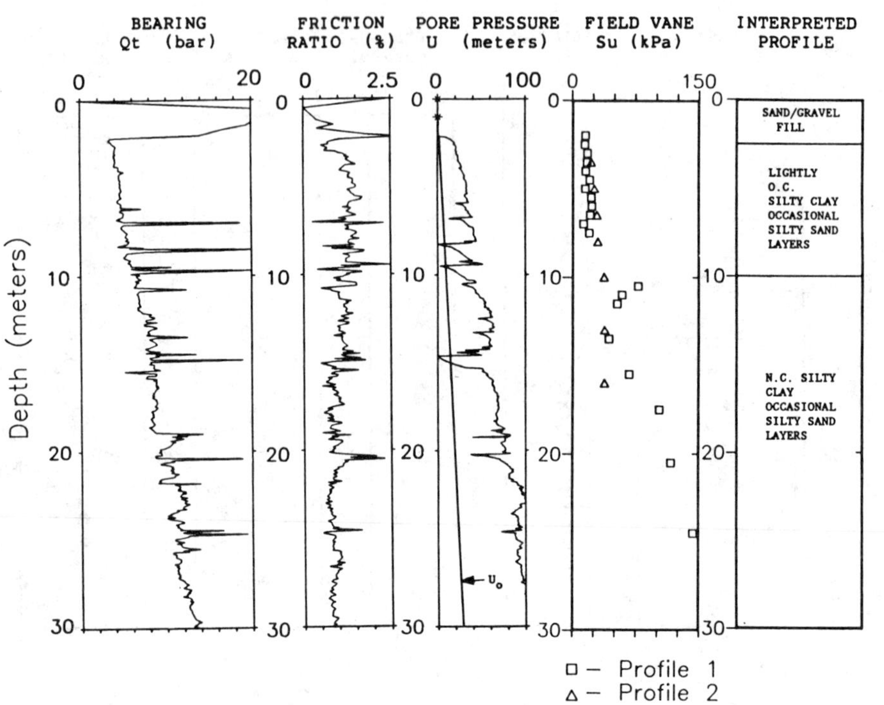

FIG. 4—*Soil profile at the Langley Railway site.*

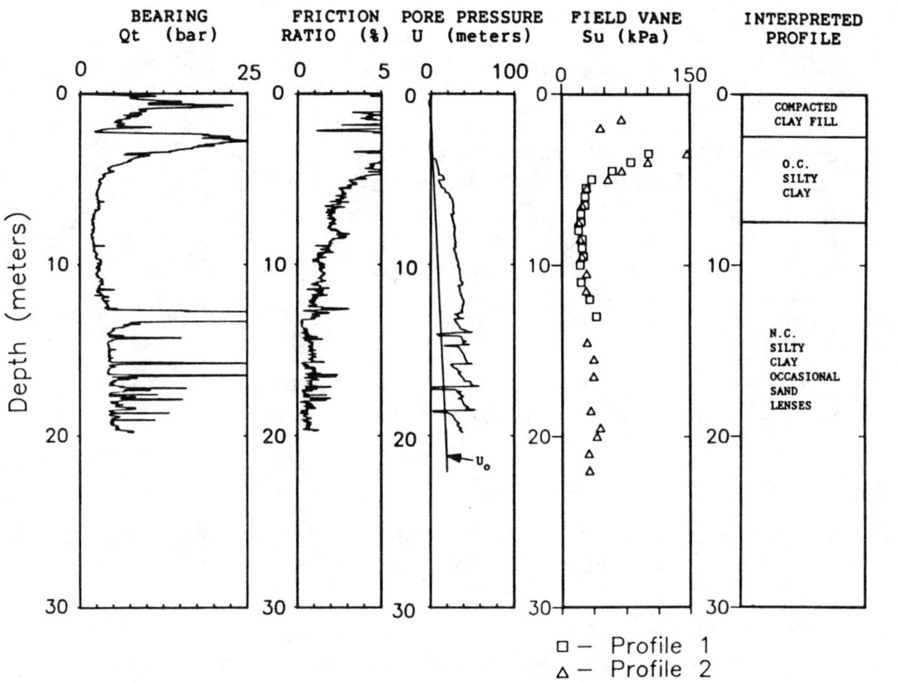

FIG. 5—*Soil profile at the Upper 232nd St. site.*

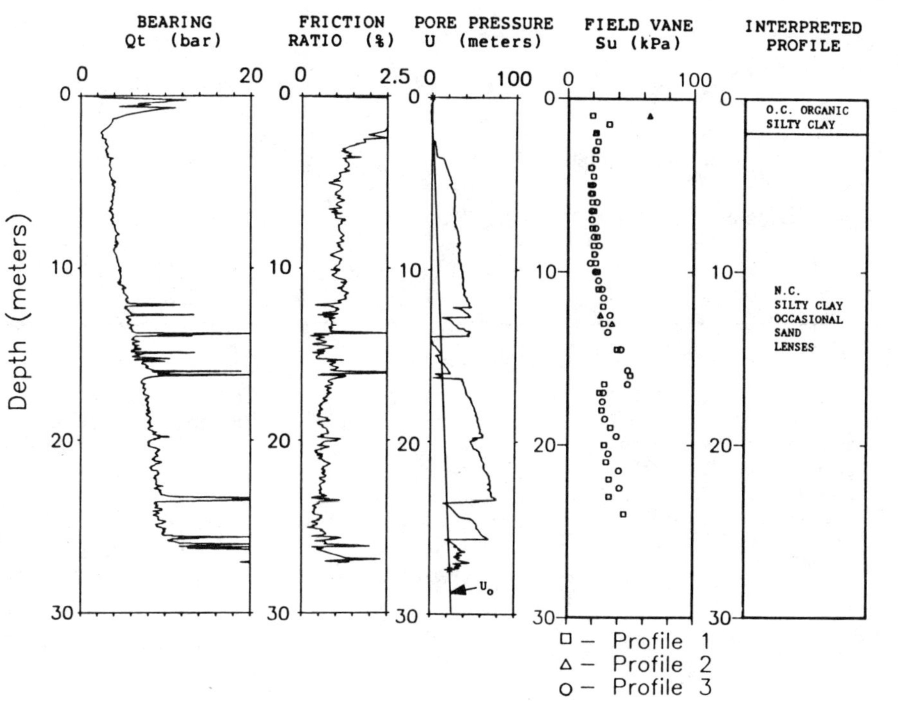

FIG. 6—*Soil profile at the Lower 232nd St. site.*

Test Equipment

The following is a list of the equipment (and their abbreviations) used through the course of this study:

1. Geonor and Nilcon Field Vane (FVST) [*2,6*].
2. Standard Marchetti Flat Plate Dilatometer (DMT) [*7,8*].
3. Double Helix Screw Plate (SPLT) [*9,10*].
4. Hughes Self Boring Pressuremeter (SBPM) [*11*].
5. Hughes Full Displacement Pressuremeter (FDPM) [*12*].
6. Roctest Pencel Probe (FDPM) [*12*].
7. UBC Piezocone (CPTU) [6,13].

Full details regarding the test equipment can be found in the references cited. A summary of the field tests conducted at each site is given in Table 3.

Analysis of Test Results

The following is a brief description of the methods used to analyze the data. A summary of the methods used is presented in Table 4.

Field Vane Shear Test (FVST)

All field vane undrained strengths were calculated using the standard expression (ASTM Method for Field Vane Shear Test in Cohesive Soil [D 2573]) for vanes with a length to diameter ratio of two

$$S_u = \frac{6T}{7\pi D^3}$$

where

S_u = undrained shear strength,
T = applied torque, and
D = diameter of the vane.

TABLE 3—*Summary of field tests conducted at the UBC research sites.*

Tests	Sites				
	McDonald Farm	Cloverdale	Langley Railway	Upper 232nd	Lower 232nd
Geonor field vane (FVST)	x	. . .	. . .	. . .	. . .
Nilcon field vane (FVST)	. . .	x	x	x	x
UBC Piezocone (CPTU)	x	x	x	x	x
Dilatometer (DMT)	x	x	x	x	x
Screw plate (SPLT)	x	x	x	. . .	x
Self-boring pressuremeter (SBPM)					
Hughes	x	. . .	. . .	. . .	. . .
Full displacement pressuremeter (FDPM)					
Hughes	. . .	. . .	x	x	. . .
Roctest pencel probe	. . .	. . .	. . .	x	. . .

NOTE: x indicates that the test was performed at the site.

TABLE 4—*Summary of the interpretation methods used.*

Method		S_u
Field vane shear test	(FVST)	$S_u = \frac{6T}{7\pi D^3}$ for $\frac{L}{D} = 2$
Dilatometer	(DMT)	S_u from program DILLY4 (empirical)
Screw plate	(SPLT)	$S_u = \frac{P_{ult}}{9.0}$
Pressuremeter (SBPM and FDPM)		$S_u = \frac{(P_1 - P_0)}{1 + \ln(G/S_u)}$
Cone penetrometer	(CPTU)	$S_u = \frac{Q_t - \sigma_{vo}}{N_{kt}}$ (for bearing) $S_u = \frac{U - U_0}{N_{\Delta U}}$ (for pore pressure behind the tip)

There has been much discussion [*1,2,6*] as to the correct interpretation of the vane test; however, most engineers appear to use the above expression. No correction factors (for example Bjerrum's [*14*] or Aas et al. [*15*]) were applied to the vane data.

Flat Plate Dilatometer Test (DMT)

The DMT data were analyzed using the standard dilatometer reduction routines, DILLY and DILLY4, supplied by GPE Inc. of Gainesville, FL. These reduction routines calculate S_u using an empirical correlation proposed by Marchetti [*7*].

Screw Plate Test (SPLT)

The screw plate data were analyzed using the method suggested by Selvadurai et al. [*10*]

$$S_u = \frac{P_{ult}}{9.0}$$

where P_{ult} = ultimate failure stress to cause plunging of the screw plate.

Pressuremeter Tests (SBPM and FDPM)

All of the pressuremeter test results were analyzed using the method developed by Gibson and Anderson [*16*]

$$S_u = \frac{(P_1 - P_0)}{1 + \ln(G/S_u)}$$

where

P_0 = lift off pressure,
P_1 = limit pressure,
G = shear modulus,
S_u = undrained shear strength, and
G/S_u = rigidity index.

In this study estimates of the rigidity index were made using the curves presented by Ladd et al. [*17*] and a knowledge of plasticity index (PI).

Cone Penetration Test (CPTU)

The cone bearing Q_t and excess pore-pressure measurements ΔU were used to estimate S_u from CPTU data. All cone bearing data were corrected for temperature and pore-pressure effects [*3,6*]. Estimates of S_u from the cone bearing were made using the cone factor N_{kt} [*18*] where

$$S_u = \frac{Q_t - \sigma_{vo}}{N_{kt}}$$

where

Q_t = cone bearing corrected for pore-pressure and temperature effects,
σ_{vo} = total vertical stress, and
N_{kt} = empirical cone factor.

Estimates of S_u from penetration pore pressures measured behind the tip were made using the pore-pressure factor [*18,19*] $N_{\Delta U}$

$$S_u = \frac{\Delta U}{N_{\Delta U}}$$

where

ΔU = $U - U_0$,
U = penetration pore pressure,
U_0 = equilibrium pore pressure, and
$N_{\Delta U}$ = empirical pore-pressure factor.

Several other methods of estimating S_u from CPT and CPTU (piezocone) have been suggested [*6,18,19,20*]; however, only the two methods described above were used in this study.

Discussion of Results

The test results for each site are presented in Figs. 7 through 11. The following is a brief discussion of the results from each research site.

McDonald Farm

The two field vane profiles (Fig. 2) are reasonably consistent with both indicating that S_u increases linearly with depth. The FVST profiles indicate a soft layer at about 20 m; however, no evidence of this could be seen in the CPTU profile. Although the shape and the trend of the DMT profile (Fig. 7) are very similar to those of the FVST, the results are consistently 20 to 30% lower. The results from the two SPLT profiles exhibit considerable scatter. However, below 18 m there appears to be a trend that is consistent with the field vane although 40 to 50% higher. The SBPM results are consistent with those from the field vane, although some low values from the SBPM were recorded between 21 and 24 m. Very

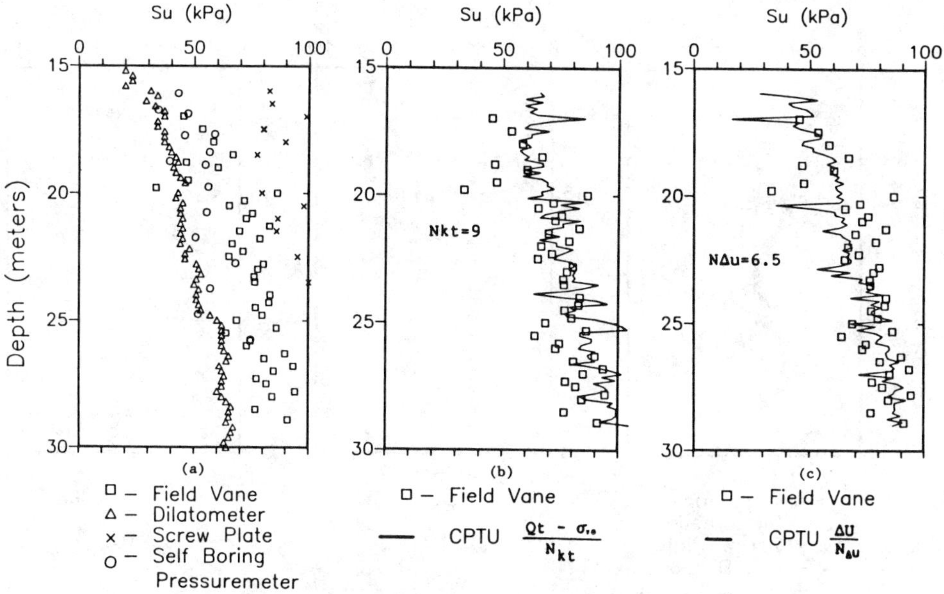

FIG. 7—*Field test results at the McDonald Farm site:* (a) *various in-situ tests,* (b) S_u *estimated from CPTU bearing* Q_t, *and* (c) S_u *estimated from CPTU pore pressures.*

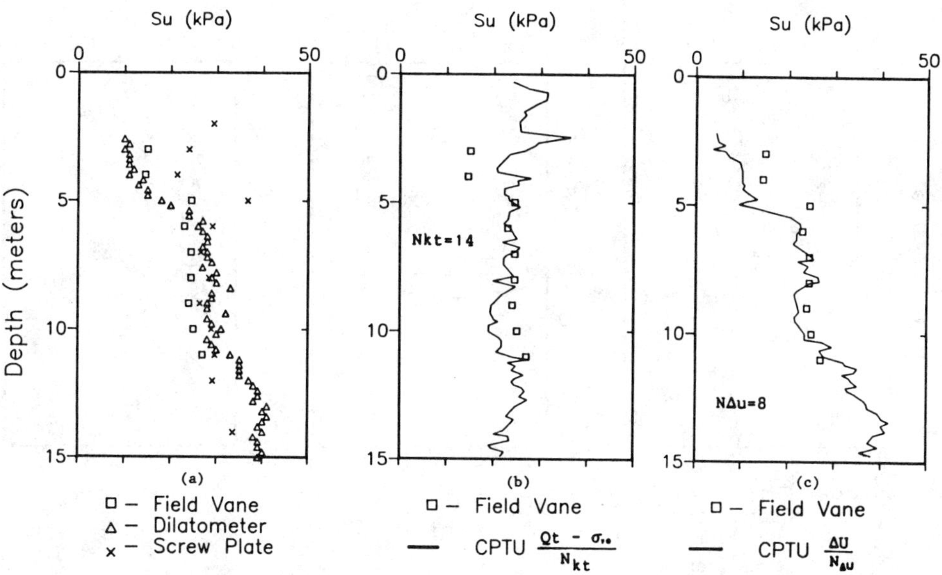

FIG. 8—*Field test results at the Cloverdale site:* (a) *various in-situ tests,* (b) S_u *estimated from CPTU bearing* Q_t, *and* (c) S_u *estimated from CPTU pore pressures.*

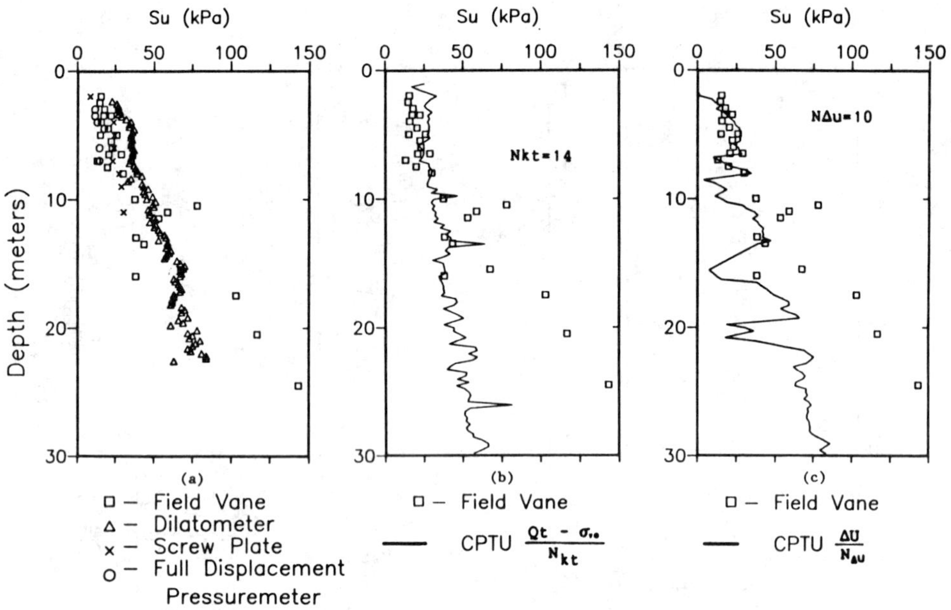

FIG. 9—*Field test results at the Langley Railway site:* (a) *various in-situ tests,* (b) S_u *estimated from CPTU bearing* Q_t, *and* (c) S_u *estimated from CPTU pore pressures.*

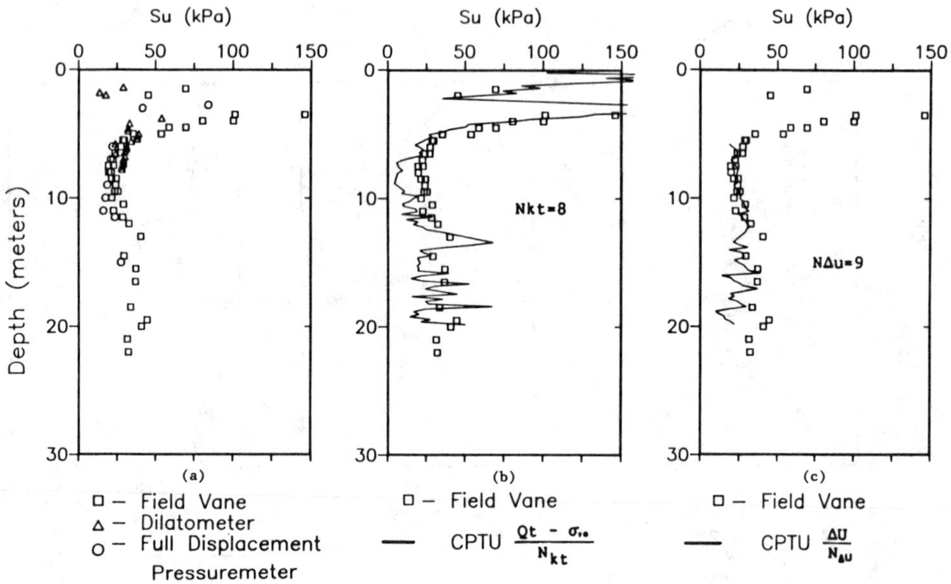

FIG. 10—*Field test results at the Upper 232nd St. site:* (a) *various in-situ tests,* (b) S_u *estimated from CPTU bearing* Q_t, *and* (c) S_u *estimated from CPTU pore pressures.*

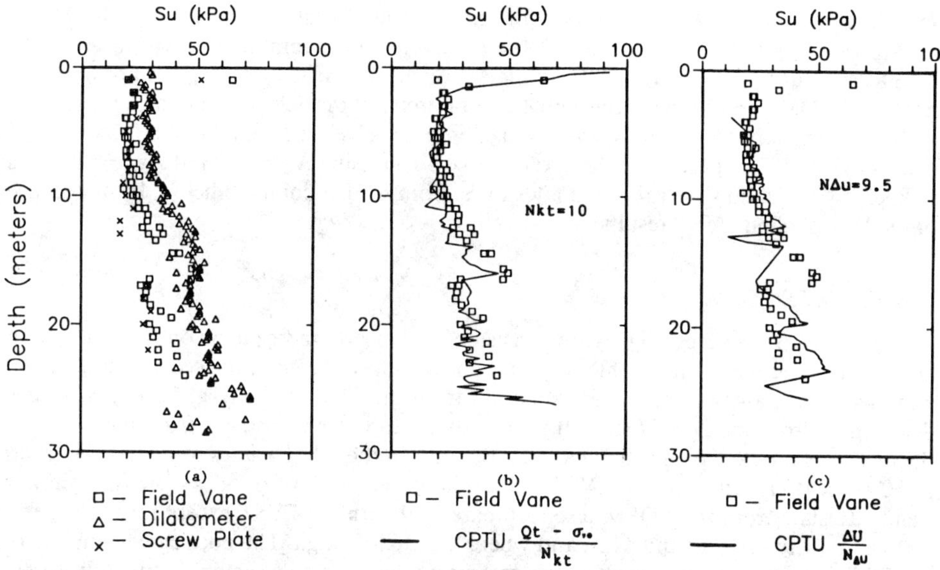

FIG. 11—*Field test results at the Lower 232nd St. site:* (a) *various in-situ tests,* (b) S_u *estimated from CPTU bearing* Q_t, *and* (c) S_u *estimated from CPTU pore pressures.*

good agreement between estimates of S_u from Q_t and FVST were obtained using a cone factor N_{kt} = 9. Some local high values in the CPTU bearing data are due to the influence of thin sand lenses.

Using $N_{\Delta U}$ = 6.5, very good agreement is observed between estimates of S_u from CPTU pore-pressure data and FVST results. Some low values are again due to the influence of thin sand lenses.

Cloverdale Site

The FVST profile (Fig. 3) indicates a slightly softer material above 5 m and a uniform deposit below 5 m that exhibits only a slight increase in S_u with depth. Both the CPTU and DMT data indicate a light overconsolidation above 10 m. The DMT S_u results (Fig. 8) follow the field vane trend very well being only slightly low above 5 m and slightly high below 5 m. The SPLT results compare well to the field vane below 5 m but are high and scattered above 5 m. Estimates of S_u from CPTU data agree well using a cone factor N_{kt} = 14 and a pore-pressure factor $N_{\Delta U}$ = 8. It is of interest to note that the shape of the DMT profile is almost identical to that from the $\Delta U/N_{\Delta U}$ profile, suggesting that the DMT predominantly measures pore pressures in soft clay [*20*].

Langley Railway Site

The two-field vane profiles shown in Fig. 4 differ considerably below a depth of 7 m. Four field vane values of S_u are significantly larger than the remaining FVST values. These high values of S_u appear to have been caused by thin sand lenses that are clearly visible from the CPTU profile in Fig. 4. A continuous borehole sample obtained at the site [*6*] confirmed the existence of frequent fine sand lenses throughout the profile. The CPTU sampling rate (25 mm) is significantly higher than the typical 1-m interval for the FVST. The results in Fig. 9 demonstrate that the relatively wide sampling intervals used in the FVST can lead to erroneous results and false conclusions if the clay layer is not homogeneous. The DMT results again reflect the general trend of the field vane but are consistently higher by approximately 30%. Very good agreement is observed with the SPLT and FDPM results. The CPTU results compare well above 16 m using N_{kt} = 14 and $N_{\Delta U}$ = 10. It is difficult to assess how well the estimates of S_u from CPTU compare below 16 m because of the lack of good FVST results.

Upper 232nd St. Site

The two FVST profiles (Fig. 5) are consistent and clearly indicate the overconsolidation above 7 m. The field vane profiles do not appear to have been influenced by sand lenses, although significant sand lenses are apparent from the CPT profile (Fig. 5). The S_u values determined from the DMT (Fig. 10) were considerably less than those from the FVST in the overconsolidated material above 5 m. Below 5 m the DMT values were on average 40 to 50 percent greater than the FVST values, however, the two profiles displayed similar trends. Results from the FDPM tests compare well with the FVST except for the slightly low values between 9 m and 11 m and where the OCR is high. The poor agreement where the OCR is high may be a result of the pressuremeter tests not reaching a true limit pressure. S_u values from the CPTU data using N_{kt} = 8 compare favorably with FVST values except between 7 and 9.5 m where the CPTU values are low. Agreement is very good where OCR is high. S_u values from ΔU show good agreement in the uniform material between 6 and 12.5 m using $N_{\Delta U}$ = 9. Below 12.5 m, there is a strong influence of sand lenses. Estimates of S_u in heavily overconsolidated materials can not be made from ΔU when the pore pressures have been measured behind the tip [*19*].

Lower 232nd St. Site

The three field vane profiles (Fig. 6) are very consistent showing a trend of S_u linearly increasing with depth except between 14 and 17 m where there appears to be a substantial increase in S_u. This change in the profile is coincidental with the sand lenses detected by the CPTU. The CPTU and FVST profiles are similar in shape. The CPTU pore-pressure profile was significantly influenced by the sand layers (Fig. 6). The DMT results (Fig. 11) are consistently higher by about 50%; however, the trend of the field vane was followed very well. Estimates of S_u from the screw plate test show very good agreement except for the low values between 12 and 13 m. S_u values obtained from CPTU results compare well in both the normally and overconsolidated regions using N_{kt} = 10. Estimates of S_u from CPTU excess pore-pressure measurements compare favorably using $N_{\Delta U}$ = 9.5.

Summary and Conclusions

This paper has presented a comparison of undrained shear strength results from the field vane shear test with those obtained by the flat plate dilatometer, screw plate, pressuremeter, and piezocone tests at five sites in the lower mainland near Vancouver, British Columbia, Canada. A summary of the results is shown in Table 5. The results clearly show that for each type of test there is no unique factor that can be used to estimate equivalent FVST S_u values for all types of clay. This is because S_u itself is not a unique parameter but depends on the type of test, the rate of strain, and the orientation of failure planes. Each of the in-situ tests used in this study shears the soil in a different manner and at a different strain rate, and can therefore be expected to produce different results.

TABLE 5—*Summary of in-situ measurements of undrained shear strength compared to the field vane shear test.*

Site	S_t[a]	OCR	In-Situ Test Method				
						CPT Using	
			DMT	SPLT	PM	Bearing Q_t	Pore Pressure Δu[b]
McDonald Farm	2 to 7	NC	20–30% lower but similar trend	40 to 50% higher	self boring: consistent	$N_{kt} = 9$ consistent	$N_{\Delta u} = 6.5$ consistent
Cloverdale						$N_{kt} = 14$	$N_{\Delta u} = 8$
<5 m	10 to 27	2 to 5+	slightly lower	high and scattered	. . .	50% higher	30% lower
>5 m	8 to 29	1 to 2	consistent	consistent	. . .	consistent	consistent
Langley Railway					full displacement	$N_{kt} = 14$	$N_{\Delta u} = 10$
<7 m	4 to 10	1 to 3	30% higher but similar trend	consistent	consistent	slightly higher	consistent
>7 m	7 to 11	NC	30% higher but similar trend	consistent	consistent	consistent	consistent where not influenced by sand layers
Upper 232nd St.					full displacement	$N_{kt} = 8$	$N_{\Delta u} = 9$
<7 m	2 to 7	1 to 10+	lower in fill at shallow depth (<3 m) then 40% higher	. . .	lower in fill at shallow depth (<3 m) then consistent	very consistent	can not use in heavily OC clay
>7 m	7 to 19	NC	50% higher (similar trends)	. . .	consistent	slightly lower	consistent except where sand had strong influence
Lower 232nd St.						$N_{kt} = 10$	$N_{\Delta u} = 9.5$
<5 m	7 to 10	1 to 7+	50% higher but similar trend	consistent	. . .	very consistent	can not use where clay is heavily O.C. or unsaturated
>5 m	10 to 19	NC	50% higher but similar trend	10 to 20% lower	. . .	consistent	consistent where not influenced by sand lenses

[a] Sensitivity from field vane.
[b] Porous element located behind tip.

Despite their differences in failure mechanism the results obtained by the in-situ methods presented in this report tend to agree fairly well with the FVST values. However, at three of the sites the DMT results did show significant error. Encouragingly though, the overall shapes of the estimated S_u profiles from the DMT were very similar to the field vane profiles. This suggests that some flexibility with respect to the input of local correlations in the DMT reduction programs is required.

The screw plate test results were considerably higher than those from the field vane only at the McDonald Farm site again showing that local correlations are often required.

The S_u results from both the self-boring and the full displacement pressuremeters agreed well with those from the field vane. However, the results were dependent on an appropriate selection of the rigidity index G/S_u.

Very good agreement was obtained using the cone bearing and various values of the cone factor N_{kt}. It is clear, however, that there is no unique value of N_{kt} for all clays. The variation in N_{kt} is influenced by such soil properties as stress history, sensitivity, and stiffness [*19*]. Increases in OCR are generally reflected in increases in N_{kt} [*6,18*]. Data from the Langley railway site [*6*] also indicate that N_{kt} increases with decreasing PI. It is essential to correct CPTU bearing values for temperature and pore-pressure effects [*3,6*] in soft clays where bearing is low and pore pressures are high.

Good agreement was obtained using CPTU excess pore pressures measured behind the tip and various values of the pore-pressure factor $N_{\Delta U}$. No unique value of $N_{\Delta U}$ was found since the generation of pore pressures is also influenced by the soil's stress history, sensitivity, and stiffness [*19*]. Reasonable estimates of $N_{\Delta U}$ can be made from the rigidity index. Estimating S_u from pore-pressure measurements is highly influenced by the location of the pore-pressure element [*19*] and the degree of saturation in the measuring system. Estimates of S_u in heavily overconsolidated clay can not be made by this method if the porous element is located behind the tip. Estimating S_u from CPTU pore-pressure data is also significantly influenced by the occurrence of sand layers.

The existence of sand lenses can significantly influence FVST results. The relatively large depth intervals used in the FVST can lead to erroneous results and false conclusions if the clay layer is not homogeneous.

The CPTU and DMT are both logging tests that economically provide near continuous data. Results from this study show that provided the locally evaluated empirical correlation factors are applied to CPTU and DMT, data near continuous estimates of equivalent FVST S_u values can be determined.

Acknowledgments

The assistance of the Natural Sciences and Engineering Research Council of Canada and the technical staff of the Civil Engineering Department, University of British Columbia, is much appreciated. The work of Don Gillespie, Bill Berzins, Ian McPherson, Clifford Tsang, and Bruce O'Neill is gratefully acknowledged. We are also indebted to John Hughes who made his pressuremeter available at several sites and assisted with the testing and interpretation.

References

[*1*] Wroth, C. P., "The Interpretation of In-Situ Soil Tests," Written Version of the 1984 Rankine Lecture, *Geotechnique*, Vol. 34, No. 4, 1984, pp. 449–489.

[*2*] Schmertmann, J. H., "Measurement of In-Situ Shear Strength," *Proceedings of the Specialty Conference on In-Situ Measurement of Soil Properties*, ASCE, Raleigh, Vol. 2, 1975, pp. 57–138.

[3] Robertson, P. K. and Campanella, R. G., "Interpretation of the Cone Penetration Test. Part I: Sand and Part II: Clay," *Canadian Geotechnical Journal,* Vol. 20, No. 4, 1983, pp. 718–745.
[4] Campanella, R. G. and Robertson, P. K., "State of the Art in In-Situ Testing of Soils: Developments Since 1978," *Engineering Foundation Conference on Updating Subsurface Sampling of Soils and Rocks and Their In-Situ Testing, Santa Barbara, California,* Engineering Foundation, 1982, pp. 254–269.
[5] Armstrong, J. E., "Post Vachon Wisconsin Glaciation, Fraser Lowland, British Columbia," *Geological Survey of Canada,* Bulletin 332, 1978.
[6] Greig, J. W., "Estimating Undrained Shear Strength of Clay from Cone Penetration Tests," M.A.Sc. thesis, University of British Columbia, Vancouver, Canada, 1985.
[7] Marchetti, S., "In-Situ Tests by Flat Dilatometer," *Journal of Geotechnical Engineering Division, ASCE,* Vol. 106, No. GT3, 1980, pp. 299–321.
[8] McPherson, I. D., "An Evaluation of the Flat Dilatometer as an In-Situ Testing Device," M.A.Sc. thesis, University of British Columbia, Vancouver, Canada, 1985.
[9] Berzins, W. E., "Determination of Drained and Undrained Soil Parameters Using the Screw Plate Test," M.A.Sc. thesis, University of British Columbia, Vancouver, Canada, 1983.
[10] Selvadurai, A. P. S. and Nicholas, T. J., "A Theoretical Assessment of the Screw Plate Test," *3rd International Conference on Numerical Methods in Geomechanics,* Aachen, West Germany, Vol. 3, 1979.
[11] Robertson, P. K., "In-Situ Testing of Soils with Emphasis on Its Application to Liquefaction Assessment," Ph.D. thesis, University of British Columbia, Vancouver, Canada, 1982.
[12] O'Neill, B., "An Evaluation of the Full Displacement Pressuremeter," M.A.Sc. thesis, University of British Columbia, Vancouver, Canada, 1985.
[13] Campanella, R. G. and Robertson, P. K., "Applied Cone Research," *Symposium on Cone Penetration Testing and Experience,* Geotechnical Engineering Division, ASCE, Oct. 1981, pp. 343–363.
[14] Bjerrum, L., "Embankments on Soft Ground," *ASCE Proceedings of the Specialty Conference on Performance of Earth-Supported Structures,* Vol. 2, 1972, pp. 1–54.
[15] Aas, G., Lacasse, S., Lunne, T., and Hoeg, K., "Use of In Situ Tests for Foundation Design on Clay," *Proceedings of the ASCE Specialty Conference IN-SITU '86 Use of In-Situ Testing in Geotechnical Engineering,* June, 1986, pp. 1–30.
[16] Gibson, R. E. and Anderson, W. F., "In-Situ Measurements of Soil Properties with the Pressuremeter," *Civil Engineering and Public Works Review,* Vol. 56, No. 658, pp. 615–618.
[17] Ladd, C. C., Foott, R., Ishihara, K., Schlosser, F., and Poulos, H. G., "Stress Deformation and Strength Characteristics," *Proceedings of the IX ICSMFE,* Vol. 2, Tokyo, Japan, 1977, pp. 421–494.
[18] Lunne, T., Christofferson, H. P., and Tjelta, T. I., "Engineering Use of the Piezocone Data in the North Sea Clays," *Proceedings of the XI ICSMFE,* San Francisco, California, Vol. 2, 1985, pp. 907–913.
[19] Robertson, P. K., Campanella, R. G., Gillespie, D., and Greig, J., "Use of Piezometer Cone Data," *Proceedings of the ASCE Specialty Conference IN-SITU '86 Use of In-Situ Testing in Geotechnical Engineering,* June 1986, pp. 1263–1280.
[20] Campanella, R. G., Robertson, P. K., Gillespie, D. G., and Greig, J. W., "Recent Developments in In-Situ Testing of Soils," *Proceedings of the XI ICSMFE,* San Francisco, California, Vol. 2, 1985, pp. 849–855.

Part VI: Field Vane Testing on Land

Vinod K. Garga[1]

Experience with Field Vane Testing at Sepetiba Test Fills

REFERENCE: Garga, V. K., **"Experience with Field Vane Testing at Sepetiba Test Fills,"** *Vane Shear Strength Testing in Soils: Field and Laboratory Studies, ASTM STP 1014,* A. F. Richards, Ed., American Society for Testing and Materials, Philadelphia, 1988, pp. 267–276.

ABSTRACT: The design of new port of Sepetiba, near Rio de Janeiro, required a detailed evaluation of in-situ behavior of the sub-soil. The subsoil at this large site consists essentially of soft clays, but with extreme variability in sand and silt content and organic matter. The construction of two large instrumented test fills provided a good opportunity to monitor the increase in undrained strength S_u with increase in effective stress. Investigations were also carried out to measure strength anisotropy using vane tests at a site nearby the test fills. The investigations suggest the following conclusions: (1) the vane strengths are very similar to those for unconfined compression and unconsolidated undrained laboratory triaxial tests, (2) the increase in vane strengths did not accompany the increase in effective stress underneath the test fills, (3) the stress ratio s_u/σ_v' after consolidation in the foundation was lower than before construction, and (4) tests with vanes of different height to diameter ratios were not successful in determining the strength anisotropy at this site.

KEY WORDS: anisotropy, Atterberg limits, in-situ testing, soft clay, test fills, undrained strength, vane tests

In view of the physical limitations for expansion of the old port of Rio de Janeiro, Brazil, studies were undertaken for construction of a new, 15-m deep berth, industrial port of Sepetiba, some 50 km south of Rio de Janeiro. The area for stockpiling coal and iron ore, essentially underlain by marine soft clays, covered an area of 2.5 million m^2. At first it was proposed to remove 3.7 million m^3 of the very soft clays by dredging and replacing them with hydraulic sand fill. Consequently a detailed program of investigations was undertaken to study the in-situ characteristics of the soft clays. As part of this investigation, two large instrumented test fills, each 65 m by 65 m in plan and 5 m high, with 3:1 slopes, were constructed. Test fill B was constructed over natural ground, while test fill D was placed over stone columns. The instrumentation for each fill comprised of 2 lines of hydraulic piezometers with 4 piezometers each at different depths, 10 settlement plates at different depths, 14 surface settlement plates, and 2 inclinometers. In order to quantify the increase of undrained strength with time, an area 5 m^2 was delineated on each test fill where vane shear tests were conducted at regular intervals (Fig. 1). Separately in the vicinity of the fills, an attempt was made to evaluate strength anisotropy using vanes of different height to diameter ratios.

This paper discusses only the results of vane tests with respect to the following aspects:

(1) comparison of vane strength with those from simple laboratory tests;

[1] Associate professor, Department of Civil Engineering, University of Ottawa, Ottawa, Ontario K1N 9B4, Canada.

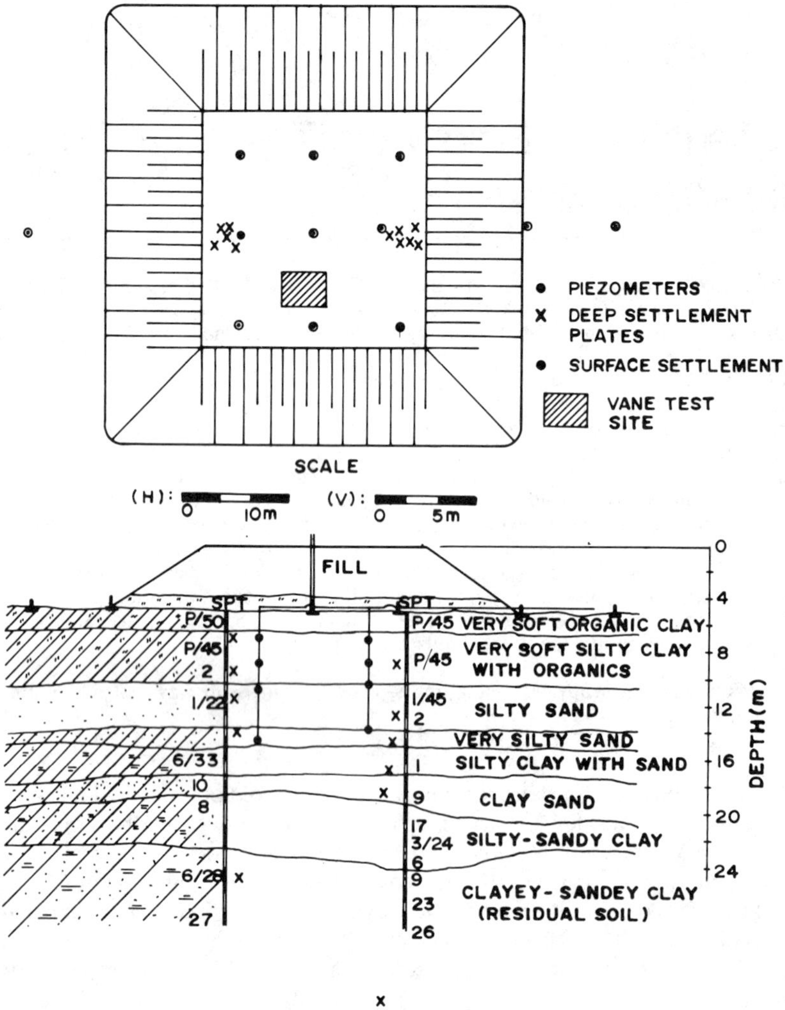

FIG. 1—*Test fill B.*

(2) the progression of measured vane strength with increase of effective stress in the foundation; and

(3) evaluation of strength anisotropy.

It should be noted that the test fill program was not of an academically oriented research nature, but was undertaken to provide answers to questions posed in the design of the works.

The Subsoil

The subsoil profile is variable, both laterally and vertically, in this area. It consists essentially of a very soft to soft silty clay with a highly variable sand content. The clay, in the upper 5 m, also contains organic matter, small soft sea shells, and other calcareous matter.

Figure 2 shows a vertical subsurface profile adjacent to the shoreline at Sepetiba Bay. The clay is extremely soft; in fact no standard penetration test (SPT) blowcounts could be measured since the spoon sampler penetrated up to 95 cm under its own weight (indicated as P/95 in the figure). It is of interest to note that even this deposit of soft clay shows clear indication of preconsolidation pressure, at least in the upper 5 m. The preconsolidation pressure σ'_{vm} indicated on Fig. 2 was determined from oedometer tests on 10-cm-diameter, thin-wall piston specimens. The variability of the site is exemplified by the vertical soil profile in Fig. 3, which is characteristic of conditions at the test fill sites. The clay has soft to medium consistency, and vane tests could not be conducted at approximately 4 to 5-m depth. This stronger clay layer had a 75% clay fraction and much higher Atterberg limits than for soils at other depths. In Fig. 3, the preconsolidation pressure σ'_{vm} as determined from oedometer tests, also reflects the stiffer nature of the clay at the 4 to 5 m depth.

Measurement of Vane Strengths

In-situ determination of undrained shear strength was carried out using a vane 7.5 cm (3 in.) diameter, 15 cm (6 in.) high, and 6.3 mm (¼) in. in blade thickness. This blade thickness was deemed to be necessary to prevent damage by sea shells. The "remolded" strength was determined after only one 360° rotation of the vane. It is appreciated that current international procedures recommend 6 to 10 rotations; hence the value remolded vane strength obtained in this study is an overestimate of this strength. An interval of

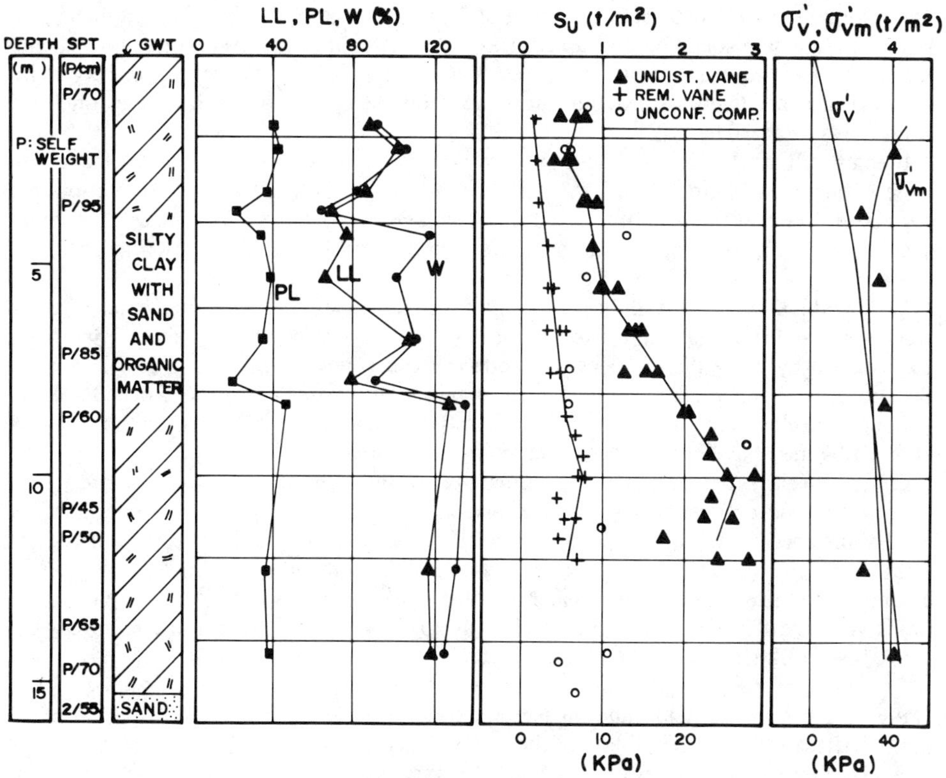

FIG. 2—*Subsoil profile at Sepetiba Bay.*

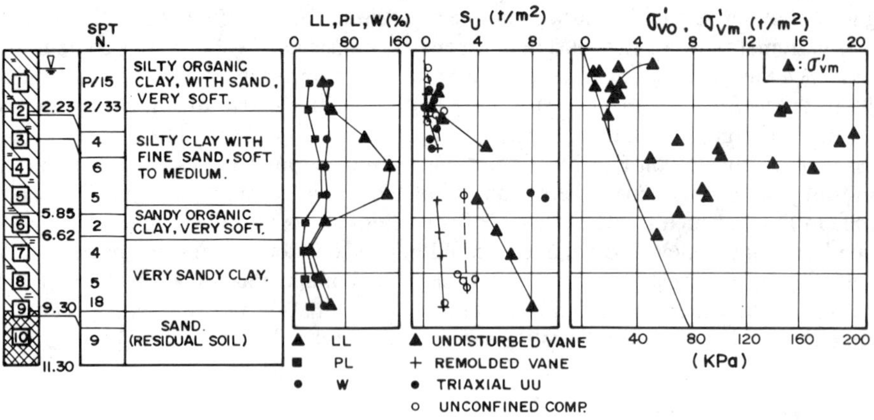

FIG. 3—*Subsoil profile at test fills.*

approximately 15 min was allowed between determination of peak and remolded strengths. The vane strengths were calculated assuming a uniform distribution of stress on top and base of the cylindrical specimen [*1*]. Figure 2 shows a comparison of undrained strengths from the vane and simple compression tests. The two results are similar at shallow depths. However the unconfined compression test yields values close to the "remolded" vane strength below a depth of 7 m. In the case of subsurface profile shown in Fig. 3, the unconsolidated undrained triaxial tests, the unconfined compression test, and the vane appear to indicate similar values of strength S_u for the very soft upper clay. With increase of consistency with depth, the vane strengths can be twice the values obtained from the laboratory tests.

Bjerrum [*2*], on the basis of field observations, proposed a correction factor for in-situ vane strengths as a function of the plasticity index. On the basis of this correlation, the correction factor at the test fill sites would range between 0.7 and 0.85. The application of such a correction factor appears to be dubious in light of the following observations:

1. The thickness of the blades has significant effect on the measured strengths. La Rochelle et al. [*3*], for example, showed that the field vane strength of sensitive Champlain clays varied by more than 25% as the blade thickness increased from 1.6 to 4.7 mm. At Sepetiba, the vane thickness was higher than that investigated by La Rochelle et al., although the clay is not sensitive. It is entirely possible that the in-situ strength measured at Sepetiba may have been underestimated by as much as 25%.
2. In some clays, the method for specimen preparation significantly affects Atterberg limits. Since the natural water content is invariably high, it becomes necessary to either first air dry the specimen, and to increase the water content in stages, or to carry out tests at intervals as the specimen dries out slowly. The significant difference in Atterberg limits caused by test procedure for the nearshore, very soft specimens at Sepetiba Bay is shown in Fig. 4. The difference in Atterberg limits was not found to be as significant for specimens from the test fill sites.

Figure 5 shows the field vane measurements at test Fill B, taken over a period of 15 months. Because of the variability of the soil, it was not always possible to obtain readings at every metre depth. A general tendency for increase of strength with time is noticed;

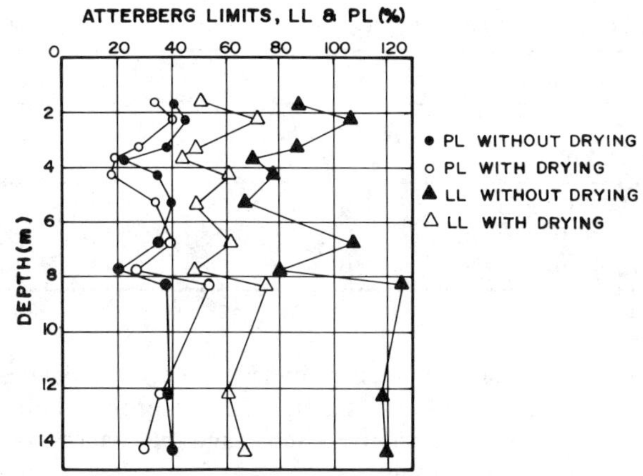

FIG. 4—*Influence of testing procedure on Atterberg limits.*

however the dispersion is too large for differences between two adjacent readings to be analyzed. Table 1 shows the average increase of effective stress $\Delta\sigma'$ at three depths at three time intervals, namely, before construction, at construction of full height of 5 m, and six months later. The values of $\Delta\sigma'$ have been evaluated directly from the pore pressure dissipation measured in the hydraulic piezometer. The best estimate of increase of undrained shear strength ΔS_u during these periods is also indicated. Clearly, the change in S_u does not reflect the change in effective stress in the soil. For this soil, the vane test cannot be used as a reliable tool to monitor the progress of consolidation in the foundation. Law [*4*] analyzed vane strength data from a number of test fills as well as triaxial-vane tests on soft to

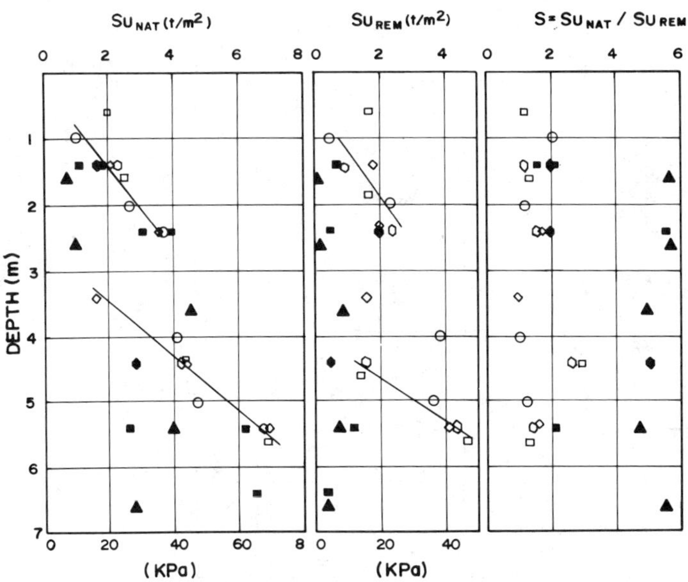

	TEST	FILL HT. (m.)	DATE
▲	SVB - I	0	22-07-76
■	SVB-III	1.9	06-01-77
●	SVB-IV	3.0	03-02-77
△	SVB-V	4.0	——
○	SVB-VI	5.0	15 18-04-77
□	SVB-VII	5.0	11 13-05-77
O	SVB-VIII	5.0	08 09-06-77
◇	SVB-IX	5.0	21-10-77

FIG. 5—*In-situ vane measurements at test fill B.*

TABLE 1—*Comparison of average increase of* ΔS_u *with change in effective stress* $\Delta\sigma'$.

	$\Delta\sigma'$, kPa, During Period		ΔS_u, kPa, During Period	
Depth, m	22.07.76 to 15.04.77	15.04.77 to 21.10.77	22.07.76 to 15.04.77	15.04.77 to 21.10.77
2	43	20	19	4
3.5	50	10.5	5	4
5.5	68	10	13	13

medium clays in Eastern Canada, and has provided conclusive evidence that increase in vane strength is noted only when there is a significant change in horizontal stress. The period of observation at Sepetiba is relatively short, and the changes in vertical and horizontal stresses are also small. The resulting changes in shear strength caused by increase in effective stress are therefore not reflected by the vane test. It will be observed that in the initial stages of loading before the height of the fill exceeded 3 m, a large scatter in vane strength measurements was obtained, as indicated by closed symbols in Fig. 5. The scatter reduces appreciably as the effective stress increases and the soil approached the normally consolidated state.

Figure 6 shows the S_u/σ_v' profiles with depth for the two fills as the consolidation proceeded. The time difference between measurements 1 to 4 is approximately five months, between measurements 4 to 6 is three months, and the last three measurements (6 to 8) at full height of the fill were made over a period of six months. It is apparent from Fig. 6 that two distinct S_u/σ_v' profiles are obtained. At low effective stresses, when the soil is still in the lightly preconsolidated state specially in the upper 3 m, the S_u/σ_v' values generally lie between 0.5 to 1.0. The S_u/σ_v' values decrease appreciably, varying between 0.2 and 0.5, as

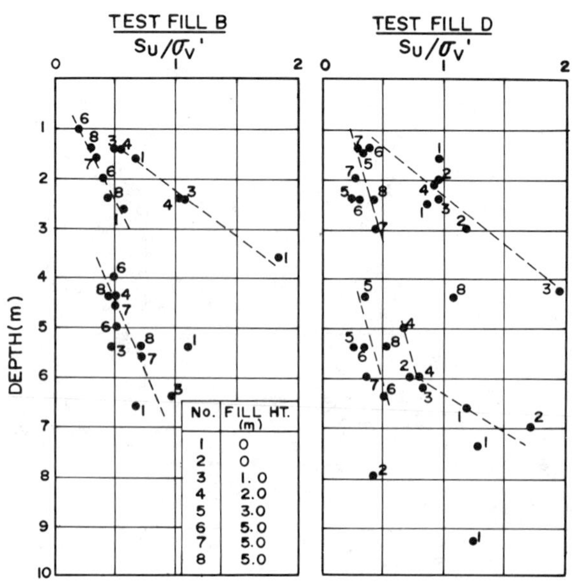

FIG. 6—S_u/σ_v' *profiles from test fills.*

the effective stress increases as indicated by measurements 5 to 8. This trend is consistent with that observed on Scandinavian clays by Aas [5], and on Indian marine clays by Mohan and Bhandari [6]. In both these studies, S_u/σ_v' increased with increasing preconsolidation. Some results from Sepetiba are plotted on the S_u/σ_v' values plasticity index plot (Fig. 7), where the S_u/σ_v' values before loading lie in the same range as observed for lightly overconsolidated clays from Tonsberg and Åserum in Norway [5]. The S_u/σ_v' values after consolidation as a result of the test fill approach the S_u/σ_v' range of 0.1 to 0.3 for the normally consolidated clays.

Investigations were also carried out to determine the strength anisotropy ratio near the test fill sites. Three vanes of different height to diameter H/D ratios (10 cm/2.5 cm [4 in/1 in.], 2.5 cm/10 cm [1 in./4 in.], and 15 cm/7.5 cm [6 in./3 in.]) were used. At each test location, vane tests were carried out in three boreholes spaced at 1.5 m in a triangle. Attempts at measuring strength anisotropy with vanes of different H/D ratios and vane shapes have been reported by Aas [5], Wiesel [7], Richardson et al. [8] and by Donald et al. [9]. S_v and S_h, the undrained strengths on the vertical and horizontal failure surfaces, respectively, are related to the overall strength S_u in the following manner

$$S_u = 0.86S_v + 0.14S_h;\ (H/D = 2)$$
$$S_u = 0.92S_v + 0.08S_h;\ (H/D = 4) \quad (1)$$
$$S_u = 0.43S_v + 0.57S_h;\ (H/D = 1/4)$$

Figures 8 and 9 show the overall strength S_u determined from the three vanes in two series of boreholes. In both cases, S_u determined from the slender vane with $H/D = 4$ is significantly larger than the corresponding values from the other two vanes. The difference in S_u from vanes with H/D equal to 2 and ¼ generally does not exceed 10 kPa (1 t/m^2). In the case of boreholes SV111, the remolded strength from the three vanes, after 360° rotation, is very similar. In contrast, for boreholes SV109, the remolded strength from vane with $H/D = 4$ was larger than the other two values. Since the slender vane with $H/D = 4$ measures strength essentially on the vertical plane, the results in Figs. 8 and 9 are difficult to explain. A plausible reason may be that the other two vanes with smaller H/D ratios suffer from progressive failure and thus yield lower strengths. However, the experience of other investigators, for example, Aas [5], suggests that long slender vanes are more likely to encounter progressive failure. The explanation for the discrepancy may lie in the fact that soils, such as those encountered at Sepetiba, are inhomogeneous containing random

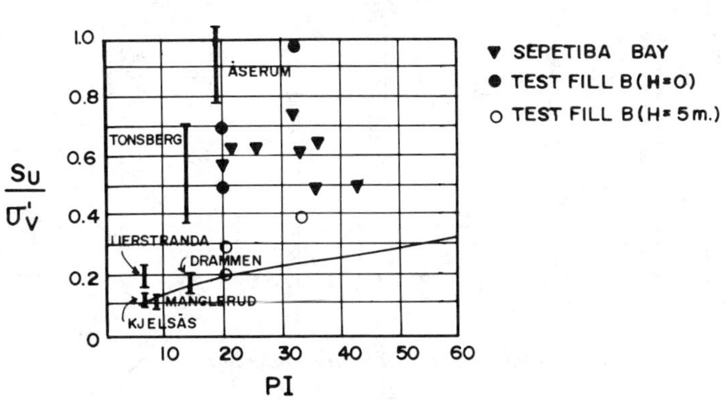

FIG. 7—S_u/σ_v' *versus plasticity index, after Aas* [5].

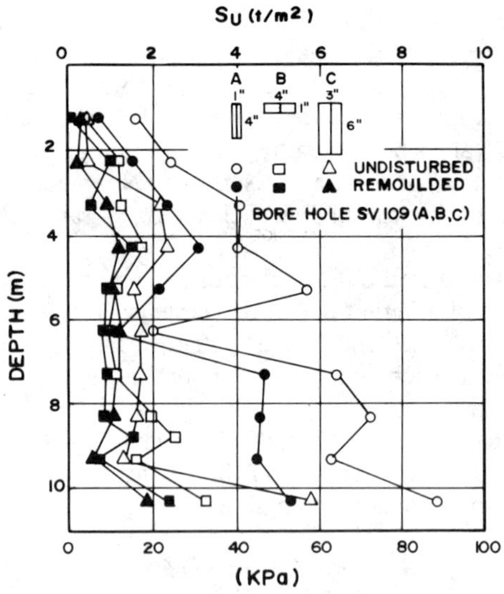

FIG. 8—*Influence of* H/D *ratio on measured strength.*

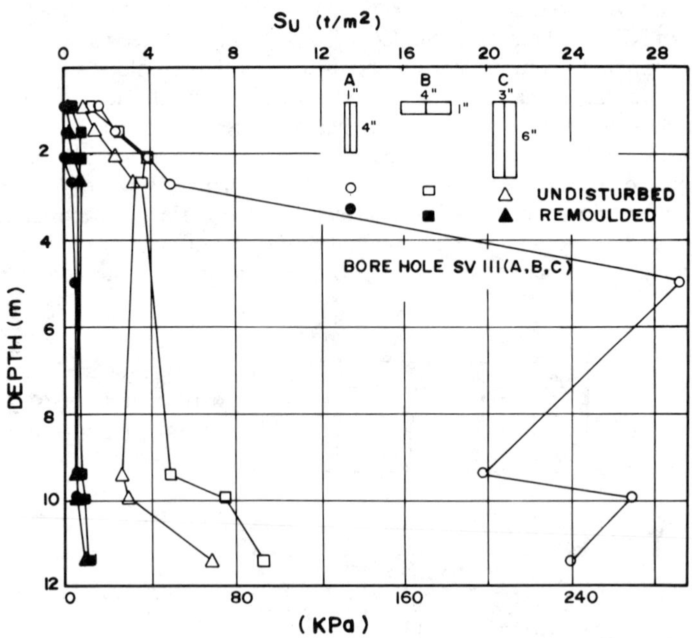

FIG. 9—*Influence of* H/D *ratio on measured strength.*

distribution of sand and silt lenses. Hence comparison of in-situ measurements, even when undertaken very close to each other, may not be very meaningful, and only general tendencies can be discerned.

The graphical representation of results from the vanes as proposed by Aas [5] was used to determine strength anisotropy S_h/S_v. In this method, results from different vanes are plotted on $T(2/\pi D^2H)$ versus ⅓DH graph (where T is the maximum torque), as shown in Fig. 10. The test data from vane with H/D = 4 could not be used for this analysis. From the measurements of the other two vanes, a value of S_v = 59 kPa, and a strength anisotropy ratio S_h/S_v = 1.2 for boreholes SV111 was obtained. In the case of boreholes SV109, a value of S_v = 29 kPa and S_h/S_v = 0.33 was indicated. The value of 0.33 is clearly unacceptable since it implies high lateral stress [5], typical of the heavily overconsolidated clays. The experience with the use of the vane test to measure anisotropy has therefore not been satisfactory. The general experience with vane testing at this site gives credence to Donald et al. [9] who note that a universally correct method for analyzing the vane test has yet to be developed.

Conclusions

The following conclusions may be drawn from the experience with vane testing at Sepetiba site:

1. S_u obtained from in-situ vane testing in soft clay is similar to the values obtained from the simple compression tests and the unconsolidated undrained triaxial tests.
2. The increase in vane strength was not proportionally reflected in the change in effective stress in the foundation.
3. As consolidation proceeded, the stress ratio S_u/σ_v' reduced to 0.2, a value significantly lower than that observed before loading.
4. S_u from vane with H/D = 4 was appreciably higher than the values from vanes with H/D = 2 and ¼.

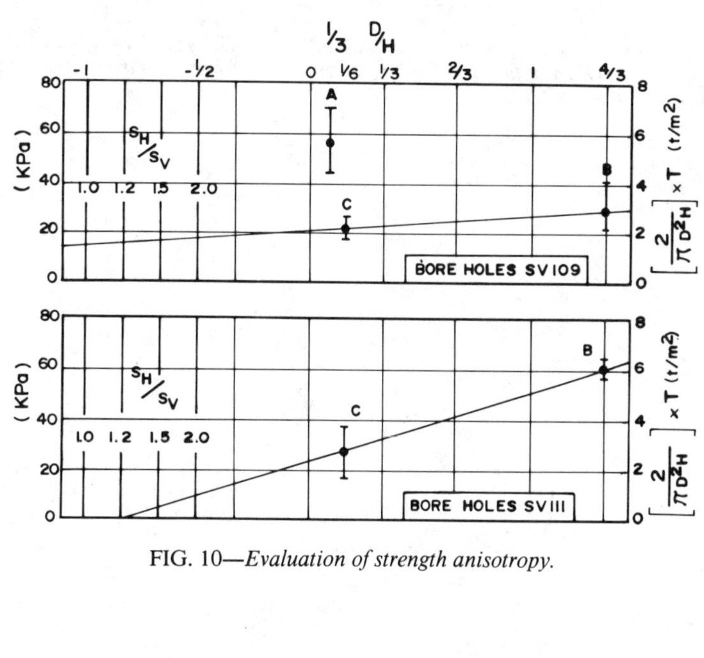

FIG. 10—*Evaluation of strength anisotropy.*

5. The determination of anisotropy ratio S_h/S_v using the method proposed by Aas [5] was not found to be satisfactory.

6. In view of the random distribution of fine sand and silt lenses in a nonhomogeneous soil, the vane test does not necessarily provide meaningful results for comparison between successive readings, and only general tendencies can be discerned.

Acknowledgments

Thanks are expressed to the consulting engineers for this project, Consorcio Planave-Planenge, for whom the author acted as a geotechnical consultant. The vane tests were performed by Romani-Gouvea Ltd. The author is deeply indebted to the late Eng. Paulo V. Carim and to Eng. Octavio Machado Villas-Boas for their collaboration on the project.

References

[1] Schmertmann, J. H., "Measurement of Insitu Shear Strength," *ASCE Conference on In Situ Measurement of Soil Properties,* Vol. II, Raleigh, NC, 1975, pp. 57–138.

[2] Bjerrum, L., "Embankments on Soft Ground," *ASCE Conference on Performance of Earth and Earth Supported Structures,* 1972, Vol. 2, Lafayette, IN, pp. 1–54.

[3] La Rochelle, P., Roy, M., and Tavenas, F., "Field Measurements of Cohesion in Champlain Clays," *Proceedings of the VIII International Conference on Soil Mechanics and Foundation Engineering,* Moscow, USSR, Vol. 1.1, 1973, pp. 229–236.

[4] Law, K. T., "Use of Field Vane Tests Under Earth Structures," *Proceedings of the XI International Conference on Soil Mechanics Tests and Foundation Engineering,* San Francisco, CA, 1985, Vol. 2, pp. 893–898.

[5] Aas, G., "Vane Tests for Investigation of Anisotropy of Undrained Shear Strength of Clays." *Proceedings of the Geotechnical Conference,* Oslo, Norway, Vol. 1, 1967, pp. 3–8.

[6] Mohan, D. and Bhandari, R. K., "Analysis of Some Indian Marine Clays," *Proceedings of the International Symposium on Soft Clays,* Bangkok, Thailand, 1977, pp. 59–73.

[7] Wiesel, C. E., "Some Factors Influencing In Situ Vane Test Results," *Proceedings of the VIII International Conference on Soil Mechanics and Foundation Engineering,* Moscow, USSR, Vol. 1. 1973, p. 475.

[8] Richardson, A. M., Brand, E. W., and Memon, A., "In Situ Determination of Anisotropy of a Soft Clay," *ASCE Conference In Situ Measurement of Soil Properties,* Raleigh, NC, Vol. 1, 1975, p. 336.

[9] Donald, I. B., Jordan, D. O., Parker, R. J., and Toh, C. T., "The Vane Test—A Critical Appraisal," *Proceedings of the IX International Conference on Soil Mechanics and Foundation Engineering,* Vol. 1, Tokyo, 1979, pp. 81–88.

Pandurang K. Nagarkar,[1] Sudhakar V. Rode,[1] Trimbak W. Shurpal,[1] and Gopal L. Dixit[1]

Vane Shear Test Apparatus: A Reliable Tool for the Soft Soil Exploration

REFERENCE: Nagarkar, P. K., Rode, S. V., Shurpal, T. W., and Dixit, G. L., **"Vane Shear Test Apparatus: A Reliable Tool for the Soft Soil Exploration,"** *Vane Shear Strength Testing in Soils: Field and Laboratory Studies, ASTM STP 1014,* A. F. Richards, Ed., American Society for Testing and Materials, Philadelphia, 1988, pp. 277–289.

ABSTRACT: There are deposits of fine clays on coastal areas in India. Those soils are soft, highly saturated, of low density, and low shear strength, sensitive, and normally consolidated. Such soils are generally termed as marine clays. Reliable undisturbed sampling of desired size in marine clays is generally difficult, and laboratory tests for determination of shear strength are also difficult for designing foundations of earth fills. In contrast, the vane shear test is a simple as well as reliable method in the hands of research workers.

Five sites for reclamation of dykes and road embankments around Bombay (India) were tackled. The height of the dykes and embankments varied from 2.5 to 7.97 m, and the depths of the marine clay deposits varied from 6 to 16 m.

During the foundation soil exploration, undisturbed soil samples were collected in open tube samplers from drilled holes. The vane shear test with its slow gear arrangement was carried out in the adjacent boreholes. The vane shear tests with vane borer (vane guard) were also conducted at one of the five sites.

Designing of the embankments, by the sliding block analysis method, was based on the average undrained shear strength determined during the vane shear tests. The settlement performance of the above embankments was studied and found to be satisfactory.

KEY WORDS: soft soils, vane shear, shear strength, stability analysis, soil instrumentation, foundations, marine clays

Soft, slushy, slightly organic marine deposits are found on coastal areas in India, such as those of Bombay. These are transported soils that are deposited due to gravitational settlement of the suspended fine clay particles from the runoffs when they are contacted by saline seawater.

Generally, these soil deposits are of recent origin and are normally consolidated. These soils contain 90 to 95% clay and are bluish grey to blackish or blue-black in color. The liquid limits are high and the character is plastic. They have low shear strengths, and generally they exhibit high compressibility. These soils bear some strength because of their peculiar structure. The soils are sensitive and their in-situ strength is considerably reduced when their peculiar structure is disturbed.

Because of the alternate drying and wetting cycles, the upper crust of these deposits becomes stiffer and comparatively stronger. This upper crust is termed, "drying crust." The thickness of "drying crust" is variable.

It is very difficult to collect undisturbed soil samples of such foundation soils for con-

[1] Chief engineer and director, research officer, scientific officer, and junior scientific assistant, respectively, Maharashtra Engineering Research Institute, Nashik 422004, India.

ducting laboratory tests, as many disturbances are caused during sampling and testing. Hence, reliable shear strength values are not available for the purpose of designing the foundation of landfills over such soils.

In such conditions, the field vane shear test is not only a quite reliable method but a quick and simple one for determining reliable values of the shear strengths of the foundation soils.

Sites Tackled

The following five sites were tackled by this Institute, for designing embankments for either reclamation dykes or road embankments. The location plan of the five sites around Bombay is shown in Fig. 1:

1. Reclamation of Land:
   (a) Wadala reclamation dyke [*1*]
   (b) Mahim reclamation dyke [*2*]
2. Road Embankments
   (a) Bandra Dharavi Link Road [*3,4*]
   (b) Bassein Creek Bridge Approaches [*5*]
   (c) Thana Creek Bridge Approaches [*6,7*]

At the first two sites reclamation bunds were constructed to reclaim land from encroachment of tidal water of the Arabian sea. The remaining three sites are approaches of road embankment for Highway bridges.

Field Work

Because of the presence of soft marine clays as foundation soil, it was found necessary in all the above cases to undertake detailed foundation soil exploration. The detailed foundation exploration comprises of collection of undisturbed soil samples for laboratory tests and conducting field vane shear tests.

Undisturbed Sampling

Undisturbed sampling was planned along the center line of the embankments. The undisturbed samples of 10-cm core diameter were attempted to be collected from boreholes of 15 cm in diameter. Because of the high water contents in these clays, the samples of smaller core diameters (for example, 5 cm) can generally be obtained. But the samples of larger core diameters are sometimes impossible to collect because of high water contents. The samples were collected in the thin seamless open tube samplers (45 cm in length) of 10-cm core diameter. The area ratio of the samplers was much less than 10%.

Field Vane Shear Test

The field vane shear tests (Indian Standard Code of Practice for In-Situ Vane Shear Test for Soils [IS 4434-1978]) were carried out by two models of the vane shear device, namely, (1) an earlier model having no vane guard and (2) a subsequent model with vane guard.

Test procedures with respect to these two types of instruments are as below.

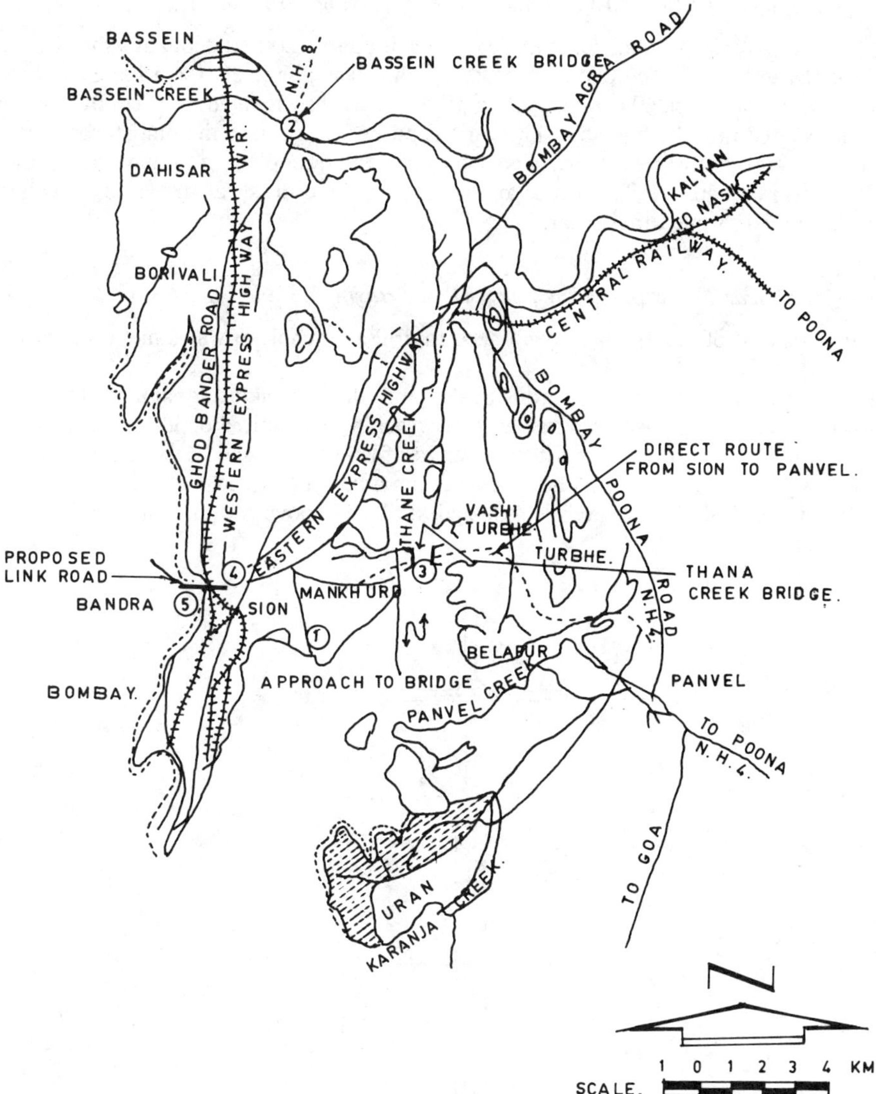

LEGEND....

1. WADALA RECLAMATION SITE.
2. BASSEIN CREEK BRIDGE.
3. THANA CREEK BRIDGE.
4. BANDRA DHARAVI LINK ROAD.
5. MAHIM RECLAMATION SITE.

FIG. 1—*Location plan of sites around Bombay, India.*

Field Vane Shear Test with a Vane Shear Device, Having No Vane Guard

A borehole of 10-cm diameter is drilled into soil up to a desired depth at which the test is to be performed. The borehole is encased with a casing pipe of 10-cm inner diameter. The standard vane (Fig. 2) connected to a pipe that can be extended in pieces by coupling is then lowered into the borehole. The vane is pushed 45 cm into the undisturbed soil at the bottom of the borehole. The vane is then rotated 90° by applying a torque that produces a strain at a rate of 0.1°/s. The torque applied is noted, and the shear strength is calculated from calibrated torque times shear.

Field Vane Shear Test with a Vane Shear Device Having a Vane Guard

With this type of device, there is no need to drill a borehole as was done in the former case.

The vane covered by a vane guard is pushed into the soil, taking advantage of the vane guard that makes this pushing possible. After reaching the desired depth, the vane guard is held in that position, and from it the vane itself is further pushed down by 45 cm. The rest of the procedure is same as the procedure previously described.

The field vane shear tests were conducted at all the above five sites with either vane

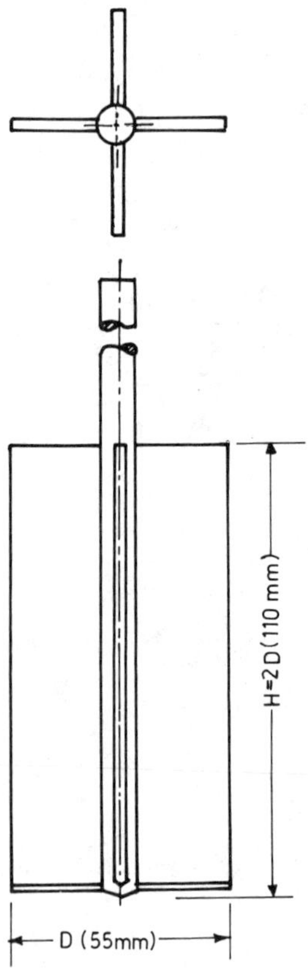

FIG. 2—*Geometry of field vane (Indian Standard 4434-1978).*

torque wrench or with slow gear arrangement in carefully predrilled holes. At one of the sites, the vane guard was used to find the shear strength of the foundation soils depthwise. The typical depth shear strength relationships for the selected sites are shown in Figs. 3 through 7. The results of the vane shear strengths determined in the experiments conducted at a site in the adjacent boreholes with the vane guard and the slow gear arrangements are shown in Fig. 8.

Laboratory Testing

Triaxial Shear Test

The vane shear device is used for "quick" test of shear strength in situ of saturated fine grained soils like marine clays. Hence, in case of laboratory test of triaxial shear (Indian Standard Determination of the Shear Strength Parameters of a Specimen Tested in Unconsolidated Undrained Triaxial Compression Without the Measurement of Pore Water Pressure [IS 2720, Part XI, 1971]), unconsolidated undrained samples are used.

For the triaxial test we used a standard Indian machine, with Indian standard test procedures. From an undisturbed sample of 3.81 cm diameter, a piece, 7.62 cm in height, suitable for the triaxial test, was cut and used.

Different soil tests were conducted for classification, identification, and studying general properties of the soils from different sites. The results are shown in Table 1. The triaxial shear tests were conducted on the undisturbed soil samples to determine shear strength. The consolidation tests were also conducted to determine compressibility.

During testing, it was very difficult to handle the soil samples as they were very sensitive

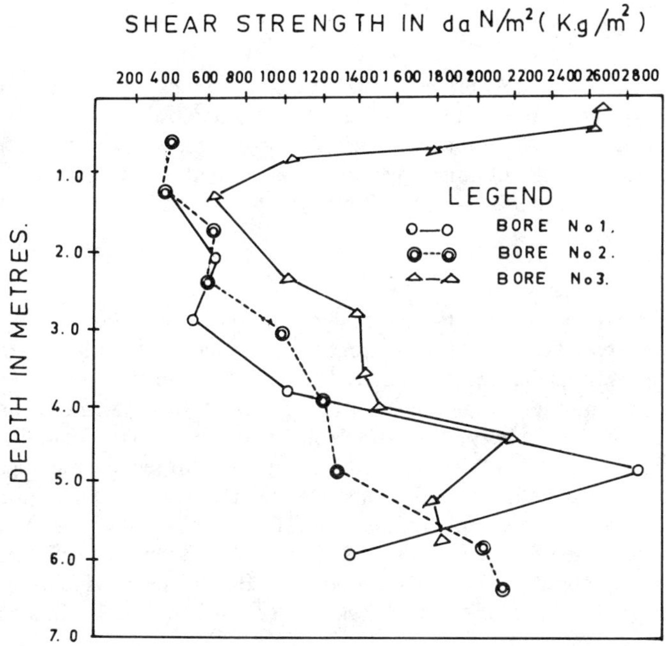

FIG. 3—*Depth to shear strength relationship for Wadala Reclamation Dyke (by torque wrench).*

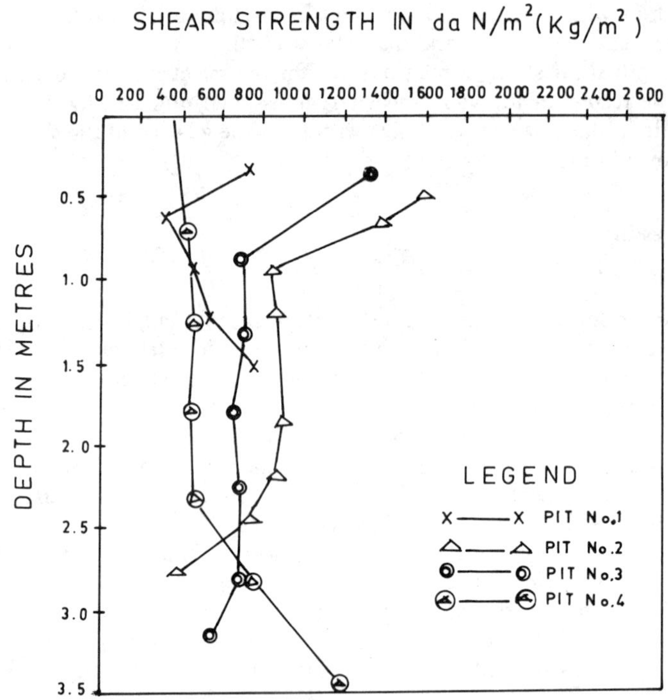

FIG. 4—*Depth to shear strength relationship for Mahim Reclamation Dyke (by slow gear arrangement).*

and contained very high moisture contents. The soil samples were disturbed during handling and testing. Therefore, the shear strength that was observed during the laboratory testing was in general less than the actual. From Table 2*a,* it will be seen that compared to the values of shear strength obtained by vane shear strength method, those determined by laboratory tests are in general lesser by about 47%.

Design Sections

All the above five embankments were designed by the "sliding block analysis method." The designs were based on minimum average shear strength of foundation soils. Sufficient care was taken to utilize lightweight material for construction of embankments to reduce total load intensity on the foundation soils. In some cases, rubble was used as an embankment material with the concept that rubble-fill (with open voids) contains 40% voids, thereby reducing total intensity on the foundation soils. Moreover, the angle of internal friction ϕ in case of rubble is 45°, which causes reduction in designed side slopes to steeper values (1:1.5). In some of the embankments, locally available lightweight murum (weathered soft rock) was used for construction, as rubble was not readily available.

The information about the designed sections for Bandra Dharavi Link Road, Bassein Creek Bridge Approach, and Thana Creek Bridge road approach embankments is as follows:

1. Bandra Dharavi Link Road: The Bandra Dharavi Link Road approach bank work is constructed along a creek-let, with maximum embankment height of 2.5 m over marine

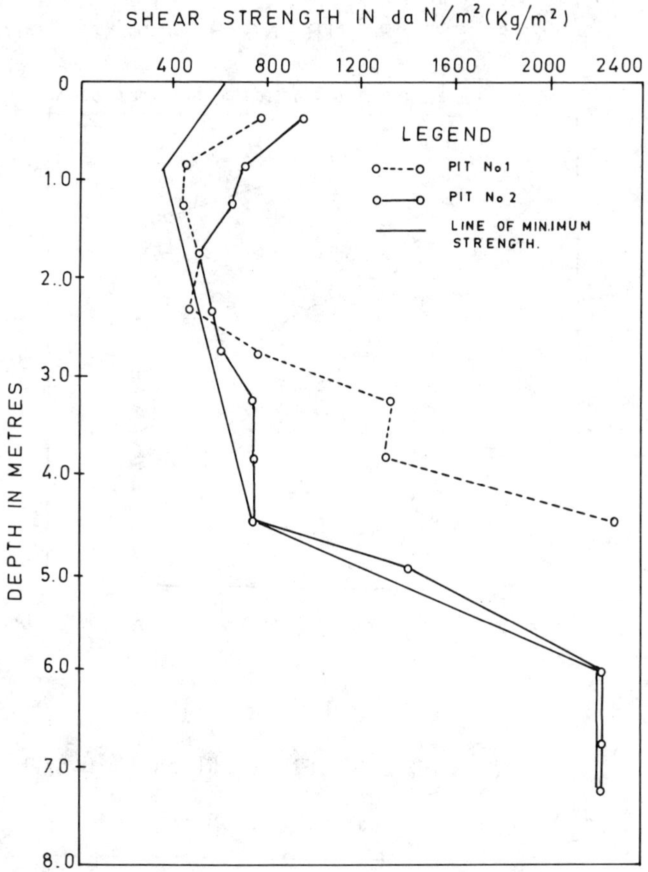

FIG. 5—*Depth to shear strength relationship for Bandra Dharavi Link Road (by slow gear arrangement).*

clay of maximum depth of 7.5 m. In this case as the wave action was not very severe, instead of rubble only lightweight murum was used as the construction material. The embankment was constructed after stripping the slushy surface to a depth of about 1 m. This excavated slush was re-utilized in construction of berms after it was sun dried.

2. Bassein Creek Bridge: In this case maximum height of embankment is 6.4 and maximum depth of marine clay is 10.5 m. The embankment is constructed of rubble with the provision of a sand trench of 1 m thickness. The loading berms were constructed of lightweight murum.

3. Thana Creek Bridge: In this case the maximum height of embankment is 7.97 m, and the maximum depth of marine clay is 16 m. This embankment of 105 m in length was constructed in two segments. First, a 90-m-long embankment away from the abutment was constructed in a single stage by using rubble with rubble fill at side berms, and second a sand mat of 0.9 m thickness was constructed below the bank work.

The portion of the embankment of 15 m length near the abutment was constructed in three steps as the strength of the soil below this portion was poor. The first stage of the embankment was constructed by murum, and the subsequent two stages were constructed

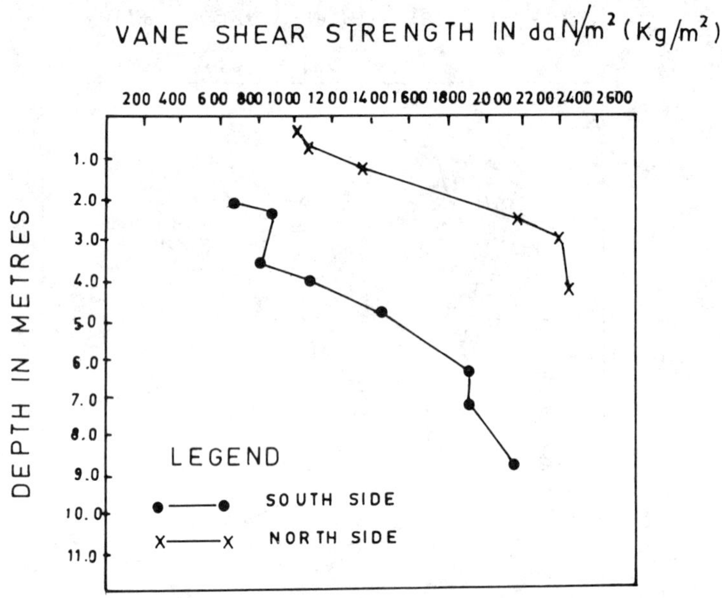

FIG. 6—*Depth to shear strength relationship for Bassein Creek Bridge Approaches (by slow gear arrangement).*

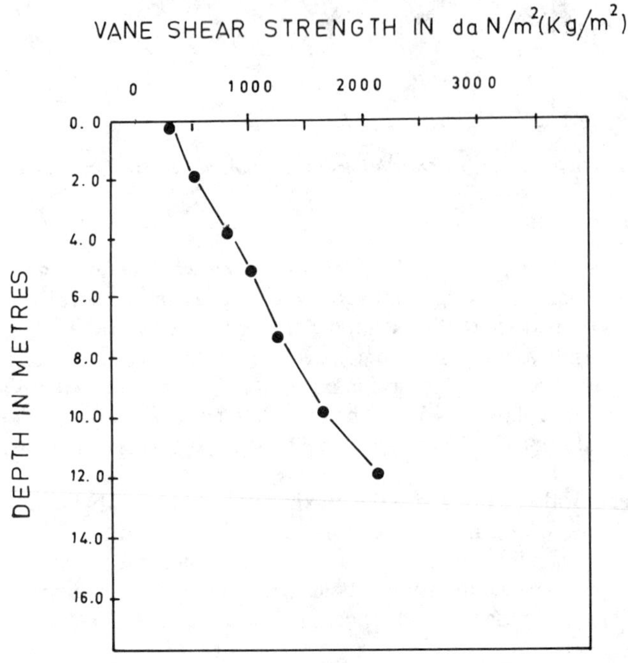

FIG. 7—*Depth to shear strength relationship for Thana Creek Bridge Approaches (by vane guard).*

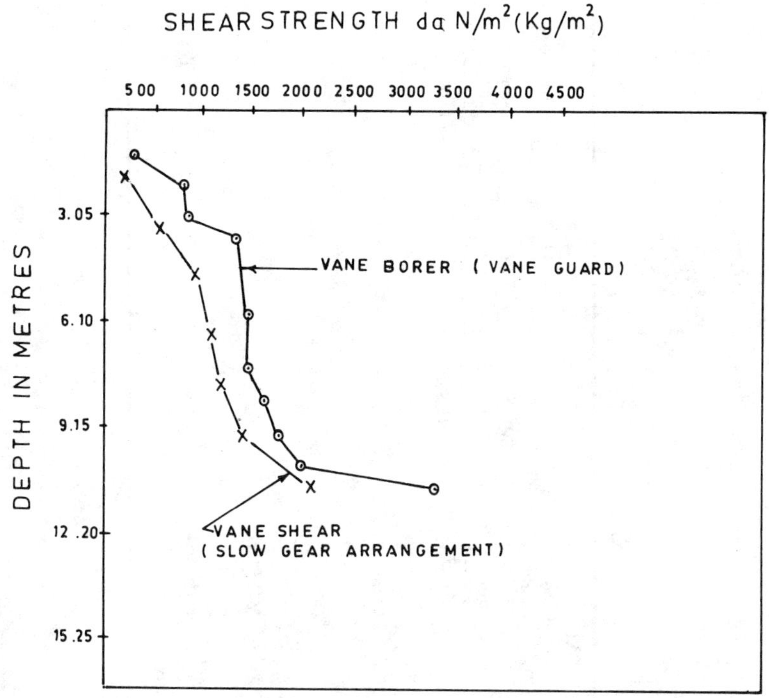

FIG. 8—*Comparison of vane shear strength by vane guard and slow gear arrangement (Thana Creek Bridge).*

by using rubble with provision of rubble loading berms. The sand mats of 0.9 and 0.3 m thicknesses were provided below the main embankment and loading berms, respectively. Vertical sand drains of 38 cm diameter with the 3-m grid in staggered layout were provided in foundation soils below the main embankment.

Instrumentation

The instruments, such as vertical settlement gages, were provided in a typical section of all the above three embankments for studying the behavior of embankments during construction as well as operation stages. The information about the theoretical and observed settlements is shown in Table 2*b*.

It is seen from the Table 2*b* that the observed settlements in the embankments were in general comparable to those anticipated theoretically.

Analysis and Discussions

Because of higher moisture contents in the foundation soil, it is possible to collect undisturbed samples of 5-cm (core) diameter and even less than that. But some times, it is very difficult to collect 10-cm (core) diameter or larger core diameter samples, which are needed for carrying out laboratory tests to find out consolidation and shear parameters.

Even when optimum precautions to minimize disturbances during boring work are taken, using the appropriate drilling bit to eject circulating water upwards and pushing the

TABLE 1—*General properties of soils.*

Name of Site	Mechanical Analysis			Liquid Limit, %	Plastic Limit, %	Laboratory Classification (Unified System)	NMC %/NDD, daN/m^3 (kg/m^3)	Sensitivity	Compression Index C_c
	Gravel, %	Sand, %	Silt + Clay, %						
Wadala Reclamation Dyke	. . .	15	85	85	46	MH–CH	95/835	1.5	0.80
Mahim Reclamation Dyke	. . .	3	97	100	35	CH	95/790	4.7	0.75
Bandra Dharavi Link Road	2	5	93	>100	38	CH	80/850	6.0	0.53
Bassein Creek Bridge Approaches	. . .	8	92	95	42	MH–CH	97/1500	2.95	. . .
Thana Creek Bridge Approaches	. . .	5	95	111	50	MH–CH	94/810	. . .	0.86

TABLE 2—*Data on designed sections and settlement embankments.*

General Information About Designed Sections, Shear Strength of Foundation Soil						Settlement of Embankments			
			Shear Strength, daN/m² (kg/m²)						
Name of Site	Depth of Marine Clay, m	Height of Earth Work, m	Field	Laboratory	% Reduction in Laboratory Value	Days	Theoretical, m	Observed, m	Remarks
Wadala Reclamation Dyke	6.0 to 10.0	2.45	2910	. . .	. . .	. . .	. . .	. . .	
Mahim Reclamation Dyke	3.5	3.66	325 to 1575	210 to 620	43.7	. . .	. . .	. . .	
Bandra Dharavi Link Road	7.5	2.5	450 to 3000	560 to 780	38.8	2040	0.71	0.72	
Bassein Creek Bridge Approaches	10.5	6.4	700 to 2160	450 to 1260	59.0	2000	1.7	1.7	single stage[a]
Thana Creek Bridge Approaches	16.0	7.97	150 to 3490	250 to 1240	40.9	1525 1200	0.82 2.10	1.05 1.90	three stage[b]

NOTE: The tests were carried out most deligently as per the Indian Standard Specifications and the data are precise.
[a] Displacement of foundation soil, when in plastic condition.
[b] Settlement in progress.

casing pipes gently by rotating to desired depths, the samples of marine clays obtained are generally disturbed. It may be observed from Table 2*a* that because of the disturbances caused in undisturbed samples while sampling and handling them during laboratory tests, in general, shear strengths are affected and are as much as 40 to 60% less than those determined by vane shear tests.

From Figs. 3 to 7, which exhibit depths of vane shear tests versus field vane shear strength, two distinct trends are generally noticed. In the drying crust at top of foundation soil, shear strengths go on diminishing to a minimum value when depth increases, and after attaining that value, the shear strengths go on gently increasing with the further increase in depth. The lower portion of foundation soils therefore indicates that the soil is normally consolidated.

From Fig. 8 it may be seen that the shear strengths determined by vane guard tests are on average 12% more than those obtained by slow gear arrangement, since in the latter case there are disturbances during the boring operations.

It is observed from laboratory tests that marine clays are highly clayey and fall in the category of silt and clay with high plasticity (MH-CH). Their natural dry density ranges from 790 to 850 daN/m^3 (kg/m^3). At the Bassein Creek Bridge approach road site, an exceptionally dry density was found, 1500 daN/m^3 (kg/m^3). Values of sensitivity vary from 1.5 to 6. As the marine clays are of low density, the values of compressibility C_c range from 0.53 to 0.86.

The designed sections of landfills at the five sites are based on minimum average values of shear strengths determined by field vane shear tests. As envisaged in Table 2*a* the shear strengths determined by field vane shear tests are higher than those obtained by laboratory tests. Therefore, designs of landfills over marine clay foundations based on field vane shear strengths have effected a considerable economy, without hampering stability.

Conclusions

1. It is uneconomical to design landfill sections on marine clay foundations by adopting shear strengths determined by laboratory tests on undisturbed soil samples. Those shear strengths are not realistic on account of disturbances caused while sampling and handling the samples during tests. The laboratory values are generally 40 to 60% lower when compared to the shear strengths obtained by field vane shear tests.

2. It is observed that the shear strengths of marine clays obtained in the field vane shear tests by using vane guard are generally more, by about 12%, when compared to those determined by using either the torque wrench or slow gear arrangements in pre-drilled boreholes. The shear strengths obtained in the latter two cases are on the lower side because of disturbances created in the foundation soils caused by the boring operation.

3. Design and construction of murum and rubble embankments on marine clays that have low density and high water contents and are sensitive to disturbances, by assuming a factor of safety as low as 1.1, in the sliding block analysis method for working out stability and adopting values of shear strengths determined by field vane shear tests (preferably with vane guards) have been found to be satisfactory. The embankments so designed are also very economical.

4. From Table 2*b*, it may be seen that the calculated and observed settlements practically tally. This fact very emphatically substantiates the assumption that the vane shear strengths determined by field vane shear tests are factual and in general acceptable. Therefore, the designs of embankments based on those values are realistic as well as economical.

5. Considering the experience and satisfactory performance of embankments designed

and constructed on soft soils onshore, namely, on marine clays, it is clear that the field vane shear test apparatus, especially with vane guard, is a very reliable tool for soft soil exploration.

Acknowledgment

The authors thank the Irrigation Department, Government of Maharashtra (India), for encouraging these research studies and acknowledge the sincere cooperation of various field officers in making available field data.

References

[*1*] "Investigation and Design of Anik Poldor (Part of Wadala Reclamation Scheme)," Maharashtra Engineering Research Institute's Technical Memorandum Number SM-89/I, Nashik, India, May 1971.

[*2*] "Design of Dykes for the Mahim Reclamation Scheme, Bombay," Maharashtra Engineering Research Institute's Technical Memorandum Number SM-67/I, Nashik, India, Jan. 1972.

[*3*] "Design of Embankment for the Bandra Dharavi Link Road, Bombay," Maharashtra Engineering Research Institute's Technical Memorandum Number SM-89/I, Nashik, India, July 1970.

[*4*] "Prototype Performance of Bandra Dharavi Link Road," Maharashtra Engineering Research Institute's Technical Memorandum Number SM-93/I, Nov. 1972.

[*5*] "Prototype Performance of South Approach of Bassein Creek Bridge, Bombay," Maharashtra Engineering Research Institute's Technical Memorandum Number SM-31/I, Nashik, India, May 1972.

[*6*] "Design of Bombay Side Approach to the Thana Creek Bridge on the Bombay Poona National Highway," Maharashtra Engineering Research Institute's Technical Memorandum Number SM-70/I, Nashik, India, July 1970.

[*7*] "Prototype Performance of Bombay Side Approach to Thana Creek Bridge on Bombay Poona National Highway Number 4," Maharashtra Engineering Research Institute's Technical Memorandum Number SM-70/I(A), Nashik, India, Jan. 1974.

Part VII: Field Vane Testing Offshore

Gary W. Johnson,[1] *Thomas K. Hamilton,*[2] *Ronald J. Ebelhar,*[3] *Jeffrey L. Mueller,*[4] and *John H. Pelletier*[5]

Comparison of In-Situ Vane, Cone Penetrometer, and Laboratory Test Results for Gulf of Mexico Deepwater Clays

REFERENCE: Johnson, G. W., Hamilton, T. K., Ebelhar, R. J., Mueller, J. L., and Pelletier, J. H., **"Comparison of In-Situ Vane, Cone Penetrometer, and Laboratory Test Results for Gulf of Mexico Deepwater Clays,"** *Vane Shear Strength Testing in Soils: Field and Laboratory Studies, ASTM STP 1014,* A. R. Richards, Ed., American Society for Testing and Materials, Philadelphia, 1988, pp. 293–305.

ABSTRACT: Site investigations were recently performed offshore at three prospective platform sites in water depths between 400 and 500 m in the Gulf of Mexico. Soil conditions at the sites generally range from very soft clays near the seafloor to very stiff clays at 120- to 150-m penetration. Analyses of stress history indicate the soils at these locations are generally normally consolidated. Laboratory tests were performed on recovered specimens to determine the undrained shear strength. Standard laboratory miniature vane shear tests and unconsolidated-undrained triaxial tests were performed in addition to consolidated-undrained tests using stress history and normalized soil engineering properties (SHANSEP) procedures. Tests performed in the field included in-situ vane and cone penetrometer. Cone factors N_k were computed using in-situ vane shear strengths as the reference strength.

This paper compares the results of consolidated-undrained (SHANSEP) laboratory and in-situ tests to determine a relationship that may be used to correlate these results. The effects of soil strength and plasticity are examined and used to correlate shear strength with the liquidity index. A comparison is also made between peak and residual in-situ vane strengths.

This paper further describes how a combination of these in-situ and laboratory tests can be used to characterize a deepwater site for foundation design. Recommendations for future site investigations are also discussed.

KEY WORDS: clays, site investigation, in-situ vane, cone penetrometer, SHANSEP, liquidity index, plasticity index, normalized shear strength

The deepwater (200 to greater 600 m) areas of the Gulf of Mexico have become the targeted "new frontier" areas for petroleum exploration and production in the last ten years. Fixed production platforms exist in water depths of 300 m, and other platforms are

[1] Geotechnical engineer, Browning-Ferris Industries, 757 N. Eldridge, P.O. Box 3151, Houston, TX 77253.

[2] Project manager, McClelland Engineers, Inc., P.O. Box 2197, Houston, TX 77210.

[3] Engineering department manager, S&ME, Inc., 225 Corp. Court, Fairfield, OH 45014.

[4] Ocean engineer, Conoco, Inc., P.O. Box 2197, Houston, TX 77252.

[5] Senior civil engineer, Shell Oil Company, P.O. Box 4480, Houston, TX 77210.

planned for depths of 500 m. Investigation of soil conditions at these deepwater sites follows an integrated approach with a number of steps [*1*] such as

- review of existing data,
- geophysical survey and data interpretation,
- soil sampling and in-situ testing,
- laboratory testing,
- data analyses, and
- integration of results.

This paper will look at results of an in-situ and laboratory testing program performed for three locations in the central Gulf of Mexico. Water depths at the three locations varied from 400 to 500 m. An analytical approach for determining a design shear strength for pile design from in-situ and laboratory results is presented.

Geology and General Soil Conditions

The three sites are located in the Green Canyon Area of the Central Gulf of Mexico, which covers an area of about 22 000 km^2. The Green Canyon Area is located on the upper continental slope, south of a large reentrant in the outer edge of the continental shelf, and southwest of the Mississippi River Delta. The topography of the Green Canyon Area is irregular because of diapirism. Average seafloor slopes in the area are less than 5%, but in areas around growth faults and diapirs, slopes may be greater than 25%. Typically, 8 to 10 m of Holocene deposits overlay several hundred feet of material deposited during the middle to late Pleistocene. Hemipelagic sedimentation is ongoing in most areas with rates of 1 to 3 cm/year.

Soil conditions in this area generally consist of normally consolidated clays with consistencies ranging from very soft ($S_u < 12.5$ kPa) at the seafloor to very stiff or even hard (200 kPa) at 150 m below the seafloor.

Generally, the upper 150 m of material may be divided into four to five depositional units. These units may differ slightly in plasticity and stress history, but the deposits are generally normally consolidated. Areas of erosion (such as previous massive landslides) will be slightly overconsolidated (overconsolidation ratio [OCR] approximately 2 to 3) while areas underlying landslide deposits may be slightly underconsolidated. The variations in plasticity characteristics and stress history result in varying rates of shear strength increase with penetration.

Geotechnical Investigation Procedures

Field Operations

Field investigation procedures included four or five borings at each site. These borings consisted of soil sampling using push or piston samplers and in-situ field vane and cone penetrometer testing. The three site investigations were carried out by drilling crews working from a dynamically positioned drill ship. Soil samples were generally obtained using 76-mm-diameter fixed piston samplers although some push samples were obtained. In-situ testing was carried out using McClelland Engineers' Remote Vane [*2*] and a standard ASTM piezocone or friction cone penetrometer. All cones have an area ratio [*3*] of 0.75 to 0.78. These sampling and testing tools were operated using one of two drill string control and reaction systems. Details of these systems may be found in other references [*4,5*].

Recovered soil samples were generally extruded in the field for on-board classification and testing by miniature vane and unconsolidated-undrained triaxial compression tests. Remaining samples were carefully wrapped in plastic film and aluminum foil to preserve moisture content and packaged for transportation to a Houston, TX, laboratory. Some samples were sealed in the stainless steel sampling tubes and later extruded. A stress history and normalized soil engineering properties (SHANSEP) [6] testing program was performed on selected piston samples from each site. Samples were consolidated to 2.0 to 2.5 times the calculated in-situ vertical stress to overcome the effects of stress relief that occurred during sampling. Samples were sheared using 18 direct simple shear tests, two K_0-consolidated, and four isotropically consolidated undrained triaxial compression tests. The K_0-consolidation phase for triaxial compression tests was conducted by applying a varying axial load to the specimen while maintaining zero lateral strain prior to the axial shear phase. Tests were run at a nominal axial strain rate of 2.5 percent per hour.

During in situ vane testing, the four-bladed vane was pushed 1 to 1.5 m into the soil below the bottom of the borehole. Details of the in situ vane tool are discussed by Kraft et al. [7] and Ehlers et al. [8]. Three different sizes of vane blades were used depending on the consistency of the soil. Length-to-diameter-ratios ranged from 2.1 to 2.7 and area ratios ranged from 15 to 23%. Tests were performed at a rotation rate of 18°/min. After completion of the test (about 2 to 3 min), the vane was pushed an additional 1 m so that another test could be performed. Undrained shear strength was computed from the maximum torque recorded during the test. At one site, a residual value (at a minimum rotation of 60°) of undrained shear strength was recorded for some tests.

Cone penetrometer testing was carried out at one site using a cone penetrometer operating through the drill string. The cone penetrometer measured cone point resistance and sleeve friction and in some cases dynamic pore pressure during the 3-m stroke. A continuous push cone penetrometer system was used at the other two sites. This system allowed continuous recording of data from the seafloor to about 30-m penetration. Data recorded included cone point resistance, sleeve friction, and dynamic pore pressure (measured at the cone tip).

Typical Soil Characteristics

Figures 1 and 2 show a soil profile for one of the deepwater sites. Index and shear strength test results are presented for this typical normally consolidated site.

Index Tests—Water contents w and liquid and plastic limits (LL and PL, respectively)

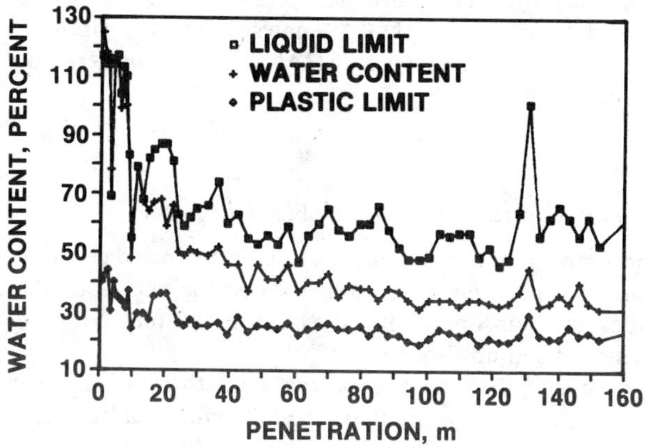

FIG. 1—*Typical deepwater soil profile of index tests.*

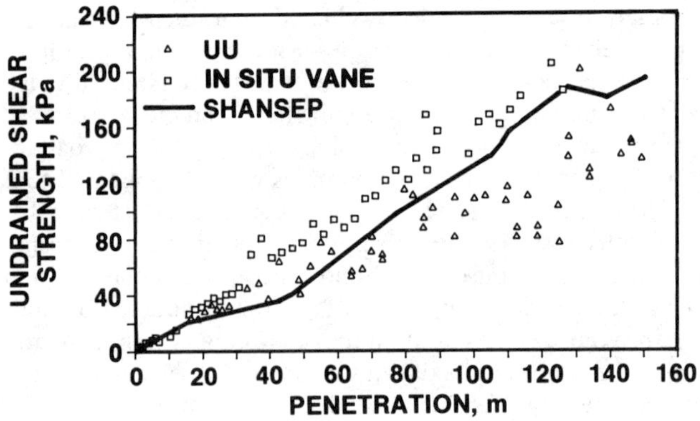

FIG. 2—*Typical deepwater soil profile of strength tests.*

were determined at selected intervals through the profile. The limit tests were performed on specimens that were initially maintained at natural water content. The liquidity index I_L for the typical site is about 1.2 at the seafloor and decreases rapidly with penetration to about 0.6 at 30-m penetration, and to about 0.25 at 150 m. Liquidity index is defined by the following expression

$$I_L = (w - PL)/(LL - PL) \tag{1}$$

The change in plasticity index I_p with penetration for this site has a curve with a shape similar to the I_L curve. The plasticity index is defined as follows

$$I_p = LL - PL \tag{2}$$

The I_p is about 80 at the seafloor decreasing to about 40 at 30 m and 30 at 150 m. Consolidation tests indicate the profile to be normally consolidated to about 75 m and slightly underconsolidated from 75 to 150 m.

Strength Results—Comparison of test results indicate the unconsolidated undrained (UU) strengths are somewhat lower than the in-situ vane strengths. The scatter in UU test data is likely due to a combination of the expansive, gassy nature of the clays and the stress relief resulting from the removal of hydrostatic pressure during sample recovery. The shear strengths determined from the SHANSEP analyses are very similar to the UU shear strengths to about 90-m penetration, where SHANSEP strengths tend to gradually increase relative to UU results.

Comparison of Results

Variation in Normalized Shear Strength with Plasticity Tests

In-situ vane shear strengths were normalized to the estimated effective overburden pressure. Since all three sites are normally consolidated (or near-normally consolidated) as determined by laboratory consolidation test results, we estimated the effective overburden pressure from measured soil unit weights.

Skempton [*9*] suggested that the normalized strength ratio of normally consolidated clays may be related to the soil plasticity by the following equation

$$S_u/\bar{\sigma}_v = 0.11 + 0.0037 I_p \tag{3}$$

where

$S_u/\bar{\sigma}_v$ = normalized shear strength ratio and
I_p = plasticity index, %.

This equation has been used by many firms and individuals in the U.S. Gulf Coast area to determine a "normally consolidated strength" for use in offshore pile design.

Figure 3 presents a plot of normalized in-situ shear vane strength versus plasticity index for these three sites. Skempton's relationship is also presented.

Bjerrum [*10*] suggested that the rate of increase in normalized in-situ vane strengths for normally consolidated clays may be a function of not only plasticity but also the age of the deposit, with secondary consolidation of the deposit over thousands of years resulting in higher shear strengths. Bjerrum presented two curves, one for "aged" clays and one for "young" clays, which are also plotted on Fig. 3. Much of the data from these three sites plot between the "aged" and "young" curves. The data tend towards the "young" curve at lower plasticities and towards the "aged" curve at higher.

The use of "aged" and "young" curves indicates that a relationship between not only plasticity but also the water content relative to the plasticity may be appropriate. While the plasticity index is considered to be a key indicator of soil behavior, the liquidity index considers both the plasticity (range) and water content (relative state) of normally consolidated soil. Plotted on Fig. 4 are the normalized in-situ strengths against liquidity index. A regression analysis was performed to develop a statistical best-fit curve of the normalized in-situ vane variation with liquidity index and a confidence band which represents the standard deviation from the best-fit curve. The resulting statistical best-fit equation is

$$S_u/\bar{\sigma}_v = 0.171 + 0.235 I_L \quad \text{for} \quad 0.2 < I_L < 1.4 \tag{4}$$

This analysis indicates a variation in normalized in-situ strength of 0.22 to 0.50 at liquidity indices of 0.2 to 1.4, respectively, for these three sites.

These data exhibit different trends from data presented by Bjerrum and Simons [*11*]. Bjerrum's data were developed for Norwegian marine clays with plasticity indices less than 35. The data for this study has plasticity indices in the range of 30 to 80 primarily, which may partially account for the variation noted.

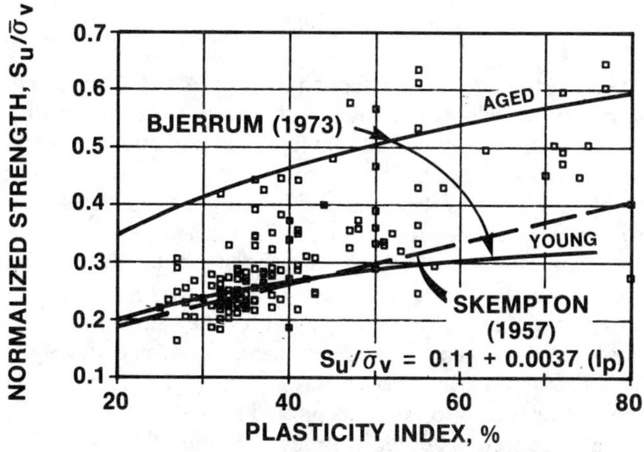

FIG. 3—*Normalized in-situ vane strengths versus index tests: a function of plasticity index.*

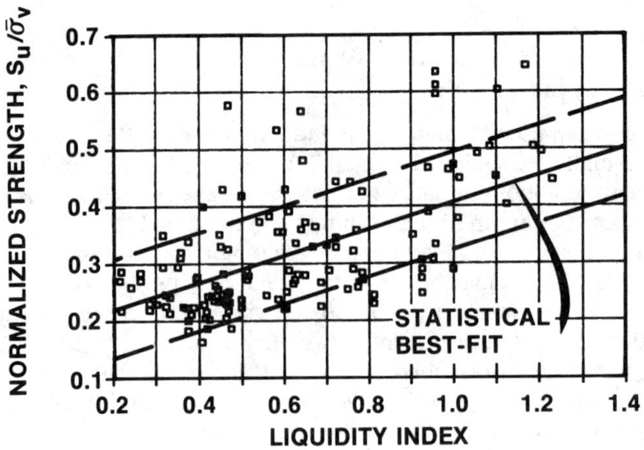

FIG. 4—*Normalized in-situ vane strengths versus index tests: function of liquidity index.*

Adjustment of In-Situ Vane Test Results

Results of in-situ vane tests generally indicate higher strengths than laboratory tests for underconsolidated to normally consolidated clays at these sites. This is due to stress relief during sample recovery, rate of shearing, sample size and handling, and a variety of reasons as discussed by Ehlers et al. [*8*]. It is generally acknowledged that in-situ vane results require some adjustment to develop parameters for most geotechnical designs. In this study, the results of in-situ vane tests are compared with laboratory tests performed on high-quality piston specimens. The goal is to develop an adjustment factor that may be applied to in-situ vane results for the purposes of determining undrained shear strengths for use in static pile capacity design procedures.

The SHANSEP approach was used to develop the relationship between strength and stress history of the materials tested, and a profile of strengths was developed. This profile was compared to the results of the in-situ vane tests. An in-situ vane adjustment factor κ was defined as

$$\kappa = (S_u)_{\text{SHANSEP}}/(S_u)_{\text{in-situ vane}} \tag{5}$$

The undrained shear strength of clays depends on the rate of loading and the anisotropy of the clay. Bjerrum [*10*] developed a correction factor μ for vane strengths based on vane tests and backfigured shear strength of several failed embankments. Bjerrum found that a correction factor μ should be applied to in-situ vane strengths to determine the shear strength that can be mobilized in the field. His correction factor μ was dependent on soil plasticity and ranged from about 1.0 to about 0.6 at plasticity indices ranging from 20 to 100, respectively. Strain rate and anisotropy effects contributed to the differences between vane and mobilized strength.

Aas et al. [*12*] suggested that the correction factor for the field vane strength was not only dependent on plasticity, but that a unique relationship between μ and $S_u/\bar{\sigma}_v$ may be found, which is valid for all normally consolidated clays, "young" and "aged." They developed a curve to represent this relationship, which is plotted on Fig. 5 along with the data from all the study sites. The same trend is noted between the Gulf of Mexico adjustment factor κ and the curve presented by Aas et al. for μ, with the Gulf of Mexico factor being somewhat lower.

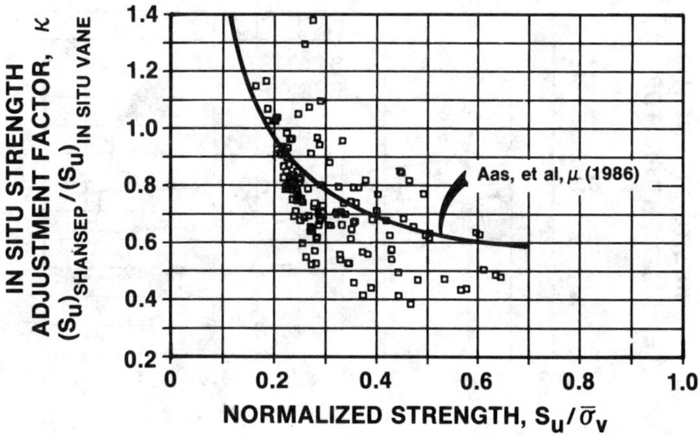

FIG. 5—*In-situ strength adjustment factor versus normalized strength.*

Variations in the adjustment factor κ are plotted on Figs. 6 and 7 as a function of index tests. Figure 6 shows the variation of the adjustment factor with plasticity index which is similar to a relationship for μ presented by Bjerrum [*10*]. Figure 7 shows the variation of the adjustment factor with liquidity index. Both figures show the results of regression analyses to develop statistical best-fit curves and a confidence band that represents the residual standard deviation from the best-fit curve. The best-fit curve for Fig. 7 is derived based on discussion in the following paragraph.

Figure 8 shows a semilog representation of the variation of shear strength with liquidity index. Independent regression analyses were performed to develop the statistical best-fit of in-situ vane and SHANSEP strengths shown on Fig. 8. The ratio of the best-fit lines was then plotted on Fig. 7 as function of liquidity index. The equations of the statistical best-fit curves, as a function of plasticity and liquidity index, are as follows

$$\kappa = 1.29 - 0.0206I_p + 0.000156I_p^2 \text{ for } 20 \leq I_p \leq 80 \tag{6}$$

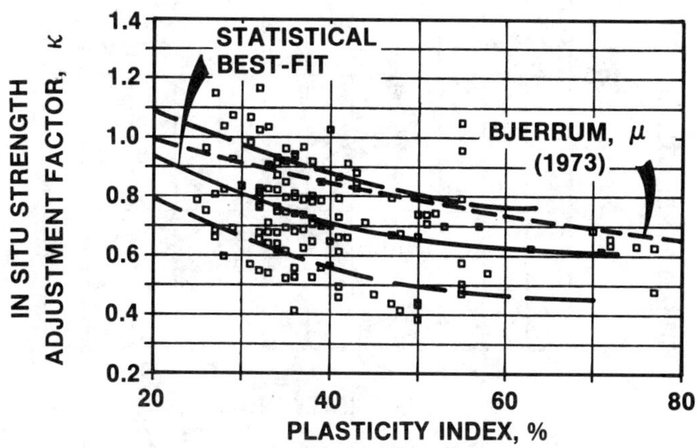

FIG. 6—*In-situ strength adjustment factor versus index tests: function of plasticity index.*

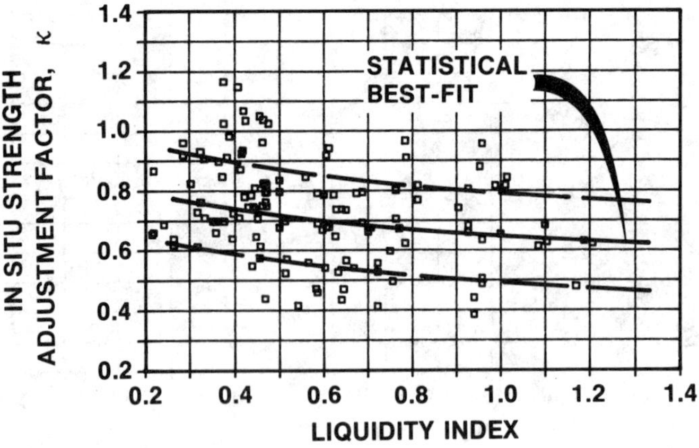

FIG. 7—*In-situ strength adjustment factor versus index tests: function of liquidity index.*

and

$$\kappa = 10^{-(0.077+0.098I_L)} \text{ for } 0.2 \leq I_L \leq 1.3 \tag{7}$$

These analyses indicate that the adjustment factor κ varies from about 0.95 to 0.6 for plasticity indices ranging from 20 to 80, respectively. The analyses show that the adjustment factor varies from 0.82 to 0.62 for liquidity indices ranging from 0.2 to 1.3, respectively. These variations are consistent in magnitude with those interpreted by Ehlers et al. [*8*] for deltaic deposits in the Gulf of Mexico.

Comparison of Peak to Residual In-Situ Vane Strength

As mentioned previously, in-situ vane tests were carried out to large rotations (>60°), so that a residual value could be determined on a number of tests at one of the sites. The

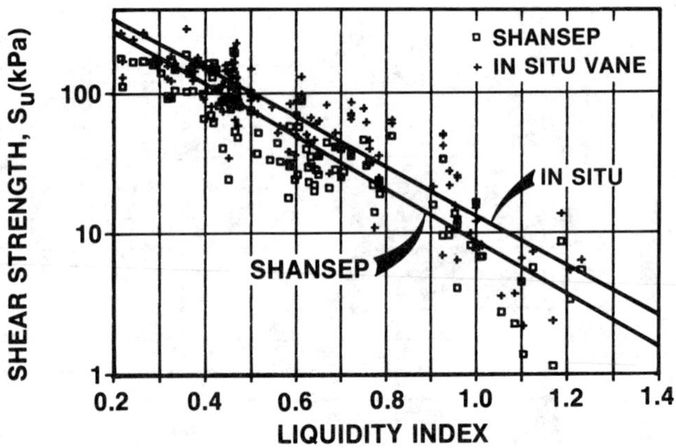

FIG. 8—*Shear strength versus liquidity index.*

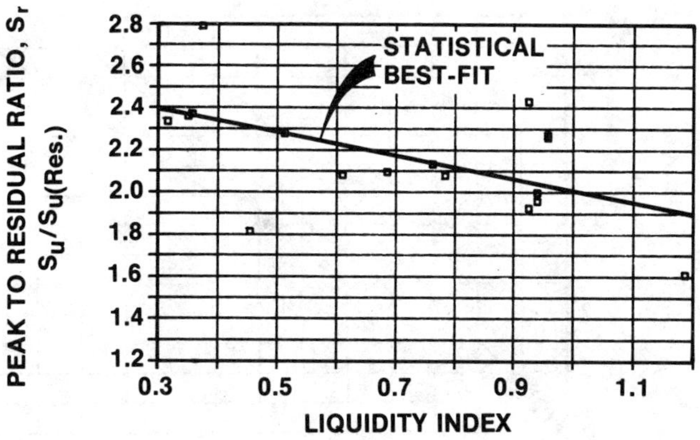

FIG. 9—*Comparison of peak to residual in-situ vane strengths.*

peak-to-residual ratio for these 17 tests are plotted on Fig. 9 as a function of liquidity index. A statistical best-fit line was determined from a regression analysis. The equation of this line is

$$S_r = 2.56 - 0.556 I_L \tag{8}$$

where S_r = ratio of peak to residual strength determined by in-situ vane.

The results suggest that clays with lower liquidity indices are more sensitive. Examination of undisturbed and remolded strength data presented by Quiros et al. [*13*] suggests a similar trend. As noted previously, the Gulf of Mexico deepwater clays generally have more structure (slickensided or platy) at deeper penetrations (lower I_L's). This may explain the peak-to-residual trend. Mitchell and Houston [*14*] present sensitivity correlations with liquidity index for Norwegian marine clays. The trends exhibited by their data are somewhat different than those observed in this study and Quiros et al. [*13*]. As mentioned earlier, there are significant differences in the range of plasticity indices for the Norwegian clays as compared to the Gulf of Mexico deepwater clays.

Comparison of Cone Point Resistance with In-Situ Vane Results

Results of cone testing indicate a linearly increasing trend of cone point resistance variation with in-situ vane strengths. Figure 10 shows the cone point resistance plotted against in-situ vane strength values (unadjusted) for all three sites. The cone point resistance was determined by subtracting the estimated total overburden pressure. The cone factor N_k was calculated according to the traditional method as defined by Terzaghi [*15*] using the following equation

$$N_k = \frac{q_c - \sigma_v}{S_u} \tag{9}$$

where

q_c = total cone point resistance, kPa,
σ_v = total overburden pressure, kPa, and
S_u = reference shear strength, kPa.

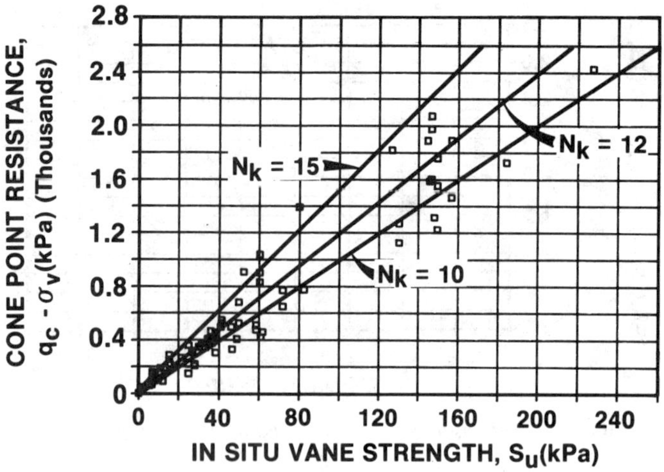

FIG. 10—*Cone point resistance versus in-situ vane strengths.*

The cone factor has been calculated by many different researchers using many different cones and reference strengths. In this paper, the cone factor is determined using a 10-cm^2 electrical friction cone with area ratio of 0.75 to 0.78 and in-situ vane shear strength as a reference strength.

Lunne et al. [*16*] found N_k to vary between 13 and 24 for Scandinavian marine clays. Quiros and Young[6] found N_k to vary from 9.5 to 15 for Santa Barbara Channel clays and silty clays. Standard 10-cm^2 cones were used, and the reference strengths were determined by in-situ vane.

The cone factors for these three Gulf of Mexico sites were found to vary between about 7 and 17.5 with the great majority of the cone factors between 10 and 15 and an average of about 12.

Variations in N_k for cohesive soils have generally been related to soil plasticity I_p. It has generally been accepted that N_k decreases with increasing I_p. Quantification of this relationship of N_k with I_p has been difficult because of use of different cones and different reference strengths. Aas et al. [*12*] recently suggested a procedure for correlating this data by correcting the cone results for dynamic pore pressure and correcting the in-situ vane data to a reference strength.

Corrected Cone Factor Correlations

Cone test results were compared with in-situ vane tests and laboratory tests to develop a method of estimating shear strengths for design purposes. Cone point resistances were first corrected for overburden pressures, as described above. Then in-situ vane strengths were adjusted using the adjustment factors κ described in a preceding section as a function of liquidity index. The cone point resistance was then divided by adjusted in-situ vane strengths to develop a corrected cone factor N_{kc}.

Regression analyses were performed to determine the statistical best-fit relationship between the corrected cone factor and liquidity index. Figure 11 presents the variation of

[6] Quiros, G. W. and Young, A. G., in this publication, pp. 306–317.

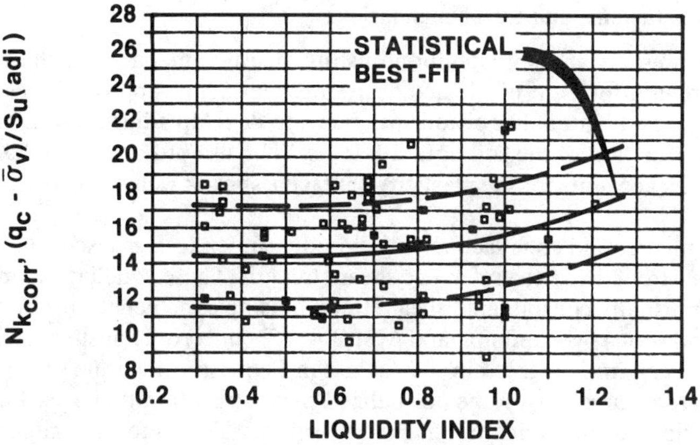

FIG. 11—*Adjusted cone factor versus liquidity index.*

corrected cone factor with liquidity index and shows the best-fit curve obtained by the regression analyses. The equation for the best-fit curve is as follows:

$$N_{kc} = 16.3 - 7.16I_L + 6.57I_L^2 \text{ for } 0.3 \leq I_L \leq 1.2 \tag{10}$$

It should be noted that piezocone data were available for only two sites; as a result, cone point resistances were not corrected for dynamic pore pressure as suggested by Aas et al. Equation 8 was used to back-calculate shear strengths from observed cone point resistances. Figure 12 shows adjusted field strengths (determined from cone and vane tests) plotted against SHANSEP shear strength results. The figure shows a reasonably good correlation between adjusted field shear strengths and laboratory strength results.

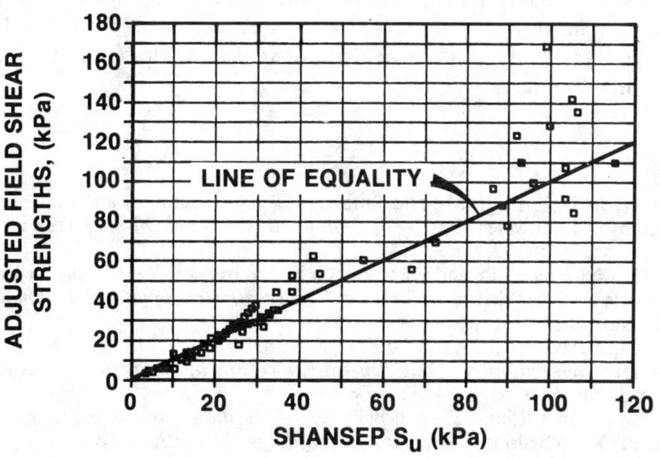

FIG. 12—*Comparison of adjusted field strengths with laboratory shear strengths.*

Summary, Conclusions, and Recommendations

This paper presents an approach for adjusting in-situ vane data to determine a SHANSEP shear strength for Gulf of Mexico deepwater clays. The approach uses the results of laboratory tests on high-quality piston specimens to develop adjustment factors for in-situ strengths as a function of the soil liquidity index. The use of liquidity index to adjust the in-situ strengths accounts for the plasticity characteristics as well as the relative state of the soil.

Variation in corrected cone factors with liquidity index tests is also examined using the adjusted vane strengths as the reference shear strength. These cone factors may be used to determine a normally consolidated shear strength using cone penetrometer data.

Recommendations by Randolph and Murphy [*17*] concerning offshore static pile design have recently been implemented in the American Petroleum Institute (API) RP 2A guidelines. These procedures suggest the normalized shear strength ratio be used in conjunction with unconsolidated-undrained triaxial and in-situ test results to determine unit skin friction for piles in clay.

In light of these recommendations and the data presented above, the authors would like to see more data comparing in-situ vane (peak and residual) results to consolidated-undrained laboratory results as well as standard test results. Comparison of in-situ, SHANSEP, and UU shear strengths at many deepwater sites (Fig. 1 and Footnote 6) indicate UU shear strengths begin deviating significantly from SHANSEP and in-situ shear strength trends at deeper depths, probably caused by stress relief during sampling. Use of SHANSEP and in-situ shear strengths can provide a means for interpreting a pile design shear strength for deepwater normally consolidated clays.

These procedures are based on a limited data base of in-situ and laboratory information. These expressions presented appear to be reasonable for the range of index, strength, and stress history conditions observed at these sites. Caution should be exercised when applying the approach to geologic and soil conditions differing from those discussed in this paper.

Acknowledgments

The authors would like to thank Shell Offshore Inc. and Conoco Inc. for permission to publish this data. The authors would also like to acknowledge Mr. John A. Focht, Jr., Mr. Charles A. Rivette, and Mr. Alan G. Young for reviewing drafts of this manuscript and making recommendations.

References

[*1*] Campbell, K. J., Dobson, B. M., and Ehlers, C. J., "Geotechnical and Engineering Geological Investigations of Deepwater Sites," *Proceedings, 14th Annual Offshore Technology Conference,* Vol. 1, Houston, TX, 1982, pp. 25–37.

[*2*] Doyle, E. H., McClelland, B., and Ferguson, G. H., "Wire-Line Vane Probe for Deep Penetration Measurements of Ocean Sediment Strength," *Proceedings, 3rd Offshore Technology Conference,* Vol. 1, Houston, TX, 1971, pp. 21–32.

[*3*] Baligh, M. M., Azzouz, A. S., Wissa, A. Z. E., Martin, R. T., and Morrison, M. J., "The Piezocone Penetrometer," *Symposium on Cone Penetration Testing and Experience,* ASCE National Convention, St. Louis, MO, pp. 247–263.

[*4*] Ferguson, G. H., McClelland, B., and Bell, W. D., "Seafloor Cone Penetrometer for Deep Measurements of Ocean Sediment Strength," *Proceedings, 9th Offshore Technology Conference,* Vol. 1, Houston, TX, 1977, pp. 471–478.

[*5*] Peterson, L. M., Johnson, G. W., and Babb, L. V., "High-Quality Sampling and In Situ Testing

for Deepwater Geotechnical Site Investigation," ASCE Specialty Conference, *In Situ Test Methods in Geotechnical Engineering Practice,* Blacksburg, VA, pp. 913–925.

[*6*] Ladd, C. C. and Foott, R., "New Design Procedures for Stability of Soft Clays," *Journal, Geotechnical Engineering Division, ASCE,* Vol. 100, No. GT7, pp. 763–786.

[*7*] Kraft, L. M., Ahmad, N., and Focht, J. A., Jr., "Application of Remote Vane Results to Offshore Geotechnical Problems," *Proceedings, 8th Offshore Technology Conference,* Vol. 3, Houston, TX, 1976, pp. 75–96.

[*8*] Ehlers, C. J., Young, A. G., and Focht, J. A., Jr., "Advantages of Using In Situ Vane Tests for Marine Soil Investigation," *Proceedings, International Symposium on Marine Soil Mechanics,* Mexico City, Feb. 1980.

[*9*] Skempton, A. W., "Discussion: The Planning and Design of the New Hong Kong Airport," *Proceedings, Institution of Civil Engineers,* London, Vol. 7, pp. 305–307.

[*10*] Bjerrum, L., "Problems of Soil Mechanics and Construction on Soft Clays," *Proceedings, Eighth International Conference on Soil Mechanics and Foundation Engineering,* Vol. 3, 1973, pp. 111–159.

[*11*] Bjerrum, L. and Simons, N. E., "Comparison of Shear Strength Characteristics of Normally Consolidated Clays," *Research Conference on Shear Strength of Cohesive Soils, ASCE,* Boulder, CO, 1960, pp. 711–727.

[*12*] Aas, G., Lacasse, S., Lunne, T., and Hoeg, K., "Use of In Situ Tests for Foundation Design on Clay," *Use of In Situ Tests in Geotechnical Engineering,* ASCE Geotechnical Special Publication, No. 6, 1986, pp. 1–30.

[*13*] Quiros, G. W., Young, A. G., Pelletier, J. H., and Chan, J. H-C, "Shear Strength Interpretation for Gulf of Mexico Clays," *Proceedings, Geotechnical Practice in Offshore Engineering,* Austin, TX, 1983, pp. 144–165.

[*14*] Mitchell, J. K. and Houston, W. N., "Property Interrelationships in Sensitive Clays," *Journal, Soil Mechanics and Foundations Division, ASCE,* Vol. 95, No. SM4, July, 1969, pp. 1037–1062.

[*15*] Terzaghi, K., *Theoretical Soil Mechanics,* John Wiley and Sons, Inc., New York, 1943, 510 pp.

[*16*] Lunne, T., Eide, O., and DeRuiter, J., "Correlations Between Cone Resistance and Vane Shear Strength in Some Scandinavian Soft to Medium Soft Clays," *Canadian Geotechnical Journal,* Vol. 13, No. 4, 1976, pp. 430–431.

[*17*] Randolph, M. F. and Murphy, B. S., "Shaft Capacity of Driven Piles in Clay," *Proceedings, 17th Offshore Technology Conference,* Houston, TX, 1985, pp. 371–378.

Gerardo W. Quiros[1] *and Alan G. Young*[1]

Comparison of Field Vane, CPT, and Laboratory Strength Data at Santa Barbara Channel Site

REFERENCE: Quiros, G. W. and Young, A. G., **"Comparison of Field Vane, CPT, and Laboratory Strength Data at Santa Barbara Channel Site,"** *Vane Shear Strength Testing in Soils: Field and Laboratory Studies, ASTM STP 1014,* A. F. Richards, Ed., American Society for Testing and Materials, Philadelphia, 1988, pp. 306–317.

ABSTRACT: A comprehensive geotechnical study was performed at a site in the Santa Barbara Channel, CA, where soils are predominately slightly overconsolidated Pleistocene clays. Undrained shear strength data at the site were acquired by (1) in-situ vane testing; (2) in-situ static cone penetration testing; (3) conventional laboratory testing consisting of unconsolidated, undrained triaxial compression and miniature vane; and (4) SHANSEP testing. The conventional laboratory and stress history and normalized soil engineering properties (SHANSEP) testing were performed on samples recovered by pushing 76-mm outside diameter (OD) thin-walled tubes. This paper compares, in detail, strength data obtained by the different testing procedures (in-situ tests and laboratory tests) and recommends the procedures that yield consistent and reliable undrained strength data for offshore foundation design.

KEY WORDS: soil investigation, soil properties, field tests, overburden, overconsolidation, preconsolidation pressure, shear strength, cohesive soils, vane shear tests, soil tests

Improvement in the state of the art of offshore foundation design requires better analytical procedures and data enhancement that better defines soil conditions. A number of groups around the world have contributed to improvement of the analytical procedures. However, since the inception of offshore site investigations, the greatest improvement has probably been achieved with data enhancement methods. During the last ten years, a number of improved data gathering tools have emerged that provide more consistent measurements of in-situ soil properties and higher quality samples, which exhibit less effects of variable sample disturbance. Yet, there has been little published data comparing the strength characteristics at offshore sites obtained by the wide variety of laboratory and in-situ testing procedures.

This paper presents strength data for a site offshore California in the Santa Barbara Channel where the following techniques were used: (1) in-situ vane testing, (2) in-situ cone penetrometer testing (CPT), (3) conventional laboratory testing consisting of miniature vane and unconsolidated-undrained triaxial compression, and (4) stress history and normalized soil engineering properties (SHANSEP) testing. The conventional laboratory and SHANSEP testing was performed on samples recovered by pushing 76-mm outside diameter (OD) thin-walled tubes. The primary purposes of this paper are to compare the strength data acquired by the different testing procedures and make recommendations relative to the procedures that best reduce the scatter in the measured strengths, thereby leading to a more consistent and reliable representation of the soil strength.

[1] Senior geotechnical engineer and vice-president, respectively, Fugro-McClelland, 6100 Hillcroft, Houston, TX 77081.

General Site Conditions

The site lies in the northwestern corner of the Santa Barbara Basin in about 365 m of water where subsurface sediments consist of clay to silty clay of moderate to high plasticity as shown in Fig. 1. An upper stratum from the seafloor to 12-m penetration is of recent Holocene origin. The more significant underlying stratum to 152.4-m penetration is Pleistocene in age and slightly overconsolidated. These soils are slightly less plastic than the soils of the upper stratum.

The clays at this site generally have material 20 to 30% finer than 2 μm. The mineralogies, as reflected by the percentage clay fraction, indicate that over half of the clay mineral is montmorillonite and the remainder is about equal proportions of kaolinite and illite. The soil activity falls between 1.5 and 1.0.

Field Investigation Procedures

Drilling, sampling, and in-situ testing at the study site were performed from a 99-m dynamically positioned drill ship. The drill ship had a 33-m 181-metric ton derrick mounted above midship and contained a 7.3- by 8.2-m moonpool that accommodated the seafloor jacking system known as Stingray [*1*]. This Stingray unit was employed to conduct push sampling and in-situ testing down to the terminal boring depth of 152.4 m, and also served as a hole reentry guide for the drill string. To reduce potential downtime caused by turbulent seas, vessel heave was isolated from the seafloor jacking system, drill string, and in-situ testing tools by means of a shipboard motion-compensation system.

Drilling and in-situ testing tools were operated through 127-mm OD, 102-mm inside diameter (ID) drill pipe. Sampling and in-situ testing were performed in a single borehole. Soil samples were generally obtained at 1.5-m intervals to about 12 m and at 3- to 6-m intervals below 12 m. The crews obtained pushed samples by latching a 76-mm OD, 72-mm ID thin-walled sampling tube into the drill bit, and then used the remote-controlled Stingray jacking unit to grip and lower the drill pipe to insert the sampler into the soil formation. Percussion samples were acquired by driving a 57-mm OD, 54-mm ID thin-walled tube into the soil using a 136-kg weight on a wire line. Most samples were recovered

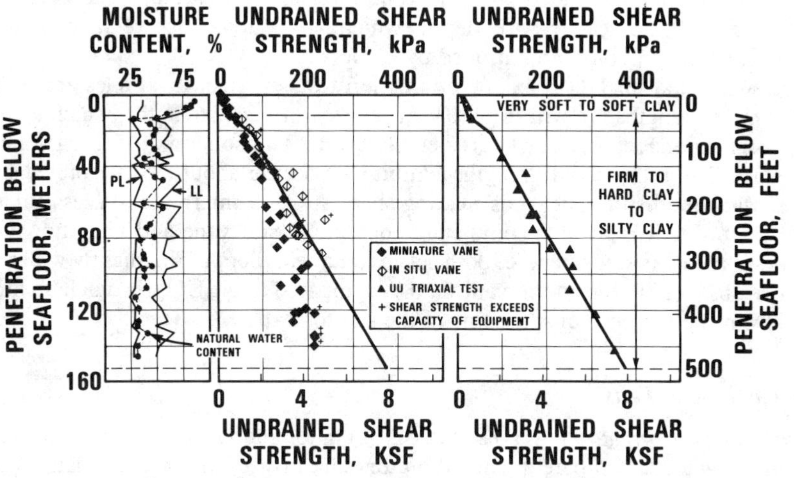

FIG. 1—*Soil conditions at Santa Barbara Channel site.*

using the push sampling technique, while driven samples were obtained at selected intervals in the boring.

In-situ strength measurements were also obtained with the wire-line operated remote vane [*2*] and the cone penetrometer [*1*]. The four-bladed vane, which is rotated by an electric motor, was pushed 1 to 1.5 m into the soil below the bottom of the borehole. Undrained shear strengths were computed from the maximum torque recorded during the test. The cone penetration test was performed by latching an electric cone penetrometer into the drill bit and advancing the cone by raising and lowering the drill pipe with the Stingray jacking unit. The cone penetrometer measured both cone point resistance and sleeve friction. Cone signals were transmitted to a recorder on board the drilling vessel by means of an electric armored cable. Because cone penetrometer tests were conducted at selected intervals between samples, cone pushes were limited to about 2 m.

Shear Strength Measurements

Laboratory Strength Measurements

The data from three different types of strength tests performed at this site are plotted on Fig. 1. An interpreted shear strength profile described in a later section is shown in the (b) and (c) parts of Fig. 1 to give a better comparison of the different types of strength data.

Miniature vane tests on pushed samples and in-situ vane tests are presented in Fig. 1b. The miniature vane strengths generally are 20 to 30% smaller than the strengths obtained with the in-situ vane down to about 60-m penetration. Below about 60 m, the strength values obtained with the miniature vane show a sharp reduction in the rate of increase with depth. The lesser values of strength obtained with the miniature vane are somewhat unexpected because, typically, in-situ vane and miniature vane measurements are consistent and agree well [*4*]. The authors believe that the lower miniature vane strength at this site probably resulted from disturbance associated with the insertion of the vane blades into very stiff to hard silty clays.

The in-situ vane strengths show a consistent rate of increase down to about 65-m depth when the strength reaches a value of about 240 kPa. The authors believe this strength is close to a maximum value for which the in-situ vane data will be meaningful because of excessive disturbance associated with the vane insertion into stronger clays. The in-situ vane data show little scatter with depth, which is an advantage relative to tests on recovered samples previously mentioned by Ehlers et al. [*3*].

The strength data for UU triaxial compression tests on pushed samples are shown on Fig. 1*c*. These strengths are within ±10% of the in-situ vane strengths, which agree with similar observations reported by Quiros et al. [*4*] and Koutsoftas and Fisher [*5*] for other marine clays. The UU triaxial strengths exhibit little scatter about the interpreted strength profile as did the in-situ vane, thus making shear strength interpretation less subjective. Another feature that is most encouraging about the in-situ vane and UU strength measurements is the consistent trend of both sets of data with depth. The slightly wider scatter that can be observed in the in-situ vane data as compared to the UU triaxial data, probably resulted from the presence of silt seams in the soil formation.

Cone Penetrometer Tests

Shear Strength Correlation—To use the CPT data for analysis purposes usually requires a correlation factor to compare it with laboratory and in-situ vane strength data. Basically, correlations have been based on an equation suggested by Schmertmann [6], whereby

$$S_u = \frac{q_c - \sigma_v}{N_k}$$

where

q_c = CPT tip resistance including total overburden,
σ_v = total overburden pressure including hydrostatic and effective overburden, and
N_k = cone penetrometer factor that theoretically corresponds to the bearing capacity factor N_c.

Figures 2 and 3 present the net cone resistance values versus undrained shear strengths measured with the in-situ vane and UU triaxial tests for pushed samples. The slope of the correlation line, forced through zero, is the factor N_k defined in the above equation. The average N_k factor based on in-situ vane strength data is 12.5 with a range in values of 9.5 to 15.0. For the UU triaxial tests, the average N_k is 12.3 with a range of 10.5 to 13.5. Thus, an N_k factor of 12 to 13 seems to be an appropriate value in obtaining estimates of undrained shear strength from CPT data in the clays at this site.

Earlier studies [7–9] comparing cone point resistance and in-situ vane strengths for normally to slightly overconsolidated clays indicate a clear tendency for N_k to decrease with increasing plasticity. The plasticity indexes of the clays at this site fall within the range of 40 to 50%. The above references indicate that soils with this plasticity generally exhibit an

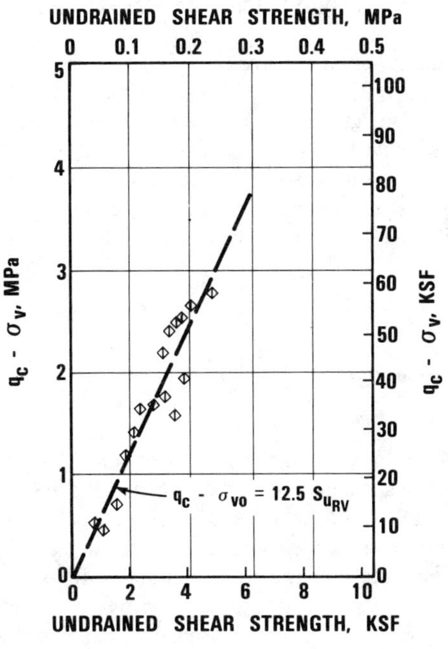

FIG. 2—*Relationship between net cone resistance and in-situ vane shear strength.*

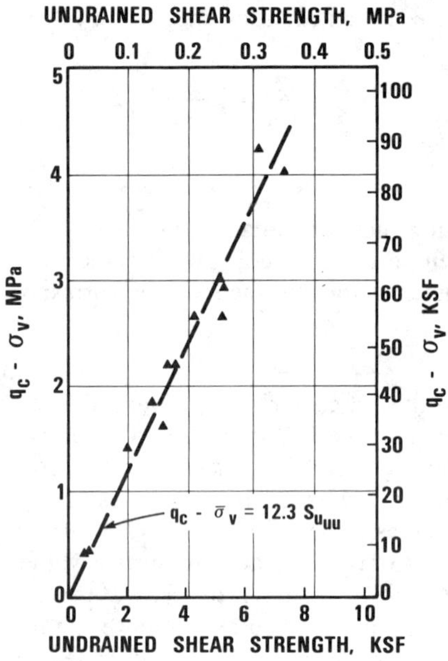

FIG. 3—*Relationship between net cone resistance and UU triaxial shear strength.*

N_k factor in the range of 10 to 15 with an average value of about 12.5. This average value agrees with the N_k factor correlated to both the in-situ vane and UU triaxial compression tests.

CPT Sleeve Friction and Remolded Strength Tests—Figure 4 illustrates there is good agreement between the sleeve friction values and the remolded strengths at the site. Marr and Endley [*10*] believe close agreement between the sleeve friction and remolded shear strength profile may also provide a good indication of the unit frictional resistance that piles will encounter during continuous driving. Hindcast drivability studies for piles previously driven at a nearby site appear to support this view. However, this approach may be relevant only in similar soil types with similar stress histories and similar pile-hammer systems.

SHANSEP Strength Measurements

General Approach

To further evaluate the reliability of the in-situ field vane strength measurements and the conventional laboratory strength tests, we used the SHANSEP method recommended by Ladd and Foott [*11*] to obtain an alternate measure of in-situ shear strength. This method requires developing normalized soil properties (such as $S_u/\bar{\sigma}_{vc}$) for a range of stress conditions that are representative of the stress history of the soil in the field. Since high-quality oedometer tests are essential, particularly in overconsolidated soils, good undisturbed samples are a major requirement. For a detailed description of the SHANSEP method of design, the reader should refer to the paper by Ladd and Foott [*11*].

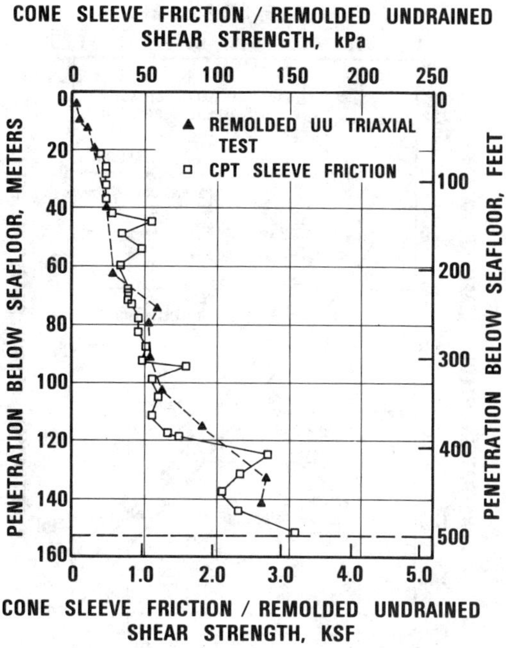

FIG. 4—*Comparison of remolded UU triaxial tests and CPT sleeve friction.*

Stress History

Consolidation Tests—Since reliable soil stress history is essential to employ the SHANSEP approach, a series of constant-rate-of-strain (CRS) consolidation tests were conducted on selected soil specimens. The maximum past consolidation pressure $\bar{\sigma}_{vm}$ interpreted from the CRS consolidation tests is shown in Fig. 5 along with the profile of computed vertical effective overburden pressure $\bar{\sigma}_{vo}$. Figure 5 illustrates that the soils below 12-m penetration are overconsolidated by a pressure ranging from about 100 to 170 kPa greater than the present vertical overburden pressure. The overconsolidated state of these sediments was also confirmed by the relatively low liquidity index values measured on the samples.

Overconsolidation of the Santa Barbara Channel soil deposit probably resulted from erosion, aging, or cyclic loading. The authors believe that overconsolidation at this site was most likely induced by overburden removal. The data in Fig. 5 suggest that as much as 22 to 31 m of sediments were eroded before the surficial 12 m of recent soils were deposited. Geologists very familiar with the site have corroborated the authors' viewpoint and have indicated that past removal of up to 30 m of sediments is not uncommon for this area.

Estimating Overconsolidation from CPT Data—This same stress history is also supported by the CPT data presented in Fig. 6. We used the q_c-depth profile to estimate overconsolidation [*6*]. As Fig. 6 illustrates, the q_c-depth profile defines an apparent linear increase of q_c with depth. Extrapolating this q_c-depth profile to $q_c = 0$ defines an intersection point on the penetration axis that can be taken as the highest probable past ground surface. Our interpretation of the data in Fig. 6 infers that about 21 to 32 m were removed in the past before deposition of the present 12 m of recent sediments. This estimate includes a correction of about -3 m of overburden that accounts for the fact that q_c was

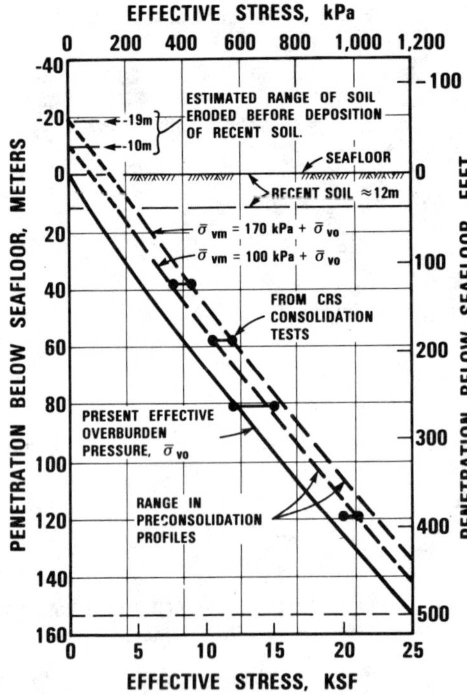

FIG. 5—*Range in preconsolidation profiles interpreted from consolidation test results.*

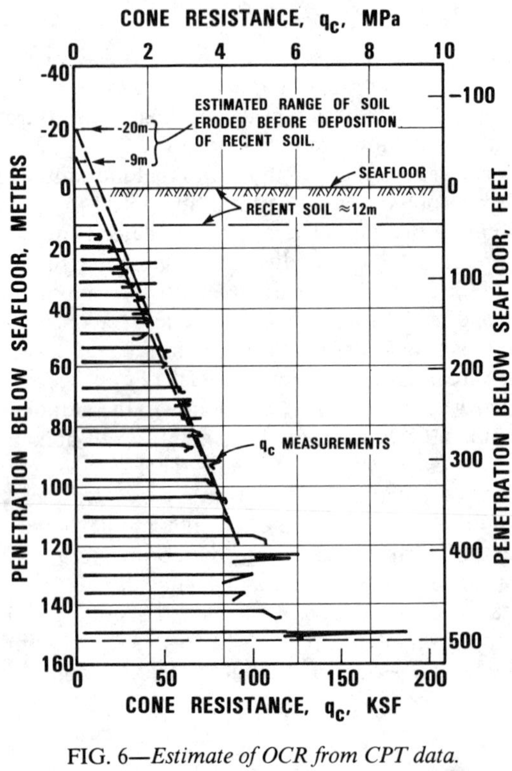

FIG. 6—*Estimate of OCR from CPT data.*

not likely to have been zero at the past seafloor surface. The surficial soil would have had an undrained shear strength of about 5 to 10 kPa.

Overconsolidation Ratio Interpretation—Based on the information presented in Figs. 5 and 6, we developed the overconsolidation ratio (OCR) profiles shown in Fig. 7. These OCR profiles reveal that the OCR of the sediments below the recent soil stratum ranges from 2.5 to 3.5 near 12-m penetration and decreases with depth.

The validity of the OCR profiles presented in Fig. 7 is supported by the results of CRS consolidation tests, in-situ vane tests, and UU triaxial tests also included in that figure. OCR values from consolidation tests were computed from the interpreted $\bar{\sigma}_{vm}$ values. OCR values from in-situ vane strengths and UU triaxial strengths were obtained by dividing the shear strength values by the corresponding $\bar{\sigma}_{vo}$ values and entering the curve in Fig. 8 developed for the Santa Barbara Channel soils. Figure 7 demonstrates that the interpreted OCR profiles agree well with the OCR data obtained from consolidation tests, in-situ vane tests, and UU triaxial tests.

Results of Laboratory Strength Tests

The suite of tests performed in this study to apply the SHANSEP method included CK_0U direct simple shear (CK_0UDSS) tests and CK_0U and CIU triaxial compression tests. Figure 9 presents the results of these tests in graphical form while Table 1 summarizes the averages.

Application of the SHANSEP Method

Undrained shear strength interpretations based on the SHANSEP concept were obtained by using the CK_0UDSS strength ratios from Fig. 8 that correspond to the OCR values computed from the consolidation test results. The interpreted strengths are presented in Fig.

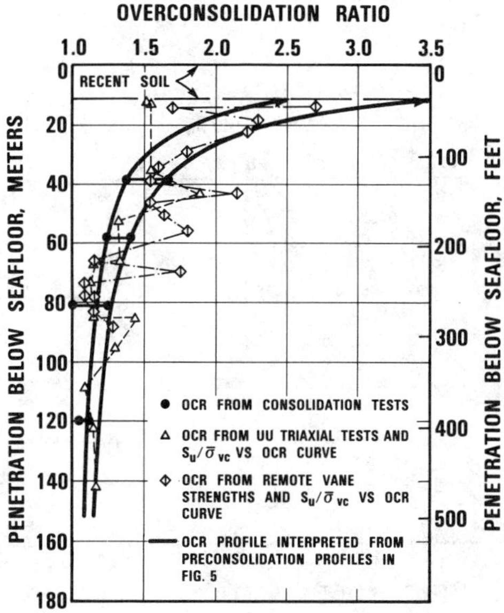

FIG. 7—*OCR data for Santa Barbara Channel site.*

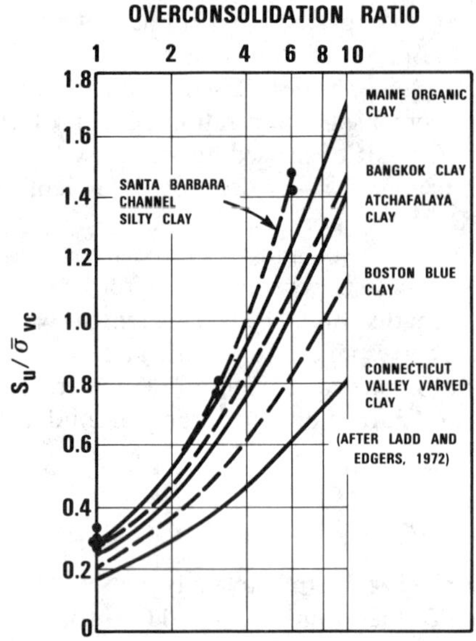

FIG. 8—*Strength ratio versus OCR from* CK_oUDSS tests.

10 along with other shear strength data measured with the in-situ vane, cone penetrometer, and UU triaxial apparatus. Figure 10 illustrates that there is good agreement between the SHANSEP strength results and the other three sets of shear strength data. Most of the data in Fig. 10 are within 15% of the interpreted shear strength profile.

It is noteworthy that, since the SHANSEP shear strength is a function of soil stress history, the good agreement between the SHANSEP strength results and the other sets of shear strength data further supports the OCR profile interpreted for the site.

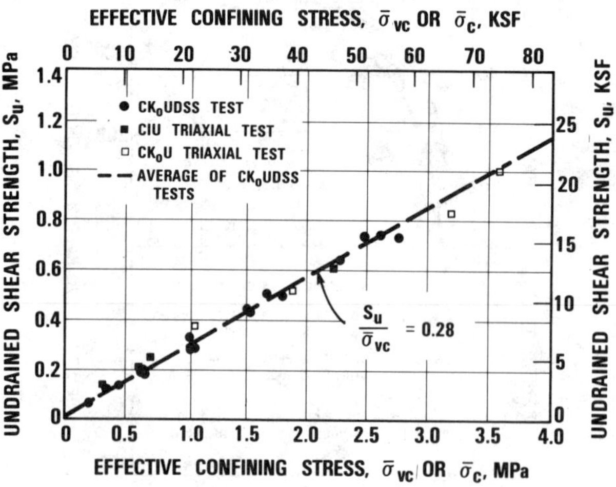

FIG. 9—*Results of laboratory strength tests.*

TABLE 1—*Summary of laboratory strength tests. Average strength ratio, $s_u/\bar{\sigma}_{vc}$ or $s_u/\bar{\sigma}_{3c}$.*

Type of Test	OCR = 1	OCR = 3	OCR = 6
CK_0UDSS	0.28	0.77	1.48
CK_0U triaxial	0.27	. . .	. . .
CIU triaxial	0.32	. . .	. . .

Comparison of Strength Measurements

The real merit of a high-quality geotechnical field program as described herein can be better appreciated by comparing the results obtained with the shear strength data secured from percussion samples. As part of this site investigation, percussion (driven) samples were acquired at selected depth intervals by driving a 57-mm OD, 54-mm ID thin-walled tube into the soil using a 136-kg weight on a wire line. Strength tests performed on the driven samples included (1) miniature vane, (2) UU triaxial, and (3) SHANSEP.

The results of the strength tests performed on these driven samples are shown in Fig. 11 along with the interpreted shear strength profile from Fig. 10. Figure 11 shows that the driven sample data fall in a range that is only 30 to 60% of the interpreted strength profile. It is also important to note the lack of consistency in the three different strengths obtained on driven samples as compared with the strengths from pushed samples and in-situ vane tests in Fig. 10. Even the SHANSEP test results from driven samples gave lower strength values than corresponding results on pushed samples because of the higher degree of sample disturbance associated with percussion sampling. As a result, consolidation tests on the more disturbed driven samples yielded preconsolidation pressures much lower than those

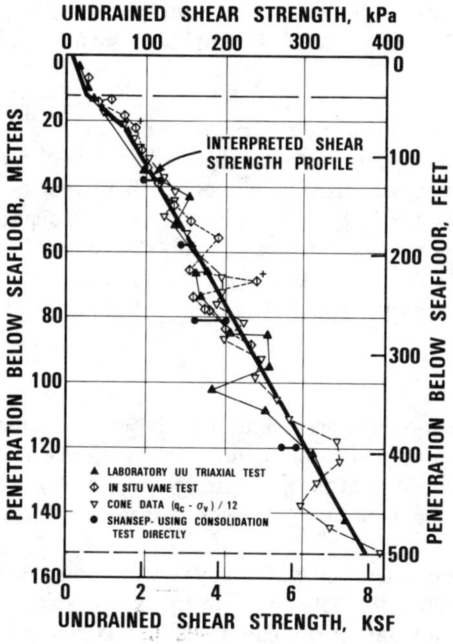

FIG. 10—*Undrained shear strength interpretation for site.*

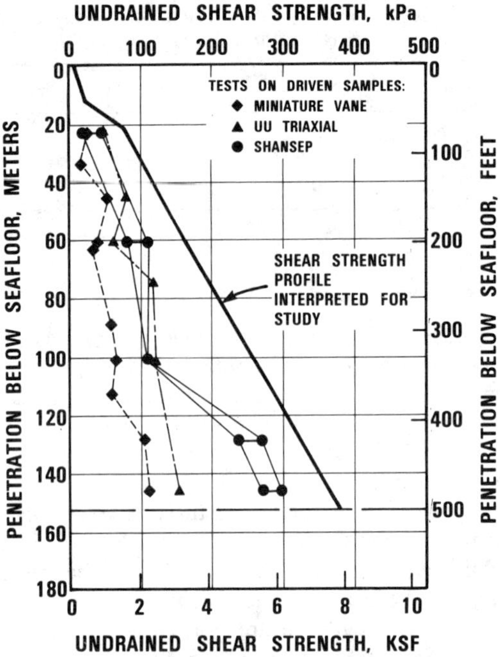

FIG. 11—*Comparison of shear strength measurements on driven samples with shear strength profile interpreted from pushed sample and in-situ test data.*

obtained for pushed samples. For additional information on the influence that sampling technique has upon disturbance effects on samples, the reader should refer to Emrich [*12*] and Young et al. [*13*].

The exercise of interpreting a shear strength profile based on the driven sample data in Fig. 11 also helps one to appreciate the importance of acquiring good quality soil samples. If only in-situ vane data and driven samples had been obtained at this site, the in-situ vane data probably would have been used only to establish a strength trend line. The resulting shear strength profile would probably have been a curve through the SHANSEP and UU triaxial strength data with values no greater than 50 to 60% of the shear strength profile actually selected for the site. Consequently, the interpreted strength for driven samples would have been very conservative.

Conclusions

A comprehensive high-quality geotechnical study was performed at a site in the Santa Barbara Channel were soils are predominantly slightly overconsolidated Pleistocene clays. Undrained shear strength data at the site were acquired by (1) in-situ vane testing, (2) in-situ CPT, (3) conventional laboratory tests consisting of UU triaxial and miniature vane, and (4) SHANSEP testing.

Miniature vanes on pushed samples generally yielded strengths about 20 to 30% less than the in-situ vane. The strengths obtained with the UU triaxial on pushed samples were within ±10% of the in-situ vane. Correlation of UU triaxial and in-situ vane strength data with the cone point resistance measurements yields an N_k factor of 12 to 13, which is consistent with other published data for normally to slightly overconsolidated clays of similar plasticity index. The sleeve friction profile from the CPT agrees closely with the remolded UU triaxial tests.

The SHANSEP testing confirmed that these soils exhibit normalized behavior. For an OCR of one, the $S_u/\bar{\sigma}_{vc}$ ratio varies from 0.27 to 0.32 depending upon the type of tests performed. The combination of stress history determined from CRS consolidation tests and $S_u/\bar{\sigma}_{vc}$ ratios from CK_0U simple shear and triaxial compression tests were used to develop an interpreted shear strength profile. It was found that the other strength data, in particular the UU triaxial and in-situ vane data, were in good agreement with the SHANSEP profile.

The close agreement between the laboratory tests on pushed samples, the CPT, and the in-situ vane provides a high degree of confidence in interpreting a shear strength profile representative of in-situ conditions. These data also underscore the value of the CPT and the in-situ vane for establishing strength trends that will permit reasoned discounting of other data (such as the miniature vane strengths in this case) that are clearly not representative of in-situ conditions. The reliability and consistency in these types of strength data help eliminate the bias towards conservatism associated with shear strength interpretation. Thus, the authors believe that a high-quality geotechnical field program of the type described here results in a better and more complete understanding of site conditions. This, in turn, results in improved reliability in foundation design.

References

[1] Ferguson, G. H., McClelland, B., and Bell, W. D., "Seafloor Cone Penetrometer for Deep Measurements of Ocean Sediment Strength," *Proceedings, 9th Offshore Technology Conference, Houston, TX,* Vol. 1, 1977, pp. 471–478.

[2] Doyle, E. H., McClelland, B., and Ferguson, G. H., "Wire-Line Vane Probe for Deep Penetration Measurements of Ocean Sediment Strength," *Proceedings, 3rd Offshore Technology Conference,* Houston, TX, Vol. 1, 1971, pp. 21–32.

[3] Ehlers, C. J., Young, A. G., and Focht, J. A., Jr., "Advantages of Using In Situ Vane Tests for Marine Soil Investigations," *Proceedings, International Symposium on Marine Soil Mechanics,* Mexico City, 1980.

[4] Quiros, G. W., Young, A. G., Pelletier, J. H., and Chan, J. D., "Shear Strength Interpretation for Gulf of Mexico Clays," *Proceedings, Specialty Conference on Geotechnical Practice in Offshore Engineering,* ASCE, Austin, TX, 1983, pp. 144–165.

[5] Koutsoftas, D. and Fischer, J. A., "In-situ Undrained Shear Strength of Two Marine Clays," *Journal, Geotechnical Engineering Division,* ASCE, Vol. 102, No. GT9, 1976, pp. 989–1005.

[6] Schmertmann, J. H., *Guidelines for Cone Penetrometer Test, Performance and Design,* FHWA TS-78-209, Federal Highway Administration, 1978.

[7] Lunne, T. O., Eide, and de Ruiter, J., "Correlations Between Cone Resistance and Vane Shear Strength in Some Scandinavian Soft to Medium Stiff Clays," *Canadian Geotechnical Journal,* Vol. 13, No. 4, 1976, pp. 430–441.

[8] Lunne, T. O. and Kleven, A., "Role of CPT in North Sea Foundation Engineering," *Proceedings, Cone Penetration Testing and Experience,* ASCE, St. Louis, MO, pp. 76–107.

[9] Sullivan, R. A., "Platform Site Investigation," *Civil Engineering,* London, Feb. 1978, pp. 26–33, 45.

[10] Marr, L. S. and Endley, S. N., "Offshore Geotechnical Investigation Using Cone Penetrometer," *Proceedings, 14th Offshore Technology Conference,* Vol. 2, 1982, pp. 783–797.

[11] Ladd, C. C. and Foott, R., "New Design Procedure for Stability of Soft Clays," *Journal, Geotechnical Engineering Division,* ASCE, Vol. 100, No. GT7, 1974, pp. 763–786.

[12] Emrich, W. J., "Performance Study of Soil Sampler Deep-Penetration Marine Borings," *Sampling of Soil and Rock, STP 483,* American Society for Testing and Materials, Philadelphia, 1971, pp. 30–50.

[13] Young, A. G., Quiros, G. W., and Ehlers, C. J., "Effects of Offshore Sampling and Testing on Undrained Soil Shear Strength," *Proceedings, 15th Offshore Technology Conference, Houston,* Vol. 1, 1983, pp. 193–204.

Joost M. Geise,[1] *John ten Hoope,*[1] and *Robert E. May*[2]

Design and Offshore Experience with an In-Situ Vane

REFERENCE: Geise, J. M., Hoope, J. ten, and May, R. E., **"Design and Offshore Experience with an In-Situ Vane,"** *Vane Shear Strength Testing in Soils: Field and Laboratory Studies, ASTM STP 1014,* A. F. Richards, Ed., American Society for Testing and Materials, Philadelphia, 1988, pp. 318–338.

ABSTRACT: The design, development, and use offshore of a commercial in-situ vane, operated from a carrier tool downhole or from a seabed jacking system, is described. This vane is inserted into soil at a rate of 20 mm/s for both deployment modes. The vane is rotated at a constant speed of either 0.1 or 0.2°/s in cohesive soils having a maximum shear strength of 200 kPa. Using the hard-tie drill string heave compensating system, the first downhole vane test may be within about 0.75 m of the borehole base in very soft soils; if the hard-tie is not used, then the first test should be about 1.5 m below the borehole base to avoid soil disturbed by drilling. Standards developed for onshore vane testing can be used for offshore vane testing with few modifications.

KEY WORDS: drilling, field tests, instrumentation, shear strength, vane testing, ocean soil, standards

This paper describes the development and subsequent use of an IN-SITU VANE designed primarily for testing clays offshore. The equipment must be capable of remote operation from either a seabed frame or in deep boreholes. It is therefore more complex than in-situ vanes used on land. An offshore vane requires the reliability and robustness of its onshore equivalent while operating under the demanding offshore conditions.

In the early 1970s a wireline in-situ vane was developed by McClelland Engineers [*1*] for commercial use in offshore site investigations. The operation of this remotely controlled vane proved to be successful, and results comparable to conventional field vane tests were obtained. At about the same time Fugro developed wireline equipment for cone penetration testing (CPT) from the drill string as well as remotely operated seabed CPT equipment [*2*]. It was clear that an in-situ vane would represent an important addition to the company's in-situ testing capabilities. In 1976 the IN-SITU VANE was used offshore for the first time. In the early 1980s the tool was redesigned for two modes of operation, namely, downhole and seabed to be fully compatible with other equipment then in use. The downhole operational mode has proved to be more popular with the first deployment from a seabed frame occurring in 1983. In recent years the demand for offshore IN-SITU VANE testing has grown worldwide as increasing numbers of site investigations have been in soft soils.

This paper describes the design considerations for the new offshore vane. A variety of national and other standards have been specified for vane testing. These were reviewed

[1] Project manager and staff engineer, respectively, Fugro Geotechnical Engineers B.V., P.O. Box 63, 2260 AB Leidschendam, The Netherlands.

[2] Staff engineer, Fugro Ltd., 18 Frogmore Rd., Hemel Hempstead HP3 9RT, United Kingdom.

before examining the practical requirements of the instrument. The second section describes the design and prototype evaluation of the new vane. Finally, a decade of vane testing is reviewed, and conclusions are drawn about offshore performance.

Required Instrument Capabilities

Vane Testing Standards

The primary requirement of the new instrument was to meet the various national and company standards. The American, British, and Norwegian standards for onshore field vanes are compared in Table 1. Although there is considerable agreement between these standards, differences exist over vane dimensions, required depth of vane insertion below the bottom of a borehole, rates of vane rotation, and the procedure for measuring remolded shear strength. There are no standards for offshore vane testing. Various Norwegian oil companies use a technical specification for vane testing that follows the Norwegian (NGF) standard with two exceptions:

1. A smaller blade, 40 mm in diameter by 1.5 mm thick, may be used in addition to the two standard blades.
2. Two or three complete revolutions are sufficient before performing a remolded test.

To meet this range of requirements, it was clear that the blades should be interchangeable. The instrument would have to operate at different speeds and be capable of multiple

TABLE 1—*Summary of National Vane standards and Fugro standard.*

Parameters	ASTM [3]	BS[a]	NGF [4]	Fugro
Vane Geometry	rectangular/ tapered	rectangular	rectangular	rectangular/ tapered
height to diameter ratio	2	2	2	2
vane blade diameter, mm	38.1/50.8 63.5/92.1	50/75	55/65	Table 2
thickness of blade, mm	1.6 3.2		2.0	Table 2
diameter of vane rod, mm	12.7	13.0	12.0	Table 2
accuracy of torque reading	±1.20 kPa	1% of range (0 to 700 Nm)	±0.5% of maximum range	±0.5% of maximum range
drive of vane	geared drive preferred	geared drive	geared drive preferred	geared drive
area ratio	<12%	<12%	<12%	Table 2
Procedure				
depth of insertion	5x hole diameter	3x hole diameter	0.5 m	>0.75 m
rate of rotation	6°/min	6 to 12°/min	12°/min	6°/min.
time of failure	2–5 min	5 min	1 to 3 min	. . .
remolded shear strength minimum revolutions	10	6	25	2 to 3
delay time	none or <1 min	5 min	<5 min?	none
Intervals between tests	>0.76 m	0.5 m	0.5 to 1.0 m?	0.5 to 0.75 m

[a] British Standard Code Practice for Site Investigation. (BS 1377 and BS 5930).

TABLE 2—*Fugro vane blade dimensions.*

	Rectangular						Tapered		
Parameters	A1	A2	A3	G1	G2	G3	A1	A2	A3
Height to diameter ratio	2	2	2	2	2	2	2	2	2
Vane blade diameter, mm	38.1	50.8	63.5	40.0	55.0	65.0	38.1	50.8	63.5
Thickness of blade, mm	1.6	1.6	2.0	1.5	1.5	1.5	1.6	1.6	2.0
Diameter of vane rod, mm	13.0	13.0	13.0	13.0	13.0	13.0	13.0	13.0	13.0
Area ratio	18.7	12.5	11.2	17.0	10.9	8.7	18.7	12.5	11.2

rotations. A torque cell was required with a performance high enough to allow for any future increase in torque capability. The system had to be capable of pushing the blades to a depth of 1.5 m below the base of a borehole. This last requirement arose from the ASTM specification combined with a nominal 300-mm-diameter hole drilled with Fugro's conventional 8-in. (200-mm) diameter drag bit. Although the standards provide for a number of revolutions required before a remolded test is performed, they do not specify at which rotation speed this should be done. For economical reasons it was decided to incorporate the ability to rotate the vane blades at a speed of 1.0°/s to reduce the time required for two or three revolutions.

Operating Requirements

In addition to meeting the standards, various other requirements were identified for the instrument as follows

1. It should be capable of operation both from a drill string and from a seabed jacking frame (Figs. 1 and 2). The first of these requirements limited the maximum instrument diameter to about 64 mm to allow for clearances inside the drill string and to be accommodated in the downhole carrier tool. In the seabed mode the vane has to operate at a substantial depth below the seabed jacking frame. Adequate torsional reaction cannot be guaranteed by the test rods under these conditions. These conditions therefore dictated that the instrument itself should provide the torsional reaction for the rotating vane blades. A substantial depth of penetration from a seabed frame also necessitated the inclusion of an inclinometer. Reliable equipment for pushing cone penetrometers into the soil from drill strings and seabed frames had already been developed by the company. It was therefore logical to make the new vane equipment compatible with the existing CPT systems.
2. Additional measures were considered necessary to minimize soil disturbance before the vane test. For both modes of deployment it was decided to jack the vane smoothly into the soil at a rate of 20 mm/s. For borehole deployment it was considered important that the drill bit should not be raised from the bottom of the borehole at any stage before the test. This should minimize the risk of borehole base failure in soft soils and prevent granular debris from higher strata clogging the base. Concern was also expressed about the difficulty of having adequate heave compensation for the drill string when drilling to the test horizon through very soft clays. This problem is discussed later in the paper.
3. Clearly the instrument had to be robust enough to survive offshore handling. Two additional requirements were specified. The instrument should be capable of testing soils with a shear strength of up to 200 kPa, and it should be able to operate in 1500 m of water.

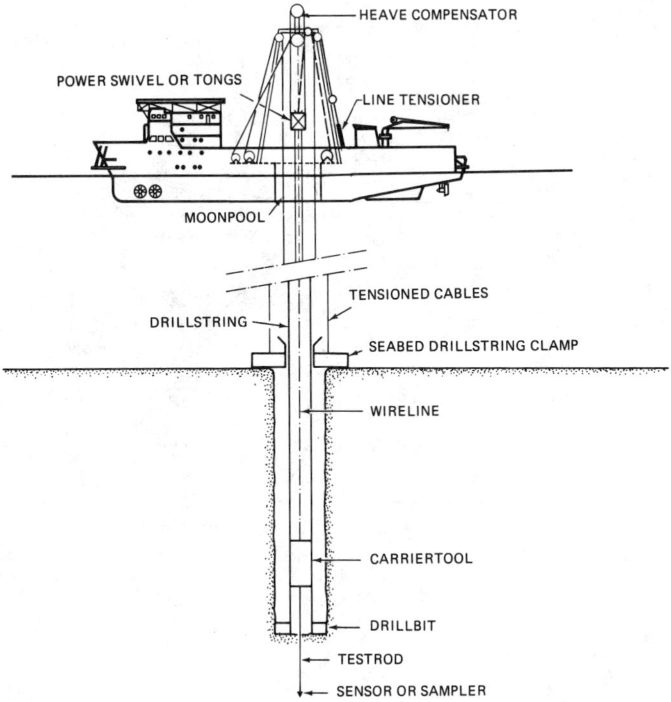

FIG. 1—*Schematic diagram of some major elements of the drill ship-wireline method of in-situ testing* [8].

Design of the In-Situ Vane

Based on these requirements, the new prototype vane was designed and built. This probe was rigorously evaluated, and after minor modifications it was ready for offshore employment.

Key Design Elements

Figure 3 shows the key elements of the vane unit, which has a total length of 1360 mm and a diameter of 64 mm. Element details follow:

1. The vane blade is interchangeable so that different types may be attached. Table 2 lists the currently available types of vane blades that conform to ASTM Method for Field Vane Shear Test in Cohesive Soil (D 2573) (Types A1 to A3, rectangular; and A1 to A3, tapered) and the NGF standard (Types G2 and G3, rectangular). Although the vane blade Types A1 (rectangular and tapered) and G1 (rectangular) have area ratios well above the recommended value of 12% by the standards, they were included in the selection because these types are specified by ASTM (both Types A1) and the clients (Type G1).
2. A reaction vane to provide the necessary torque reaction while performing a test. The four reaction blades form an integral part of the probe body and have a total length of 475 mm. This is about 3.6 times the height of the largest vane blade, Type G3 (rectangular).

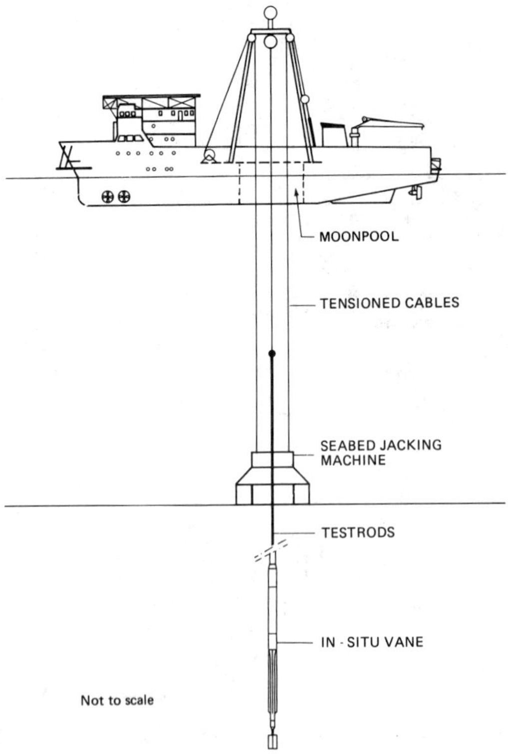

FIG. 2—*Schematic diagram of some major elements of the seabed in-situ testing system* [8].

The reason for such long reaction blades is to minimize rotation to mobilize the torsional reaction.

3. The torque measuring element is a torque cell with strain gages. The technique to measure load using the strain gage technique has been used by Fugro for many years, especially in electric cone penetrometers [5].

4. To rotate the vane blade, a small electric motor was chosen with sufficient power to test soils having an undrained shear strength of 200 kPa when using the smallest vane blade. Since such a small electric motor does not provide the required torque of 50 Nm at a rotation speed of 0.1 or 0.2°/s, a gear box is used to allow the motor to run at a higher speed.

5. To ensure that the rotation speed is kept constant, a tacho generator is used. The output from the tacho generator is compared to a reference signal by an electronic control system. This monitors the difference between the two signals and automatically adjusts the power supply to the motor to keep the rotation speed constant. The output of the tacho generator is also used to determine the rotation of the vane blades during a test. The system design prevents errors in the rotation measurement caused by internal slippage.

6. An inclinometer was considered essential for the seabed mode of operation based on Fugro's cone penetration testing experience. When pushing the IN-SITU VANE to a substantial depth below the seabed, the tool may be deviated from the vertical by changes in soil consistency, or by the presence of sand or stiff clay layers in a soft clay stratum.

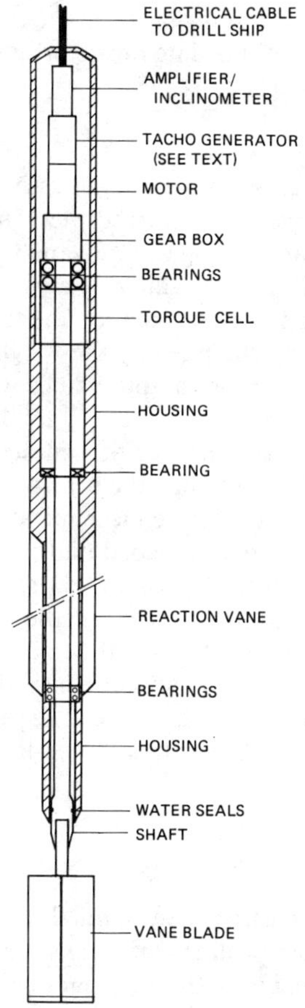

FIG. 3—*Key elements of the Fugro in-situ vane.*

7. The measured signals of the torque cell are transmitted from the IN-SITU VANE to the drill ship. Transmission lines become very long with increasing water depth and thus are likely to influence the quality of the signals. To overcome this problem the signals are amplified in the IN-SITU VANE before they are transmitted to the surface.

A control unit was designed and built so that the vane could be remotely operated from the drill ship. With the unit the test speed can be selected and maintained throughout the test. Displays are available for torque readout, angle of rotation, rotation speed, and inclination.

Data Acquisition

Data results are displayed on the remote control unit and are recorded by a mini-computer and on dual-pen strip-chart recorders. Two recorders are used, one recording the torque and rotation speed versus time and the other the torque versus rotation. The mini-computer displays the shear strength versus rotation. The digitized measurements are stored on magnetic media (either disk or tape) by the mini-computer system. This facilitates easy processing and editing of the data onsite or in the office.

Calibration

The torque cell calibration was performed on an assembled vane, which was fixed horizontally. A circular plate on a steel bar replaced the vane blade. Attached to the circumference of the plate was a wire and a weight. The vane was then switched on at a rotation speed of 0.1°/s and allowed to lift the weight. The signals of the torque cell were recorded continuously on a *x-y* plotter and after 90° rotation the test was stopped. The mean output was then compared with the torque applied by the weight (weight × radius of the plate). By using different weights a set of mean output values was obtained. The calibration line was computed by linear regression.

The mechanical friction of a vane may be of significance when testing very soft sediments. The mechanical friction is caused by the bearings of the driving shaft and the seals used to keep the instrument water tight. The mechanical friction was checked through performance of a vane test with the tool unloaded and in a vertical orientation. The result prompted a reduction in the number of O-ring water seals from three to one and replacement of the original bearings with low friction needle bearings. The mechanical friction was thus reduced to a value between 0.8 and 1.0 Nm.

Tests were included during the initial calibration period to check the mechanical stiffness and strength of the rotating parts from the top of the system (motor) to the bottom (vane blades). It was concluded from these tests that the design objectives had been met and no further modifications were required.

Accuracy

The accuracy of the torque measurement, and thus the shear strength measurement, depends mainly on the bearings, sealings and transmission of the torque by the vane's drive shaft. The torque cell properties are important as well, but the inherent inaccuracy can be kept small if a transducer of sufficient quality is used. Such a transducer will be very stiff, have minimal hysteresis, have a high output to load factor, and will be fully temperature compensated over a range of 0 to 30°C. In addition the excitation voltage must be precisely controlled. The major uncertainties in the system can be measured during the calibration of the assembled tool. They consist of an error at zero load, a calibration error, and a load error. The mechanical friction (zero load) should be determined in the field before a test or a series of tests. This is accomplished by rotating the vane blades in the water close to the drill bit when deploying the vane downhole, or just above the seabed in the seabed mode. The calibration error is the standard deviation from the calibration line. The load error is visualized during calibration as the fluctuation around the mean signal. The maximum error of the torque measurement was calculated at 0.5% of the maximum torque of 50 Nm. Independent measurements were made of the vane's angular rotation speed. These showed that the speed could be maintained constant to within 4% over the full range of speed settings with torques varying up to 50 Nm.

Pressure Testing

Upon completion of the calibrations, the entire unit was placed in a water filled pressure vessel. The water pressure was maintained at 3 MPa for a period of 48 h. The vane was occasionally switched on during this period and allowed to rotate for periods of about 10 min. The instrument was then stripped down and inspected. No evidence of leakage past the static or rotating seals could be found.

Onshore Tests

The offshore vane lends itself also for use on a cone truck, and this ability was used to run an extensive onshore test program with the prototype instrument. The selected site comprised a layer of loose to medium dense sand of about 1 m thickness overlying soft clay. Tests in the clay were carried out between a depth of 1.5 and 5 m, and a typical result is presented (Fig. 4). During the test program, no operational problems were found, and hence the tool was ready for offshore use.

Offshore Experience

In this section a description is given of the operation of the equipment. The experience gained over a decade of testing in a variety of offshore soils is then reviewed.

Operational Procedures

Operational procedures were written to ensure proper offshore vane shear strength testing by field crews. These procedures comprise handling procedures and testing procedures. The handling procedures, besides providing instructions on how to handle the equipment, include essential operations that have to be performed to record the data necessary for the interpretation of the vane test, and also provide a means for the field crew to assess the operational condition of the IN-SITU VANE. The testing procedures comprise the instructions for an equipment operator to perform the tests. The standard test procedure is to rotate the vane blades at 0.1°/s until the required rotation is achieved. When a remolded test is not required, the torque is relieved from the blades by turning the vane blades in the opposite direction for a maximum of 5°. If a remolded test is required, then at least two full rotations are made at 1.0°/s, before the remolded test is performed at 0.1°/s.

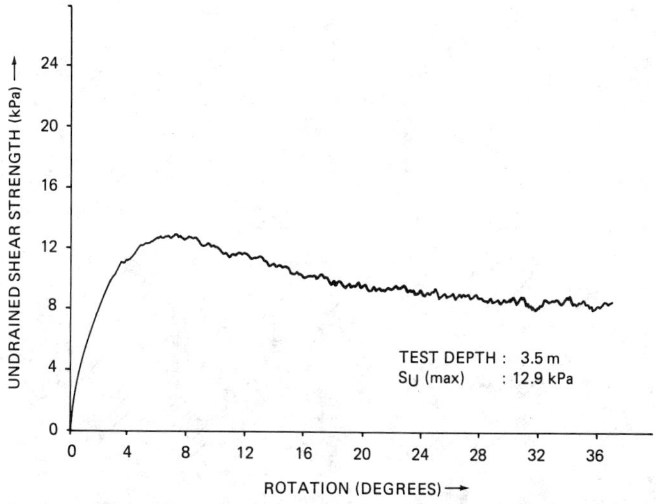

FIG. 4—*Onshore test result.*

Downhole Mode

The IN-SITU VANE can easily be connected to a carrier tool for downhole testing (Fig. 5). The drill bit is set on the bottom of the borehole, and the drill string immobilized at seabed using the SEACLAM (Fig. 1). This will minimize lateral movement of the vane when it is being pushed into the soil. Then the carrier tool is lowered on an umbilical cable through the drill string down to the high latching position (Fig. 5*a*). The vane blades are rotated for about 5 min with the vane held in this position. Next the carrier tool is lowered to the drill bit, where it automatically latches into the grooves of a special drill collar, which is called the bottom hole assembly (Fig. 5*b*). Then penetration to the first test depth begins, and the vane is pushed in a controlled fashion out of the carrier tool and into the soil beneath the bottom of the borehole at a rate of 20 mm/s (Fig. 5*c*). The maximum insertion depth is 1.5 m (Fig. 5*d*). Ample reaction for pushing the vane into the soil is provided by the SEACLAM. Subsequently a test is performed following the testing procedures as described above.

The maximum available thrust is 80 kN, and the operator obtains a continuous readout

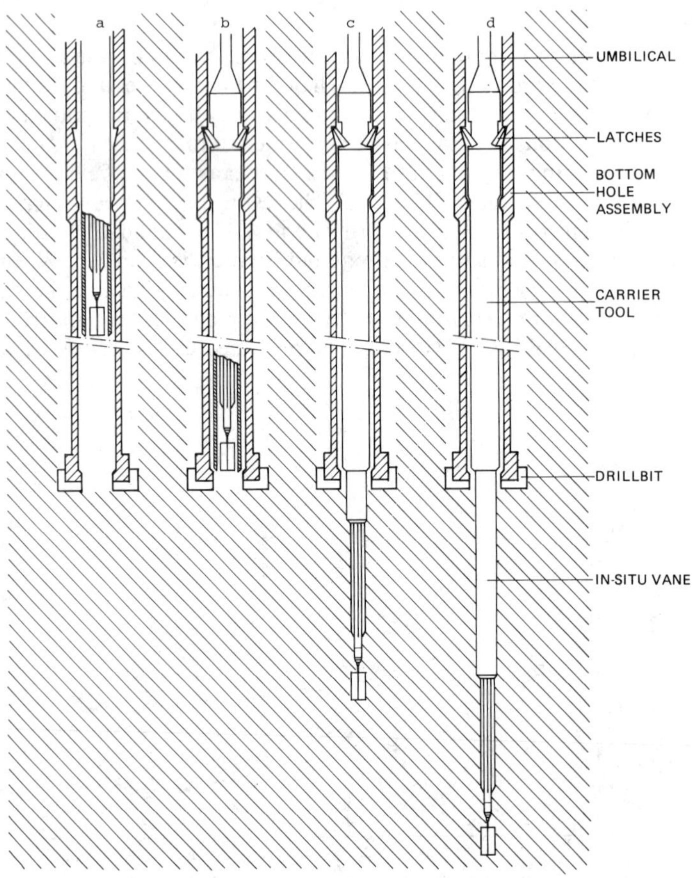

FIG. 5—*Schematic diagram of the four steps in downhole vane shear testing, described in the text.*

of the applied thrust by means of the oil pressure measured in the carrier tool. The carrier tool also measures very accurately the distance of penetration of the vane blades. With these two measurements the operator has excellent control over the insertion of the vane blades into the soil.

Seabed Mode

In IN-SITU VANE may be connected to a string of test rods and pushed into the seabed using the SEACALF (Fig. 2) or other seabed jacking devices. The vane design accommodates a penetration of about 25 m in normally consolidated clay. The limiting factor for greater depths and for other soils is the buckling potential of the test rods caused by insufficient lateral support. The available thrust of a seabed jacking device, which is generally about 100 to 200 kN, will thus not be utilized entirely.

When the test rod string has been assembled and the vane connected to the rods, the seabed jacking frame is lowered to the seabed. The vane blades are rotated for about 5 min before reaching the seabed to determine the zero reference. The seabed frame is then set on seabed, and the tilt of the frame can be observed from the dual-axis tilt meters. When the frame sits sufficiently level (less than 5° tilt), the vane is pushed into the soil to the required test depth. Upon reaching the test depth a vane test is performed. Vane tests can be performed at any desired depth through a repetition of the pushing and testing cycles.

More recently, in 1987, a small, lightweight seabed template, called SEABED VANE, has been introduced. The 1.2-m^2 template shown in Fig. 6 clamps the IN-SITU VANE for testing at a selected penetration below its base, the maximum penetration being in the order of 4 to 5 m. The unit uses a solid state memory and a programmable logic controller to perform an undisturbed and a remolded test at a location. Data from up to nine locations can be stored into memory before the unit has to be brought back to the surface.

Control of Test Quality

The measured data from the IN-SITU VANE as well as the applied thrust and the penetration of the vane blades are monitored in real time onboard the drill ship for both modes of operation except when used with the SEABED VANE. This means that the quality of the test can be continuously controlled. For example, the applied thrust may be compared with CPT data from adjacent locations to control and optimize the penetration of the vane blades and ensure it is placed in the desired soil layer. Also, from observation of the torque-rotation data, it can be judged immediately whether or not the test result agrees with expectations (again based on previous CPT or vane tests, or sampling, and so forth). It can be then decided whether a test may be terminated prematurely and a repeat test is desirable. This degree of control ensures that a high success rate, that is, number of good tests over total number of tests, can be achieved.

Offshore Results

The use of the IN-SITU VANE was mainly in the Gulf of Mexico until the early 1980s. In recent years, it has frequently been deployed in other areas, especially the North Sea and Arabian Sea offshore India. This increased use of the vane shear strength test may be explained by the fact that new oil and gas discoveries in the North Sea were located in deep water where the seabed soils consisted of very soft clays in the upper 25 to 50 m. In India, there has been a growing interest in in-situ shear strength testing, because of the presence of thick seabed strata with very soft to stiff silts and clays at the newer oil and gas fields.

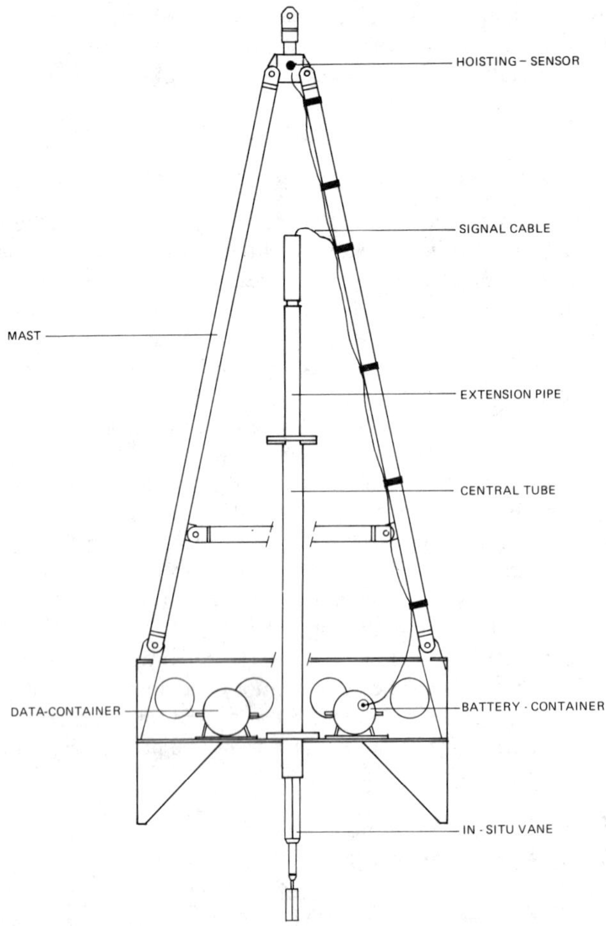

FIG. 6—*Seabed vane.*

Downhole Vane Shear Strength Testing

The majority of the tests performed have been made deploying the vane in its downhole mode. Water depths at the test sites have varied between 10 and 340 m, while the shear strengths of the soils tested have varied between about 2 and 100 kPa. Frequently clients specified in their sampling and testing schedule that vane tests were to be performed at two depths below the drill bit. These depths are generally 0.75 and 1.5 m below the bit. This is in conflict with the earlier stated requirement from the ASTM standard. This problem is discussed in a following section of this paper. Typical results are shown in Fig. 7. These tests followed the procedures as described earlier, using a rectangular vane blade with a 63.5 mm diameter (Type A3) and a rotation speed of 0.1°/s. The determination of the mechanical friction resulted in average value of 0.8 Nm (Fig. 7), which has been taken into account when computing the undrained shear strength. The computed shear strengths were 19 and 22 kPa for Tests 1 and 2, respectively.

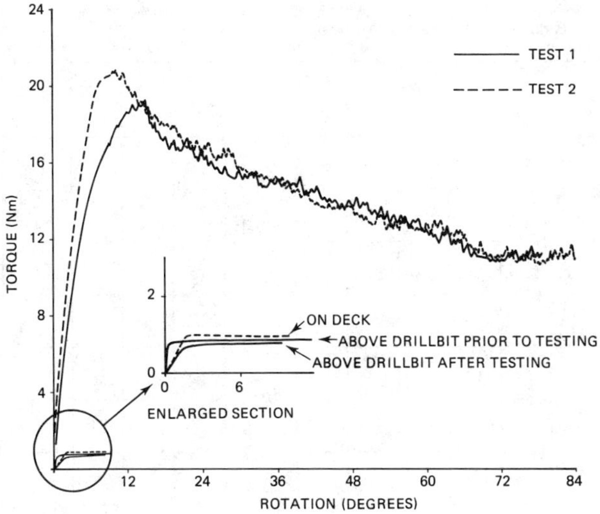

FIG. 7—*Downhole test result: drill bit at 12.6 m and test depths of 13.4 and 14.1 m for Tests 1 and 2, respectively.*

Seabed Vane Shear Strength Testing

The first commercial vane test from the seabed using a remotely operated seabed jacking device was performed in 1983. The water depth at the North Sea site was about 340 m, and the seabed soils consisted of very soft to soft silty clays to a depth of 25 m (Fig. 8), making the site ideal for vane tests to be performed from the seabed. The SEACALF [*6*] pushed the IN-SITU VANE to a final penetration depth of 22 m at the first location and 16 m at the second location. The distance between the two locations was 400 m. At location one, tests were performed every 0.5 m between about 2 and 11 m penetration and every 1 m thereafter, whereas at location 2 all tests were performed at 1 m intervals. Occasionally a remolded test was carried out at both locations. Testing at both locations began at a 2-m penetration depth below the mudline as the seabed frame was fitted with 1-m skirts for to ensure stability. In total, 43 undisturbed and 8 remolded tests were performed. Figure 8 presents a summary of the vane tests at locations one and two plotted as undrained shear strength versus depth. All tests were performed using a rectangular vane blade with a 50.8 mm diameter (Fugro Type A2) and a rotation speed of 0.1°/s. Before a remolded test was carried out, the vane blades were rotated over 720° using a rotation speed of 1.0°/s. The remolded test itself was performed at a rotation speed of 0.1°/s. The computed sensitivities S_t from the undisturbed and remolded tests performed at the same depth have been listed alongside of the vane data in Fig. 8. Correlation lines have been computed for the undisturbed and remolded test data, and the resulting equations have been included in Fig. 8 where z represents the depth in metres and s_u the undrained shear strength in kPa. The results of a SEACALF piezocone test (PCPT) about midway between the two seabed vane test locations is included in this figure. The results of the PCPT show that the cone resistance, sleeve friction, and pore pressure profiles increase linearly with depth. No distinctive soil or strength changes can be observed. The homogeneity of the clay layer was also confirmed by sampling in the boreholes and other SEACALF PCPTs carried out at the same

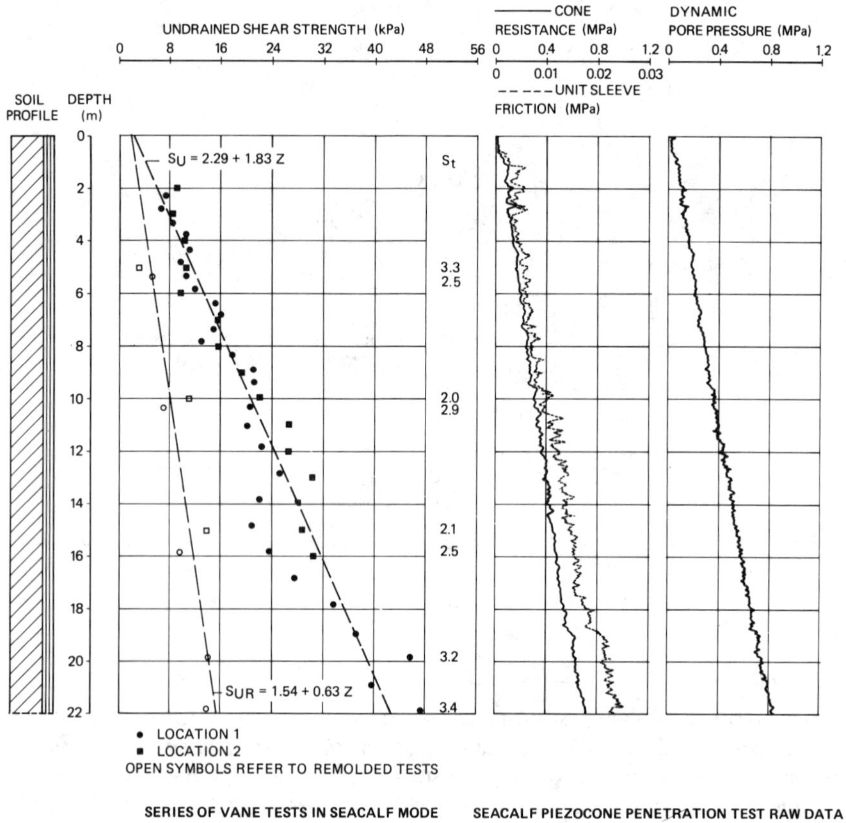

FIG. 8—*Summary of seabed vane and PCPT tests.* S_t = *sensitivity, see text.*

site. The reduction in vane shear strength between 13 and 18 m at Location 1 may partially be attributed to the slightly lower liquid limits measured at this depth on adjacent borehole specimens. The influence of the physical properties is investigated in more detail in a companion paper by Kolk et al.[3]

Soil Disturbance

There are many factors affecting the quality of in-situ testing. The obvious problems center around disturbance to the soil caused by drilling and also by in-situ testing itself [*7*]. The following discussion is based on these factors in light of Fugro's experience.

The disturbance caused by drilling will cause a degradation of the soil strength. Depending on the extension of the disturbed zone below the drill bit the vane shear test result will be affected. To minimize this disturbance of the soils directly below the drill bit, the North Sea type of drill ship is equipped with heave compensators [*8*].

To activate the drill string heave compensator and have it operating smoothly a certain bit load is required. This load is typically between 5 and 10 kN depending on the design and condition of the heave compensating system. If drilling disturbance is to be mini-

[3] Kolk et al., in this publication, pp. 339–353.

mized, the bit load must not exceed the bearing capacity of the soil. In soft soils the bearing capacity is inadequate, and consequently the drill bit will move in a slurry. The soil disturbance has been studied by the authors through a correlation of the ratio of the peak and post-peak shear strength measured at different depths below the borehole base in a normally consolidated clay. The peak shear strength is defined at the highest shear strength determined while the post-peak shear strength is the lower post-peak shear strength measured over a constant interval of approximately 10° rotation after the vane has rotated not less than 40°. The vane tests have been divided into two categories. The first covers the zone of soil less than 1 m below the drill bit, where tests generally were performed about 0.8 m below the bit. The second category comprises tests in excess of 1 m below the drill bit, with tests generally performed at 1.5 m below the bit. Figure 9 shows the ratio of peak to post-peak strengths for both categories of tests. In the lower part of Fig. 9 the ratios are plotted against depth. The vane tests performed at more than 1 m below the bit exhibit appreciably higher peak/post-peak strength ratios than their shallower counterparts for depths less than 30 m. The same data are plotted in the upper part of Fig. 9 against peak shear strength, showing that a peak shear strength of about 20 kPa was required before the ratio for the two categories of tests was similar. It may therefore be concluded that the soil in the zone within 1 m below the bit is subject to appreciable disturbance for peak shear strengths less than about 20 kPa.

It is likely that the disturbance is primarily caused by the heave of the drill bit and is dependent on the soil strength. Thus drilling in very soft soils using a drilling system even with a good heave compensator will cause a degradation of the soil strength caused by the

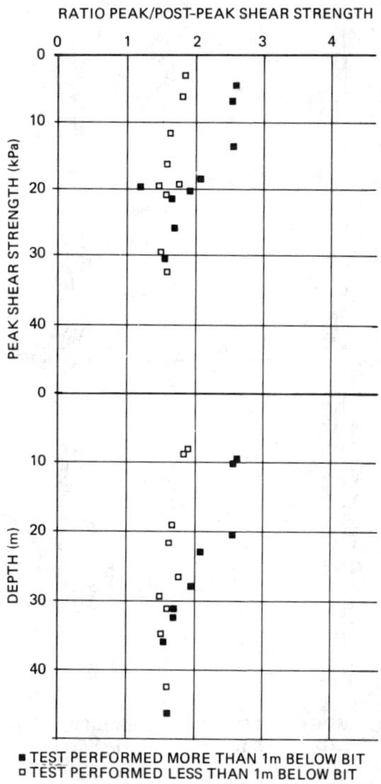

FIG. 9—*Analysis of disturbance caused by drilling. See text for discussion.*

drill bit heave, which results in a lower vane shear strength. Figure 10 (left) shows the principles of a conventional heave compensation system including the seabed frame (SEA-CLAM). To ensure a properly working drill string heave compensator in soft soils, Fugro engineers in 1983 invented the hard-tie system (Fig. 10, right) to overcome the problem of insufficient drill bit reaction [*9,10*]. In this method, the drill string and seabed frame are "tied" together ensuring that the drill string heave compensator will be activated and the motion of the respective heave compensators (drill string and seabed frame) will be synchronized in the same direction. By this means, drilling is performed in a displacement-controlled mode rather than the usual force-controlled mode.

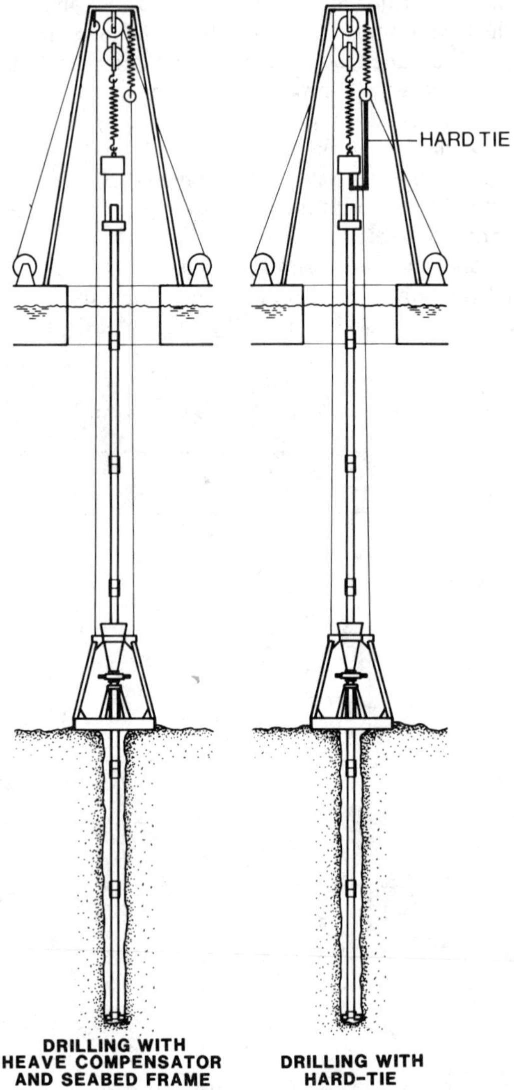

FIG. 10—*Conventional heave compensation* (left) *and the hard-tie system* (right). *See text for discussion.*

To investigate the reduction of drilling disturbance through use of a hard-tie system, the authors have made a similar correlation using vane test data from a site where the boreholes were drilled employing the hard-tie system. It must be noted that sea states at this site were generally poorer than those found at the site of previous correlation. Again the ratio of the peak and post-peak shear strength has been computed for the two above described categories. Figure 11 shows that no clear distinction can be made between tests from either category. It may therefore be concluded that drilling disturbance in very soft soils is minimized through use of a hard-tie system, and that when such a system is used that high-quality vane tests may be performed in the zone of soil less than 1 m below the bit. As high-quality vane tests can be performed in this zone one may assume with reasonable confidence that high-quality soil specimens can also be obtained in this zone when using the hard-tie system.

Further study of first data set provided some interesting aspects with respect to the influence of the drilling disturbance on the vane shear test. The authors suspected that the shape of the torque-rotation curve could also be affected by the drilling disturbance. A reaction vane located in the disturbed zone would require more rotation to mobilize the torsional

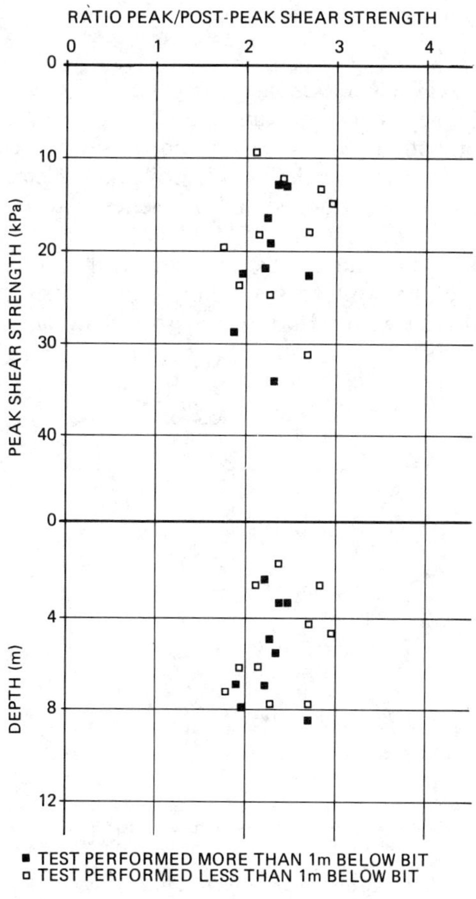

FIG. 11—*Analysis of reduced drilling disturbance through use of the hard-tie system. See text for discussion.*

reaction than if it were placed in an undisturbed zone. The reaction vane provides the reference point for the rotation measurement. Consequently the initial slope of the torque-rotation curve will be smaller because the peak value will be measured at a greater rotation value. Figure 12 shows the initial slope of the torque-rotation curve plotted against depth. Below about 30 m, the initial slopes for both categories are comparable. This confirms the conclusion from the ratio of the peak to post-peak shear strength; namely that below this depth the soil within 1 m of the bit has not severely been disturbed by the drilling. It is clear that the use of a hard-tie system will also ensure that comparable initial slopes of the torque-rotation curves are obtained for vane tests performed less than 1 m below the drill bit and tests performed more than 1 m below the bit.

So far torque-rotation curves have not been used for engineering purposes because they do not represent a stress-strain relationship similar to the result of a triaxial test. In this test the soil stiffness may be computed from the initial slope of the stress-strain curve. This is not the case for the torque-rotation curve. Therefore it is considered uneconomical at the present time to engineer a modification of the present in-situ vane design to allow the vane to clamp itself to the drillstring to provide the torsional reaction and thus to reduce unwanted relative rotation of the vane with respect to its reference point.

Downhole versus Seabed Testing

Not many data were available to compare vane tests performed in both downhole and the seabed modes, since the two methods are generally used as alternatives. At one site in the North Sea, however, both modes were used to perform tests in a clay layer occurring between about 2 and 10 m. Three boreholes were drilled in which 22 undisturbed and six remolded downhole tests were performed and one series of seven undisturbed and two remolded tests were performed using the SEACALF. The hard-tie system was employed while drilling the boreholes to minimize disturbance caused by drilling. This made the site ideal for the evaluation of the two modes, since the seabed tests should not be affected by drilling disturbance. All tests were performed using a rectangular vane blade of 50.8 mm

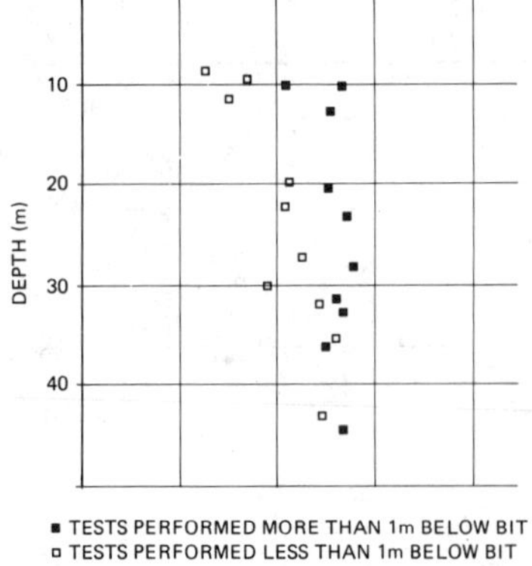

FIG. 12—*Influence of drilling disturbance on the initial slope of the torque (shear strength) rotation curve.*

diameter (Type A2) and a rotation speed of 0.1°/s. The remolded tests were made similar to the remolded tests described previously. The sensitivities derived from the undisturbed and remolded tests performed at the same depth are tabulated in Fig. 13. They vary between 2.5 and 4.8. The undrained shear strengths measured in the vane tests are also plotted in Fig. 13. No significant differences between the results from the two modes can be observed at this location. This demonstrates again the significant improvement by the hard-tie system on the drilling disturbance since downhole vane tests performed in the zone of soil less than 1 m below the bit, in this particular case at 0.75 m, yielded similar results as the seabed tests and the downhole tests performed 1.5 m below the bit.

Recommendations Relative to the ASTM Field Vane Standard

1. Particular precautions are required if high-quality results are to be obtained in soft to very soft cohesive soils when a conventional heave compensation system is used. For soils with shear strengths of less than about 20 kPa, insufficient bit reaction is available to allow proper heave compensation of the drill string during drilling. The authors recommend that in such soils a hard-tie system be used for drilling. Vane tests can then be performed at 1.0 and 1.5 m below the drill bit. Although the test at 1.0 m is in conflict with the ASTM standard, the authors are confident that with the hard-tie system high-quality tests can be performed at that penetration depth.

2. In the event a hard-tie system is not used for drilling, vane tests should only be performed at 1.5 m (five times the borehole diameter of 300 mm when using a conventional 8-in., 200-mm, diameter drag bit) below the bit for soils having a shear strength of less than 20 kPa. Occasionally a vane test may be performed at a smaller distance from the bit. The

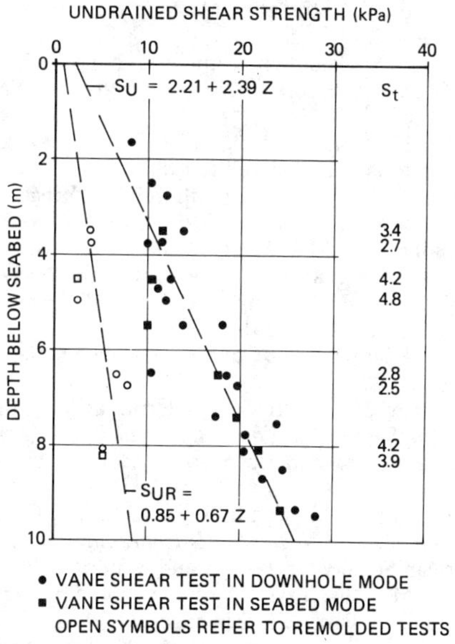

FIG. 13—*Comparison of vane tests performed in the downhole mode and the seabed mode.*

result of such a test may be used to evaluate the quality of the soil specimens obtained in the zone directly below the bit.

3. For remolded tests offshore, two, or possibly three, revolutions at a rotation speed of 1.0°/s are considered sufficient for economic reasons. For the practice of commercial vane testing downhole or using seabed jacking machines, this procedure can be expected to yield a sufficiently low remolded shear strength for a reasonable estimate of soil sensitivity.

4. The use of vane blade Types A1 (both rectangular and tapered), which conform to the ASTM onshore vane standard, is not recommended in view of the high area ratios of both vane blade types. The same applies to vane blade Type G1 (rectangular). The authors recommended that these blades be redesigned to meet the requirement by all three standards [*3,4*] that the area ratio be less than 12%. It should, however, be noted that these blades most often will be used in very stiff clays and that sufficient strength and stiffness should be incorporated into a new design.

5. Finally, ASTM D 2573 field vane standard for onshore testing is satisfactory for offshore testing without further modification at this time. The procedures for quality offshore vane testing, however, should conform to the above recommendations whenever they are applicable.

Acknowledgments

We appreciate having the permission of ONGC and Statoil to use certain data contained in this paper. A. F. Richards, H. J. Kolk, and N. J. Withers reviewed the manuscript and made a number of constructive comments. D. Eijmaal drafted the new illustrations.

References

[*1*] Doyle, E. H., McClelland, B., and Ferguson, G. H., "Wire-Line Vane Probe for Deep Penetration Measurements of Ocean Sediment Strength," *Offshore Technology Conference Proceedings,* Vol. 1, Paper 1327, 1971, pp. 21–32.

[*2*] Zuidberg, H. M., "Use of Static Cone Penetrometer Testing in the North Sea," *European Symposium on Penetration Testing Proceedings, Stockholm,* Vol. 2.2, National Building Researching, Stockholm, Sweden, 1976, pp. 433–436.

[*3*] *1986 Annual Book of ASTM Standards,* Vol. 04.08, Soil and Rock; Building Stones, American Society for Testing and Materials, Philadelphia, 1986.

[*4*] Norsk Geotekniske Forening, "Veiledning for Uflførelse av Vingeboring," NGF Medling, No. 4, 1982 (available from the Norwegian Geotechnical Institute, Oslo, Norway).

[*5*] Schaap, L. H. J. and Zuidberg, H. M., "Mechanical and Electrical Aspects of the Electric Cone Penetrometer Tip," *Penetration Testing, Proceedings of the Second European Symposium on Penetration Testing,* A. Verruijt, F. L. Beringen, and E. H. de Leeuw, Eds., Vol. 2, A. A. Balkema, Rotterdam, The Netherlands, 1982, pp. 841–852.

[*6*] Zuidberg, H. M., "Seacalf: A Submersible Cone-Penetrometer Rig," *Marine Geotechnology,* Vol. 1, No. 1, 1975, pp. 15–32.

[*7*] Richards, A. F. and Zuidberg, H. M., "In-Situ Determination of the Strength of Marine Soils," *Strength Testing of Marine Sediments: Laboratory and In-Situ Measurements, STP 883,* R. C. Chaney and M. R. Demars, Eds., American Society for Testing and Materials, Philadelphia, 1985, pp. 11–40.

[*8*] Richards, A. F. and Zuidberg, H. M., "Sampling and In-Situ Geotechnical Investigations Offshore," *Marine Geotechnology and Nearshore/Offshore Structures, STP 923,* R. C. Chaney and H.-Y. Fang, Eds., American Society for Testing and Materials, Philadelphia, 1986, pp. 51–73.

[*9*] Moeyes, G. and Hackley, M., "Soil Investigations in the Troll Area," Offshore Northern Seas Advanced Projects Conference, Paper T6, Stavanger, Norway, 1983, 37 pp.

[*10*] Zuidberg, H. M., Richards, A. F., and Geise, J. M., "Soil Exploration Offshore," *Field Instrumentation and In-Situ Measurements Proceedings, Fourth International Geotechnical Seminar,* Nanyang Technological Institute, Singapore, 1986, pp. 3–11.

Discussion

N. J. Withers[1] *(written discussion)*—The conclusions arising from the papers in Part VII of this STP, the panel discussion at the symposium, and this paper have much in common. It is perhaps constructive to list the items that need to be addressed by an offshore vane standard. They are offered to stimulate discussion and are based on Fugro's experience with its offshore vane, which is shown in following conclusions.

Experience has shown that repeatable, accurate vane data can be obtained with the following.

- Well designed equipment which
  (1) is operationally convenient
  (2) meets defined specs for accuracy/resolution
  (3) can be deployed in a controlled manner (penetration rate and thrust measured)
  (4) can be used at speeds of 0.1°/sec to 0.2°/sec for tests and 1.0°/sec for 3 revolutions before remoulded test
  (5) can accommodate blades compatible with various standards/specs.

- (For downhole mode) Adequate heave compensation of drillbit using a hard-tie system.

Standard needs to specify acceptable

- Vane dimensions for ranges of C_u, 0 to 50, 50 to 100, and 100 to 200 kPa
- Vane shape factor ($H/D = 2$)
- Vane blade thickness (area ratio $< 12\%$)
- Method of calibration of torque, rotation, rotation speed
- Internal friction w.r.t. full-scale torque (2% FSO)
- Accuracy, resolution of torque (0.5% FSO), rotation, rotation speed (4% FSO)
- Angular velocity of blade (0.1°/s to 0.2°/s for $D = 50 - 65$ mm)
- Maximum rotation or (tip displacement) during pre-post peak test (100°) remolded test
- Definition of peak torque, post peak torque, remolded torque
- Number of revolutions before remolded test (3)
- Speed revolutions before remolded test (1°/s)
- Maximum time to maximum rotation for undrained conditions throughout
- Depth of test (1.5 m unless hard-tie used in downhole mode)
- Test interval (>0.7m)
- Maximum bit pressures, mud pressures as function of expected soil strength to avoid unacceptable disturbance (these can only be advised).
- Parameters to be measured during deployment
  internal friction (a must)
  thrust (desirable)
  penetration rate (desirable)
  penetration (a must)

[1] Fugro Geotechnical Engineers, P.O. Box 63, 2260 AB Leidschendam, The Netherlands.

Internal friction should also be measured after each test
(downhole mode) or series of tests (seabed mode)

- Waiting times for start of test and for between post-peak and remolded phases (<5 min in both cases)
- Standard for presentation of data
- Guidance for interpretation of data (including limitations, corrections factors, and so forth)

Harry J. Kolk,[1] *John ten Hoope,*[1] *and Brian W. Ims*[2]

Evaluation of Offshore In-Situ Vane Test Results

REFERENCE: Kolk, H. J., Hoope, J. ten, and Ims, B. W., **"Evaluation of Offshore In-Situ Vane Test Results,"** *Vane Shear Strength Testing in Soils: Field and Laboratory Studies, ASTM STP 1014,* A. F. Richards, Ed., American Society for Testing and Materials, Philadelphia, 1988, pp. 339–353.

ABSTRACT: An evaluation is presented of results from about 450 tests obtained with an offshore in-situ vane during geotechnical investigations in the North Sea and off the west coast of India. The soils tested are normally consolidated and included silts and clays with differing carbonate contents. The strength values of the in-situ vane are compared with strength results from triaxial tests, laboratory vane, and Torvane tests. These data are also correlated with in-situ test results obtained with the piezocone penetrometer. The in-situ vane shear strength data show a small scatter, and good agreement was found with unconsolidated undrained (UU) triaxial test results. The rate of post peak reduction in soil strength, measured on the vane shear strength-rotation curve, appears to depend on plasticity. Based on this study it is recommended that the vane be rotated to a minimum of 40° for very plastic clays to approximately 150° for clays with low plasticity, to determine post-peak strength behavior with great accuracy. It is suggested that the undrained shear strength values determined with the in-situ vane may be of greater value than other undrained shear strength measurements for axial pile design purposes. This study also yielded encouraging estimates of the coefficient of lateral stress at rest K_0 based on in-situ vane and triaxial test data.

KEY WORDS: carbonate soils, clays, coefficient of lateral stress at rest, drilling, in-situ testing, in-situ vane, laboratory vane, ocean soils, shear strength, silt, triaxial tests

This paper summarizes the results of a study which was made to (1) compare typical results of in-situ vane and cone penetrometer with laboratory tests, (2) assess the applicability of the results to geotechnical design, and (3) investigate if and how vane test procedures and standards could be improved. The investigation included correlations of typical vane test results with laboratory test results (laboratory vane, Torvane, unconsolidated undrained (UU) and consolidated isotropic undrained (CIU) triaxial tests, and Atterberg limits) and other in-situ tests (cone penetrometer and piezocone penetrometer). About 450 in-situ vane tests were performed with four borings at one location (A) and five borings at another location (B) off the west coast of India, and one location (C) in the North Sea (Fig. 1). The vane tests were performed either in a seabed mode or a downhole mode. These modes are described in a companion paper by Geise et al.[3]

The vane tests were made in cohesive soils with varying grain size distributions, Atterberg limits, and carbonate contents (Table 1). The classification and nomenclature follow ASTM Classification of Soils for Engineering Purposes (D 2487) and Practice for Description and Identification of Soils (Visual-Manual Procedure) (D 2489). It may be concluded

[1] Manager of engineering and staff engineer, respectively, Fugro Geotechnical Engineers B.V., P.O. Box 63, 2260 AB Leidschendam, The Netherlands.

[2] Senior engineer, Fugro Ltd., 18 Frogmore Road, Hemel Hempstead HP3 9RT, United Kingdom.

[3] Geise, J. M., Hoope, J. ten, and May, R. E., in this publication, pp. 318–338.

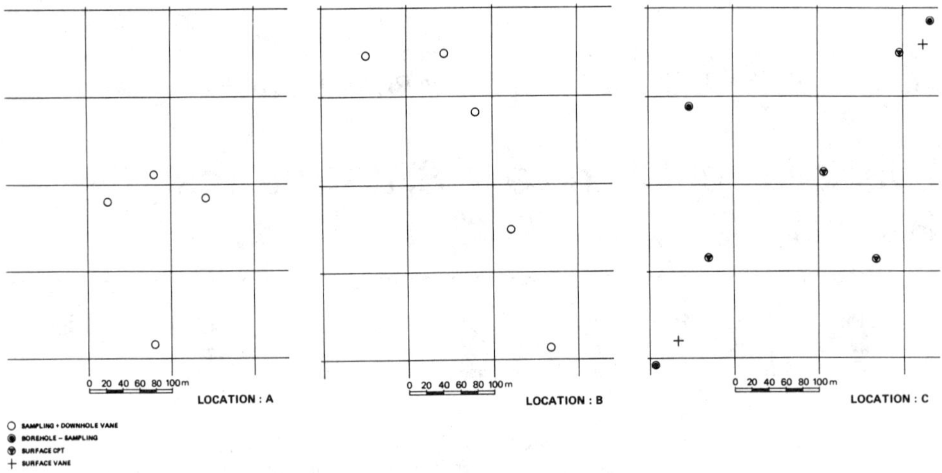

FIG. 1—*Relative position of the boreholes of the investigated sites.*

from the plasticity chart (Fig. 2) that there are three different types of soil, although all are classified as elastic silts to fat clays (Table 1).

In-Situ Vane Testing and Sampling Methods

At Locations A and B samples were taken and downhole in-situ vane tests were performed alternately in the same boreholes. The majority of samples from these locations were taken using percussion samplers, while the remainder were taken using push samplers, which produced a higher sample quality. In all cases thin-walled sample tubes ("shelby") were used. At Location C, the in-situ vane tests and piston sampling were done in separate boreholes. Vane tests also were performed from the seabed. A conventional drill string heave compensating system was used at Locations A and B. At Location C, a

TABLE 1—*Physical properties of the investigated soils.*

Parameters	Location A	Location B	Location C
Soil description	Elastic silt to fat clay (MH to CH)	Elastic silt (MH)	Fat clay (CH)
Grain size distribution			
sand, %	. . .	. . .	5
silt, %	60	71	51
clay, %	40	29	34
Carbonate content, %	18 to 27	48 to 57	<5
Plastic limit, %	36 to 49	23 to 38	22 to 23
Water content, %	55 to 110	54 to 94	49 to 53
Liquid limit, %	107 to 127	72 to 96	55 to 75
Plasticity index, %	58 to 89	35 to 58	33 to 52
Specific gravity, –	2.54[a]	2.79[a]	2.65 to 2.80

[a] Only one test.

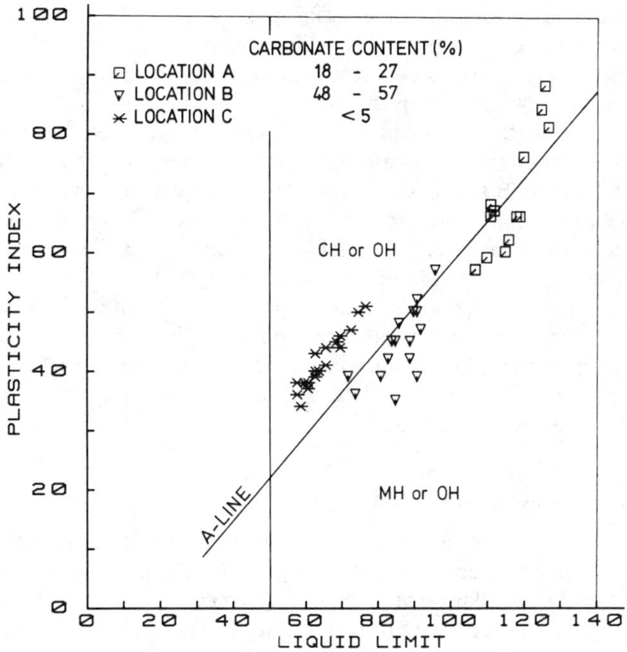

FIG. 2—*Plasticity charts of investigated locations.*

hard-tie heave compensator system was used in addition. The differences between these two systems are discussed in the companion paper by Geise et al.[3] The field testing program, the sampling methods, and the heave compensator system used at Locations A and B restricted the number of samples suitable for triaxial testing.

Table 2 summarizes the standards and procedures followed for the in-situ and laboratory tests. The vane shear strengths were calculated according to ASTM Method for Field Vane Shear Test in Cohesive Soil (D 2573), which assumes that the distribution of shear strength is uniform across the ends and around the perimeter of a cylinder enclosing the blades of the vane. The error introduced by this simplification results in a conservative shear strength [*1*].

TABLE 2—*Standards and procedures for tests.*

Test Name	Standard or Reference for Procedure	Notes
In-situ vane	ASTM D 2573 Geise et al. (Footnote 3)	vane blade dimensions 63.5 by 127 mm; rotation speed 0.1 °/s.
Cone penetrometer	ISSMFE [*10*]	. . .
Piezocone	Senneset and Janbu [*4*]	. . .
Penetrometer	Zuidberg et al. [*11*]	
Laboratory vane	Wilson [*12*]	vane blade dimensions 12.7 by 25.4 mm; motorized rotation speed 0.33°/s
UU triaxial test	Bishop and Henkel [*13*]	. . .
Atterberg limits	ASTM D 4318	one point method, wet preparation

The borehole vane tests generally were performed at depths of 0.75 and 1.5 m below the drill bit. To prevent the influence of drilling disturbance in the soil affecting the results, only the vane tests at a penetration depth of 1.5 m below the drill bit have been selected for the correlation studies reported in this paper.

The typical shear strength versus rotation curves have been schematized to three typical forms (Fig. 3). The vane tests at Location A generally exhibited Curve 1 type behavior, and the tests at Locations B and C showed strain softening as illustrated by Curves 2 and 3. The following terminology is used to describe three types of shear strength deduced from these curves. The peak shear strength is defined as the highest shear strength measured. The post-peak shear strength is the lowest shear strength measured over an interval of approximately 10° rotation after a minimum of 40° rotation from the beginning of the test. The remolded shear strength of the soil is measured by the in-situ vane after remolding the soil by two or three complete revolutions of the vane blade at a higher rotation speed, following the procedure given in ASTM D 2573. A least squares regression method was used to derive the line best fitting the data.

Variation of Shear Strength with Depth

The measured shear strengths are plotted against depth in Figs. 4 to 6. The shear strengths based on the Atterberg limit data using a correlation proposed by Skempton [*2*] are presented in Fig. 7. The relation between the corrected cone resistance q_{net} and the shear strength ($q_{net} = N_k s_u$) [*3*] is presented in Fig. 8. The soils at the locations were found to be

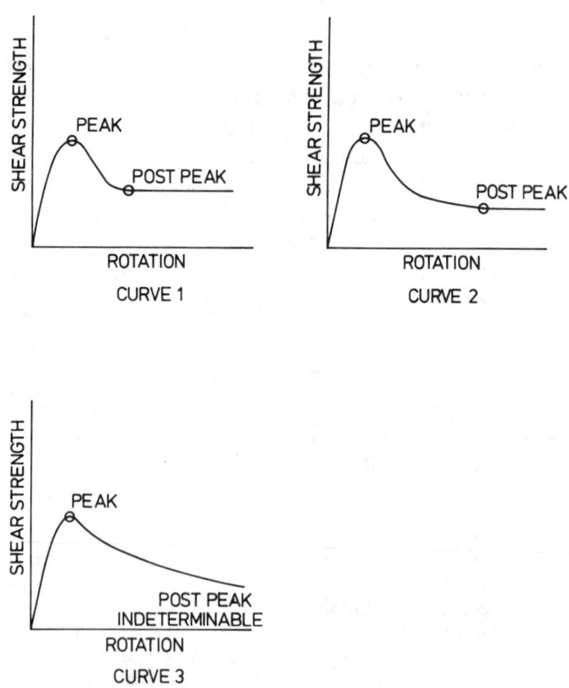

FIG. 3—*Definition of peak, post-peak, and remolded shear strength and schematic shear strength versus rotation curves.*

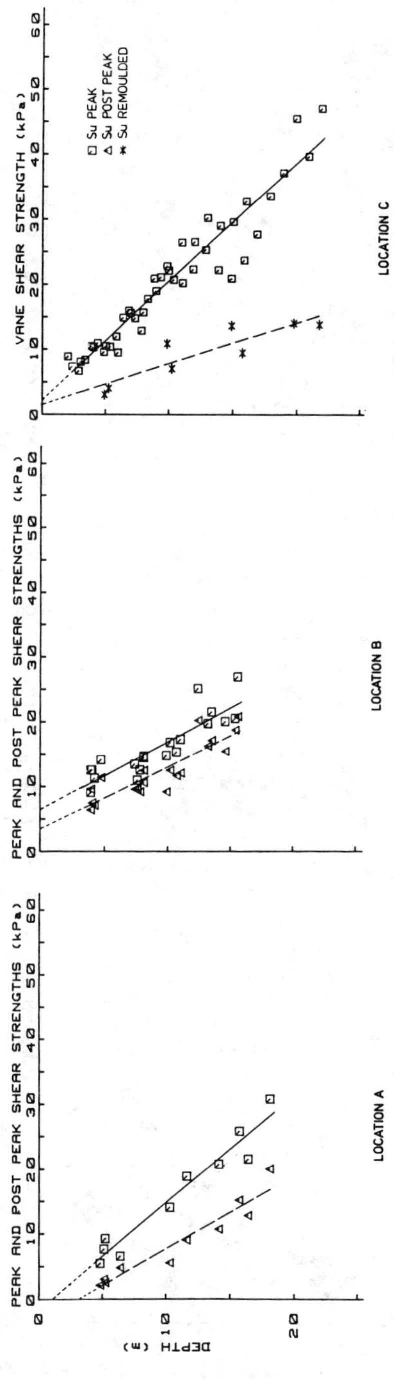

FIG. 4—*In-situ vane peak and post-peak shear strength versus depth. The symbols refer to data from all three locations.*

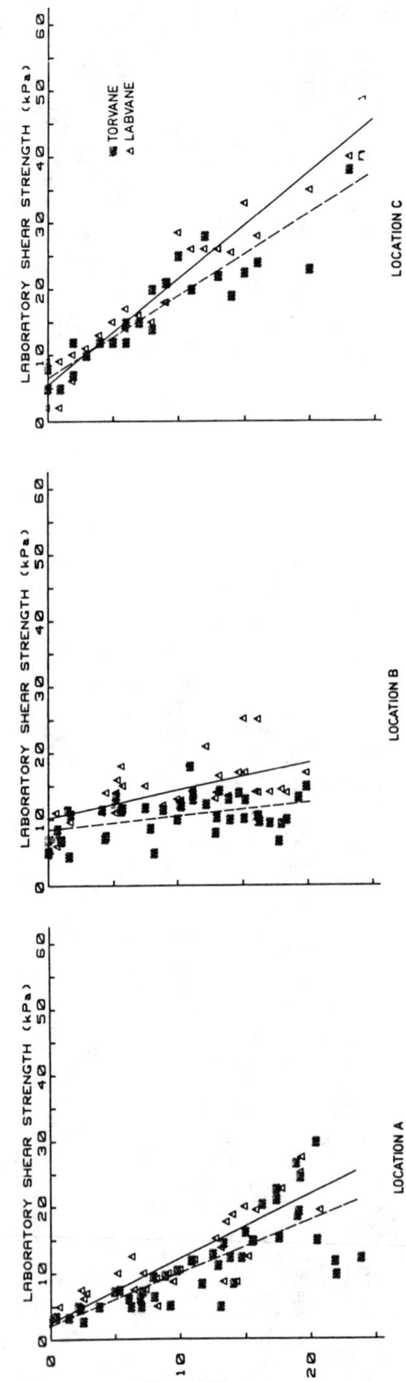

FIG. 5—*Laboratory shear strength of Torvane and laboratory vane versus depth. The symbols refer to data from all three locations.*

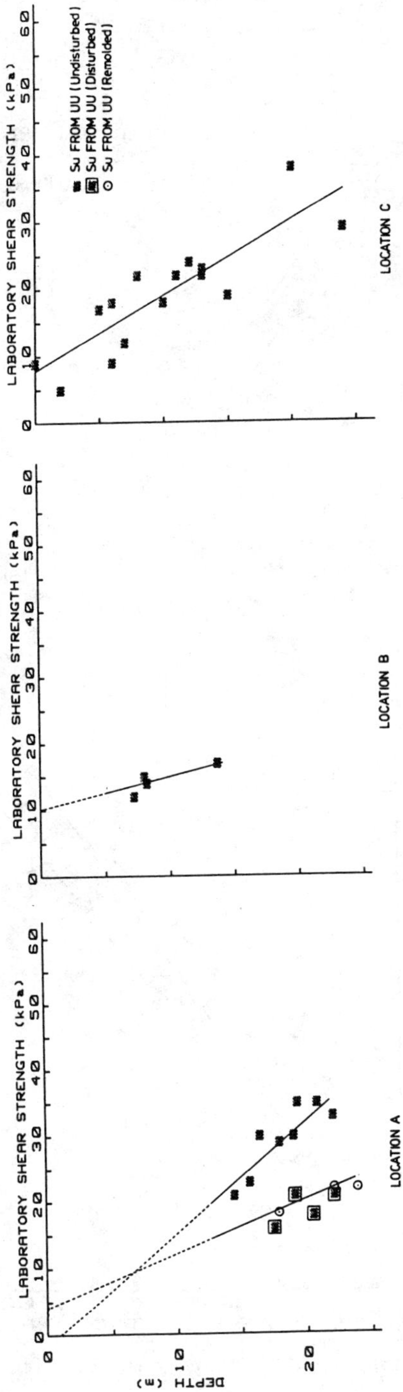

FIG. 6—*Laboratory shear strength of triaxial tests versus depth. The symbols refer to data from all three locations.*

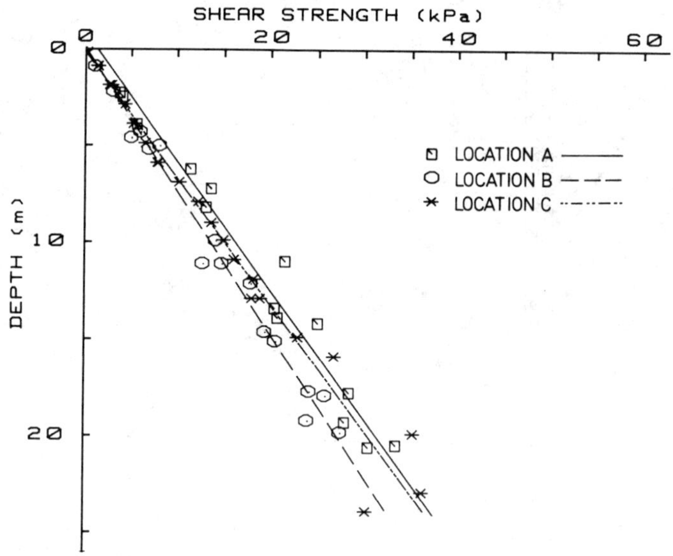

FIG. 7—*Shear strength according to the Skempton relationship* [2] *for normally consolidated clays.*

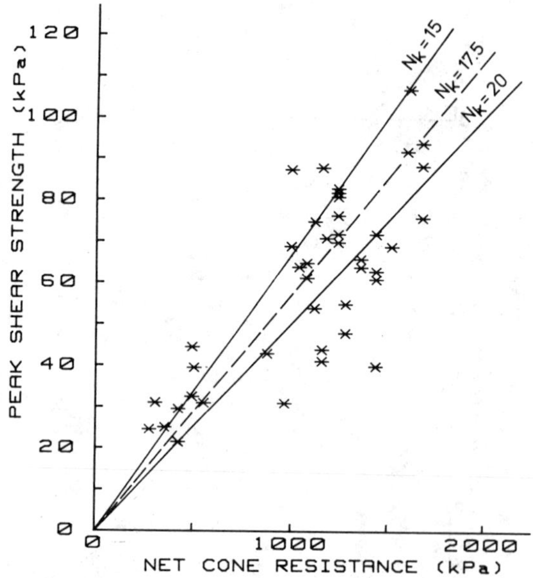

FIG. 8—*Relationship of peak shear strength versus net cone resistance.*

geologically normally consolidated to slightly overconsolidated (Figs. 4 to 6). Geotechnical data from the B location had considerably more scatter than the data from the other two locations. It is suggested that the higher carbonate content of Location B is associated with cementation and that this is represented by more scatter in strength measurements. Specific comments on the correlations between the different strength measurements follow.

In-Situ Vane Strength Data

The peak and post-peak shear strength data generally show a small scatter, which is reflected by the correlation coefficients of 0.77 to 0.94, respectively (Table 3). The largest scatter was found for Location B. Remolded vane tests were performed only at Location C.

The peak, post-peak, and remolded shear strengths show a linear increase with depth (Fig. 4). For Location A, the peak and post-peak regression lines diverge, while for Location B the lines are parallel. However, it should be noted that the post-peak strengths at Location B are not so clearly defined as at Location A. The peak and remolded strength regression lines for Location C also diverge. Post-peak values have not been measured at Location C, where the vane rotation did not extend beyond 40°.

The regression lines fitting the data from Location A suggest a small negative shear strength at the mudline. This is an artifact produced by extrapolation of the regression lines based on the available data. As no data are available from the top soil the correctness of the derived regression line cannot be ascertained in this zone.

Torvane and Laboratory Vane Data

The Torvane and laboratory vane test data show a large scatter, especially for Location B soils (see Table 3 and Fig. 5). The laboratory vane tests give higher shear strength results

TABLE 3—*Statistical relationships of shear strengths and depth.*

	Regression Coefficients[a]								
	Location A			Location B			Location C		
Source of Shear Strength	A, kPa	b, kPa/m	r^2, kPa/m	A, kPa	b, kPa/m	r^2, kPa/m	A, kPa	b, kPa/m	r^2, kPa/m
Field vane peak	−1.71	1.66	0.94	6.4	1.0	0.77	2.19	1.83	0.91
Field vane post-peak	−3.60	1.13	0.92	3.5	1.0	0.78	0.25	1.83	0.90
Field vane remolded		no data			no data		1.54	0.63	0.80
Torvane	2.0	0.8	0.61	8	0.2	0.21	6.52	1.25	0.87
Lab vane	2.5	2.0	0.72	10	0.4	0.33	5.17	1.63	0.94
UU undisturbed	−2.0	1.7	0.76		not relevant		6.4	1.3	0.77
disturbed	4.03	0.83	0.81		no data			no data	
Plasticity index [2] ($s_u = (0.0037I_p + 0.11)\sigma_{vo}'$	1.18	1.51	0.98	−0.68	1.36	0.99	−0.36	1.54	0.97

[a] Linear Regression line of the form $s_u = A + b(z)$ fitted through data on a plot of shear strength s_u against depth z.

than the Torvane tests (s_u laboratory vane = 1.2 to 1.5 s_u Torvane). Extrapolation of the regression lines fitted to the shear strength values measured by both methods indicates that the soil probably has measurable strength at the mudline.

Triaxial Test Data

Unconsolidated undrained (UU) triaxial tests were performed on undisturbed, disturbed, and remolded specimens. Disturbed samples are in this case defined as samples of which the consistency has clearly been affected by the drilling and sampling process. Remolded samples are samples fully remolded in the laboratory and recompacted to the estimated in-situ density. The results of these tests are presented in Fig. 6. About 95% of the UU triaxial tests were performed offshore. These tests show a linear relationship with depth. The regression lines through the data points have correlation coefficients of 0.76 to 0.81. A clear distinction can be made between the tests on undisturbed specimens and those on remolded or disturbed specimens. As the results from the tests on both disturbed and remolded specimens from Location A show a good correlation, they have been combined in the regression analyses and further comparisons. The regression lines for the remolded shear strengths and the undisturbed shear strengths from the UU tests for Location A intersect owing to the lack of data in the top of the soil unit.

Shear Strength Determined from the Skempton Relationship

Shear strength values were computed from Atterberg limit data and estimated effective overburden pressure using the relation proposed by Skempton [2]

$$\left(\frac{s_u}{\sigma_{vo}'}\right)_{NC} = 0.0037 I_p + 0.11$$

where

s_u = undrained shear strength,
σ_{vo}' = effective overburden pressure, and
I_p = plasticity index.

The resulting shear strength values are shown in Fig. 7. Linear regression analyses for the three locations show high correlation coefficients of 0.97 to 0.98. The regression lines for the three locations plot close together. These shear strength values compare well with the in-situ vane and UU triaxial results (Table 3). If anything, the comparison is better for the carbonate clays (Sites A and B) than for the low carbonate clays (Site C). This is remarkable since Skempton's relation was developed for the latter type of clays.

Relationship Between Cone Resistance and Shear Strength from In-Situ Vane

Cone penetrometer and piezocone penetrometer test data were only available for Location C. At this location, the in-situ vane strengths have been plotted against cone resistance measured at approximately the same depth (Fig. 8). In this plot, the net cone resistance q_{net} is the measured cone resistance q_c, which has been corrected for pore-pressure effects and the overburden influence [4].

Although there is lateral variation in soil strengths at all of the borehole and test sites at Location C, the trend of vane strength and cone resistance with depth is generally linear.

The relationship between vane strength and cone resistance is similar to that between triaxial compressive strength and cone resistance; the ratio q_{net}/s_u is in the range of 15 to 20 [2] with an average of 17.4 (Fig. 8).

Summary of Test Data

The general patterns shown by the different shear strength data can be summarized as follows:

1. The shear strengths have some finite value at mudline and increase linearly with depth. The mudline shear strengths range from zero to about 20 kPa, based on the extrapolation of the linear regression lines fitted to the test data.
2. The linear increase of shear strength values with depth suggests a normally consolidated soil.
3. The scatter in the shear strength data, as measured by the regression coefficients, increases with increasing carbonate content.

Discussion

Comparision of Torvane and Laboratory Vane with In-Situ Vane

The Torvane and laboratory vane shear strength measurements show a larger scatter than the other tests (Figs. 4 to 6). There are various reasons for this. First, the volume of soil specimen tested is relatively small. Second, the test results are susceptible to differing operator techniques.

The higher strength measurements for the laboratory vane might be explained by the fact that the test is more controlled and takes place within the specimen away from its edges. Therefore the results are probably less affected by specimen disturbance. Both the Torvane and laboratory vane data plot below the peak shear strength of the in-situ vane, except for Location C where the laboratory vane and in-situ vane data are similar.

Influence of Angular Shear Velocity

The shear strength deduced from vane tests depends on the shear velocity of the vane blade edges. Results from different vanes can be compared properly if they are normalized to a standard angular shear velocity [5]. The laboratory vane was rotated at 0.33°/s in conjunction with 12.7- by 25.4-mm (0.5- by 1-in.) vane blades, which yields an angular shear velocity of 37 μm/s. An angular shear velocity of 55 μm/s was derived for the in-situ vane (0.1°/s with a 63.5- by 127-mm vane blade). Perlow and Richards [5] indicated an increase of shear strength of 1.8 to 2.6 kPa when the angular shear velocity was increased by 1 mm/s, for MH to CH soil types. The difference between the angular shear velocity of the in-situ vane and the laboratory vane is only 18 μm/s. The effects of this difference are considered to be minimal, implying that a comparison of laboratory vane and in-situ vane test data can be made without further normalization of the shear strengths.

Comparison of UU and In-Situ Vane Data

Comparison of UU triaxial test data and in-situ vane data shows that the shear strength s_u from the UU tests on undisturbed specimens is nearly the same as the peak shear strength from the in-situ vane tests. Accordingly, it can be concluded from the triaxial tests that these soils require no correction factors to convert vane shear strength to UU triaxial shear strength. Correction factors were initially proposed by Bjerrum [*6*] and recently have been re-evaluated by Aas et al. [*7*]. The latter suggest a factor of approximate unity for young normally consolidated clays with similar plasticity indexes as these clays, which agrees with the observed data.

Comparison of Post Peak and Remolded Shear Strength

It can be seen from Figs. 4 and 6 that the post-peak shear strength of the in-situ vane is similar to the remolded UU shear strength s_{ur} at Location A. Unfortunately no comparisons could be made for the other locations because of a lack of data.

The highly plastic clays at Location A reached the post-peak shear strength after approximately 40° rotation (Fig. 3, Curve 1). However, this level was not reached within the rotation range of up to 90° for the less plastic clays at Locations B and C (Fig. 3, Curves 2 and 3).

Relationship of Vane Shear Strength with K_0

A discussion concerning the interpretation of in-situ vane strengths has been presented by Wroth [*1*]. Based on a consideration of stress paths and the experimental evidence from Law [*8*], Wroth suggested that the nondimensional vane strength might be expressed by

$$\frac{\tau_{\text{vane}}}{\sigma_v'} = K_0 \sin \phi_{ps}$$

or

$$\frac{\tau_{\text{vane}}}{\sigma_v' \sin \phi_{ps}'} = K_0 \qquad (1)$$

where

τ_{vane} = peak shear stress of vane, kPa,
σ_v' = effective overburden pressure, kPa,
K_0 = coefficient of lateral stress at rest, (−),
ϕ_{ps}' = effective angle of internal friction, deg, for plane strain and $\phi_{ps}' = \frac{9}{8} \times \phi_{\text{triax}}'$ [*1*], and
ϕ_{triax}' = effective angle of internal friction (deg) from CIU test.

Values of K_0 deduced from the peak vane strengths of the Eq 1 have been plotted in Fig. 9 against the K_0 estimated using empirical methods proposed by Brooker and Ireland [*9*]. It can be seen that there is an encouraging agreement between the independently determined values of K_0. Hence it would appear that it may be possible to estimate K_0 from in-situ vane tests, in combination with triaxial data.

Application of Vane Shear Strength Data to Design

The in-situ vane test gives highly reproducible results, which are reflected by the high correlation factors (Table 3); consequently, the selection of shear strength parameters for engineering purposes can be rather precise. The data presented in this paper were obtained from sites containing a wide variety of predominantly normally consolidated soils. From

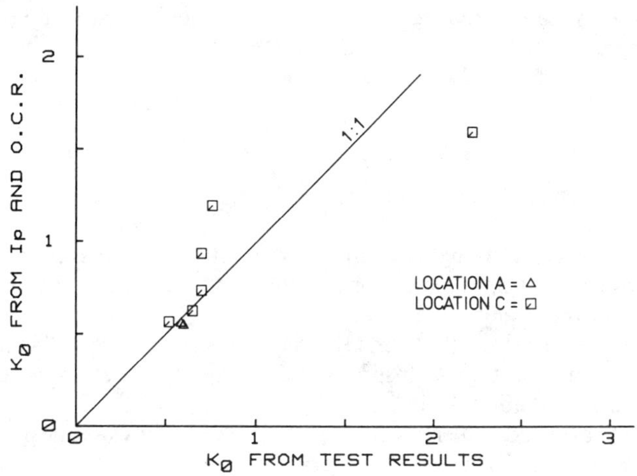

FIG. 9—*Relationship of* K_o *according to Brooker and Ireland* [9], *and Wroth* [1] *for data from Locations A and C.*

an examination of the various correlations presented it appears that no correction factor is required for these soils to obtain shear strengths similar to those measured in the laboratory under triaxial compression conditions. However, the correlation between the best estimate of K_0 and the nondimensional vane strength would suggest that the vane strengths would be indicative of strengths measured in simple shear tests on vertical slices rather than the conventional horizontal slices. Care must therefore be exercised when selecting the shear strength appropriate for design; for example, in-situ vane shear strength data may be highly relevant and perhaps preferable to conventional strength measurements on specimens selected from conventionally cut specimens for the design of axially loaded piles. However, corrections should be considered when the main load components are vertical compression or horizontal shear or both (for example, gravity structures).

Suggestions to Improve Test Procedures and Standards

The experience gained from the 450 tests, which form the basis of this paper and other published data [*5,6,7*] suggest the following:

1. The in-situ vane should be rotated sufficiently to measure the post-peak degradation of the shear strength with blade edge displacement. A suggested range of minimum rotation is from 40° for very plastic clays to perhaps as much as 150° for clays with low plasticity.
2. The shear strength determined from vane tests is dependent on the angular shear velocity of the tests. Higher angular shear velocities result in higher shear strengths. A standard angular shear velocity is recommended to produce results that can be compared with other available data. The selected angular shear velocity should meet the following requirements:

   a. sufficiently low to limit its influence on the undisturbed or peak shear strength [*5*] and
   b. sufficiently fast to test the soil in an undrained condition.

To meet both these requirements, an angular shear velocity of about 55 μm/s is recommended. This agrees with a rotation speed of 0.1°/s (as recommended in ASTM D 2573) in conjunction with 63.5- by 127-mm vane blades.

Conclusions Concerning In-Situ Vane Data

The following conclusions have been derived from this study:

1. The peak vane shear strengths compare well with the UU triaxial shear strengths. Correction factors are not required to convert the vane shear strengths to laboratory UU triaxial strength for these soils.
2. The remolded shear strengths from UU triaxial tests compare well with the post-peak vane shear strength at Location A. Insufficient data exist to allow similar comparisons to be made for Locations B and C.
3. The ratio of corrected cone resistance q_{net} to peak in-situ vane shear strength for this study is in the range of 15 to 20, with an average of 17.4.
4. The Torvane and laboratory vane generally resulted in lower shear strength measurements than the in-situ vane. This is probably a result of sample disturbance of very soft soils. Better results were obtained at Location C, where piston sampling and the hard-tie heave compensator system were used to obtain specimens of higher quality.
5. The in-situ vane test gives highly reproducible results, and as such they are of great value for establishing with great accuracy the trend of strength variation with depth.
6. Encouraging estimates of the coefficient of lateral stress at rest K_0 were obtained from in-situ vane shear strength data together with triaxial test results.

Acknowledgments

We appreciate having the permission of Oil and Natural Gas Commission (India) and Statoil (Norway) to use certain data contained in this paper. A. F. Richards, J. M. Geise, and N. J. Withers reviewed the manuscript and made a number of suggestions for its improvement.

References

[*1*] Wroth, C. P., "The Interpretation of In Situ Soil Tests," Géotechnique, Vol. 34, No. 4, 1984, pp. 449–489.

[*2*] Skempton, A. W., "Discussion: The Planning and Design of the New Hong Kong Airport," *Proceedings of the Institute of Civil Engineers,* Vol. 7, 1957, pp. 305–307.

[*3*] De Ruiter, J., "The Use of In-Situ Testing for North Sea Soil Studies," *Offshore Europe '75 Conference Papers,* Aberdeen, Paper 219. Spearhead Publications, London, 10 pp.

[*4*] Senneset, K. and Janbu, N., "Shear Strength Parameters Obtained from Static Cone Penetration Tests," *Strength Testing of Marine Sediments: Laboratory and In-Situ Measurements, STP 883,* R. C. Chaney and M. R. Demars, Eds., American Society for Testing and Materials, Philadelphia, 1985, pp. 41–55.

[*5*] Perlow, A. M. and Richards, A. F., "Influence of Shear Velocity on Vane Shear Strength," *Journal of the Geotechnical Engineering Division, Proceedings of the ASCE,* Vol. 103, No. GT1, Jan. 1977, pp. 19–32.

[*6*] Bjerrum, L., "Problems of Soil Mechanics and Construction on Soft Clays," *Proceedings of the Eighth International Conference on Soil Mechanics and Foundation Engineering, Moscow,* Vol. 3, 1973, pp. 111–159.

[*7*] Aas, G., Lacasse, L., Lunne, T. and Høeg, K., "Use of In-Situ Tests for Foundation Design on Clay," *Use of Insitu Tests in Geotechnical Engineering,* S. P. Clemence, Ed., Proceedings of In Situ '86, ASCE, New York, pp. 1–30.

[8] Law, K. T., "Triaxial Vane Tests on a Soft Marine Clay," *Canadian Geotechnical Journal,* Vol. 16, 1979, pp. 11–18.

[9] Brooker, E. W. and Ireland, H. O., "Earth Pressure at Rest Related to Stress Theory," *Canadian Geotechnical Journal,* Vol. II, No. 1, 1965 pp. 1–15.

[10] "Report of the Subcommittee on Standardization of Penetration Testing in Europe," Appendix 5 of the minutes of the Executive Committee held in Tokyo, *Proceedings of the Ninth International Conference on Soil Mechanics and Foundation Engineering,* Vol. 3, Tokyo, Japan, 1978, pp. 95–152.

[11] Zuidberg, H. M., Schaap, L. H. J., and Beringen, F. L., "A Penetrometer for Simultaneously Measuring of Cone Resistance, Sleeve Friction and Dynamic Pore Pressure," *Penetration Testing, Proceedings of the Second European Symposium on Penetration Testing,* A. Verruijt, F. L. Beringen, and E. H. de Leeuw, Eds., Vol. 2, A. A. Balkema, Rotterdam, The Netherlands, 1982, pp. 963–970.

[12] Wilson, N. E., "Laboratory Vane Shear Tests and the Influence of Pore-Water Stresses," *Laboratory Shear Testing of Soils, STP 361,* American Society for Testing and Materials, Philadelphia, 1964, pp. 377–385.

[13] Bishop, A. W. and Henkel, A. J., *The Measurement of Soil Properties in the Triaxial Test,* Edward Arnold, London, 2nd ed., 1962, 227 pp.

Armand J. Silva[1] *and Robert M. Wyland*[2]

Autonomous Seafloor Strength Profiler: Comparison of In-Situ and Core Results

REFERENCE: Silva, A. J. and Wyland, R. M., "**Autonomous Seafloor Strength Profiler: Comparison of In-Situ and Core Results,**" *Vane Shear Strength Testing in Soils: Field and Laboratory Studies, ASTM STP 1014,* A. F. Richards, Ed., American Society for Testing and Materials, Philadelphia, 1988, pp. 354–371.

ABSTRACT: The autonomous seafloor strength profiler (ASSP) has been developed to determine the strength profile of the upper 1.5 m of marine sediments. The system can operate in water depths to 6000 m and is designed to remain in a dormancy state of up to one year. The instrument consists of four integrated packages: controller and data acquisition system, mechanical system, volume/pressure compensator, and power supply. This paper describes the ASSP system and presents results of in situ vane tests in a deep sea (5845 m) illitic clay and shallow water (60 m) clayey silty fine sand. The in-situ shear strengths of the deep sea clay are 26% higher than those of core samples. The shear strength profile for the shallow water sediment is quite different from that of a nearby core.

KEY WORDS: in situ, shear strength, vane shear, comparison, autonomous

The analysis of most geotechnical engineering problems requires information on the shear strength of the natural sediment. Because of the myriad of problems associated with obtaining undisturbed samples of marine sediments, especially in deep water, there has been a growing interest in conducting in-situ strength tests.

The vane shear method is one of several that can be used, depending on the particular situation and sediment type. The vane test is intended for use in fine-grained (cohesive) material and will not give meaningful results for anything coarser than a very fine sand or silt. The cone penetrometer method is better suited for the coarser grained cohesionless materials. Therefore it is advantageous to have both the vane and cone methods available.

One of the advantages of the vane method is that it measures strength more directly, since the sediment is sheared on a well-defined surface. With reasonable assumptions regarding stress distribution, it is a simple matter to obtain a relationship between the torque required to produce failure and the shear strength. If torque versus rotation angle is monitored through the full test range, it is possible to estimate the shear modulus and to determine residual strength. The remolded strength and sensitivity can also be determined if another test is conducted at the same site after full rotation, to assure complete remolding.

Although the system described here was designed primarily for use with a vane probe, it also obtains a record of penetration resistance, and a cone probe can be substituted for the vane blades. The purpose of this paper is to (1) describe the autonomous seafloor

[1] Professor of ocean and civil engineering, University of Rhode Island, Kingston, RI 02881.
[2] General engineer, Naval Post Graduate School, Oceanography Department, Monterey, CA 93943.

strength profiler (ASSP) for obtaining a profile of strength of the upper 1.5-m sediment, using either a vane or cone probe, (2) present data obtained in both deep sea (5800 m) and shallow water (60 m) sediments, and (3) compare results of in-situ and core data.

The URI/ASSP System

General Requirements

A more detailed description of the development history of the instrumentation can be found in recent papers [*1–4*]. The motivation for the ASSP system came from the United States Subseabed Disposal Project (SDP), which is a study to determine the feasibility of burying solidified high-level nuclear wastes within certain geologically stable deep-sea sediments [*5*]. The plan is to conduct a long-term in-situ heat transfer experiment to determine the thermal, geochemical, and geotechnical response to a heat source in the sediment. This experiment is to be carried out over a one-year period in 5800 m water depth in the north central Pacific Ocean. In addition to the ASSP, various components (for example, thermal sensors, piezometers, and pore-water sampler) will be mounted on a tubular structure approximately 4.3 m long and 2.3 m high, and a 400-W isotopic heat source will be implanted into the sediment to a depth of 1 m below the seafloor. The vane shear measurements will be made at the end of the one-year experiment with the vane passing within 20 mm of the heater. Comparisons of in-situ measured responses with the predictions of numerical models for thermal, mechanical, and chemical behavior will be used to evaluate the applicability of the techniques being developed in the SDP [*6*].

The present configuration of the system has in part been dictated by the needs of the SDP, but the same basic design can be used for other geotechnical applications. Based on preliminary studies, the basic design requirements for the present version of the system were defined as follows:

- Water depth (pressure), 6000 m (600 bar).
- Sediment strength, 70 kPa.
- Profile depth, 1.5 m.
- Time on bottom, 1 year.
- Sediment temperature, 1.5 to 300°C.

The system is autonomous with its own power, controller, and data acquisition system, so it can be used in other seafloor studies. In the present configuration, the system can be mounted on a bottom-supported platform or used from a submersible. Also, a cone probe can be substituted for the vane.

Prototype (Model A)

Because of the complexity of the long-term in-situ experiment and the fact that most of the instrumentation was entirely new, a short-term (30 days) scale model (0.287 scale) laboratory experiment was conducted in the 3-m-diameter pressure vessel at the Naval Ship Research and Development Center, Annapolis, MD. The test bed for this simulation experiment consisted of a 1-m-diameter by 1-m-deep tank of saturated, reconstituted, reconsolidated north Pacific illite [*2*]. A tank of seawater was mounted above the test bed, and the entire apparatus was pressurized to 550 bars and maintained at a temperature of 4°C for a period of 30 days.

The system configuration was modified for use in this simulation experiment because of space limitations. This prototype (Model A) was shortened to a 0.6-m penetration and consisted of three major integrated packages:

- mechanical system (with internal volume compensator),
- electronic control and data acquisition system, and
- power supply (external).

The mechanical system was contained in an oil filled pressure-compensated housing, with pressure equalization accomplished by two compensator tubes equipped with floating Teflon® pistons to equalize internal and external pressures.

The Model A system was used successfully during the 30-day simulation experiment at a pressure of kPa bar, and three strength profiles were obtained:

(1) unheated profile before heating and pressurization,
(2) heated profile near the heater (within 20 mm) after 30-days heating under kPa bar pressure, and
(3) unheated profile (within 70 mm of heater) after sediment cool-down and depressurization.

Comparison of data from the pretest unheated and heated profiles indicates that a ten- to twelve-fold increase in shear strength occurred at the midplane depth of the heater in the 200°C temperature zone.

Present System (Model B)

The device that was fabricated for the Subseabed Disposal Project (Model B) consists of four integrated packages (Fig. 1):

- mechanical system,
- electronic controller and data acquisition system,
- volume compensator, and
- power supply.

A detailed description of the present system (Model B) is given by Silva et al. [*3*] and Silva [*4*].

Both the mechanical system and volume compensator are oil filled and designed to operate at deep-ocean pressures. The power supply and controller operate at atmospheric pressure and are thus contained in high-pressure housings. As in Model A, cabling between these packages is oil filled to provide volume compensation and special marine connectors are used. The controller can be programmed with fixed operation sequences. The present configuration allows for 22 vane test sites within a depth of 1.5 m below seafloor.

Information gained from the simulation experiment with the prototype (Model A) was used to design the present system (Model B) with several improvements. For example, it was determined that purchased internal components used in Model A were capable of operating at ambient deep-ocean temperature and pressure after a dormancy period of one month. Some details of the mechanical unit are shown in Fig. 2. Nonmetallic components are incorporated in less critical areas to reduce the weight, and the volume-compensating bladder is mounted externally. The weight of the entire system with the stainless steel hous-

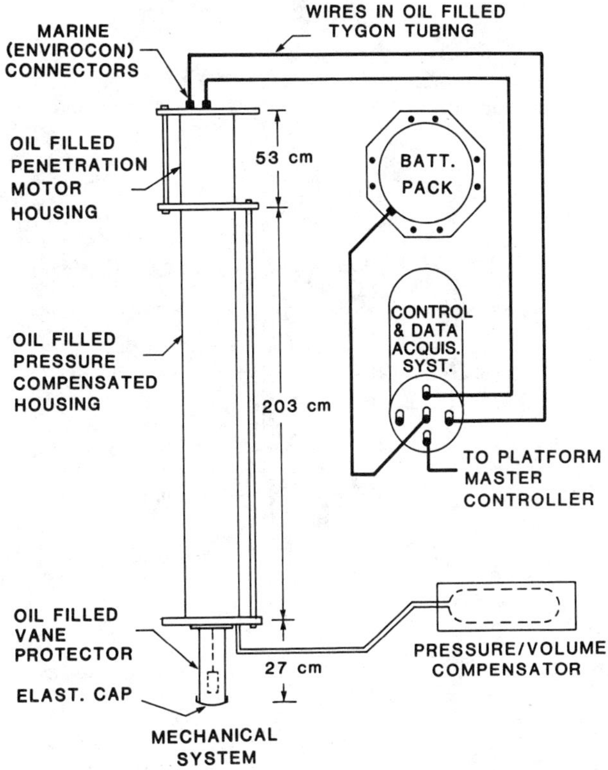

FIG. 1—*Schematic diagram of autonomous seafloor strength profiler (ASSP).*

ing and oil is 3.0 kN in air and approximately 1.4 kN in water. With the polyvinyl chloride (PVC) housing, these are reduced to 2.6 and 1.1 kN, respectively. Another new feature of the design is a self-pressure compensating vane protector housing with a neoprene cap to isolate the vane and seal this critical zone from seawater until just before penetration.

The vane itself is not the standard rectangular shape but rather has a diameter of 20 mm, height of 30 mm, and a rounded insertion end with a small needle-like tip. This rounded design is an effort to keep the vane from dragging down objects during penetration, such as manganese nodules found on the North Central Pacific Ocean Floor. The vane blades are tapered toward the side and bottom edges to give it a streamlined shape. The ratio of the cross-sectional area of the vane blades to that of the sheared area is 8.5%. The pointed tip assures that the temporary neoprene cap is easily punctured during initial penetration.

The heart of the electronic system is an Intel 8751 microcontroller (Fig. 3), which has 4K of Eprom program memory, 128 bytes at random access memory (RAM), a full duplex serial port, two 16-bit counter/timers, and two external interrupts. System configurance characteristics, such as the number of sample sites, their locations, gear ratios, output data format, and so forth, are stored in a standard 2716 Eprom, which allows the user to easily modify the system configuration to meet changing requirements. Interfaced to the 8751 microcontroller chip are a Sensotec Model 41 force transducer to measure the penetration force, a Lebow Model 2120 torque transducer to measure the torque, a Datel LPS-16 cas-

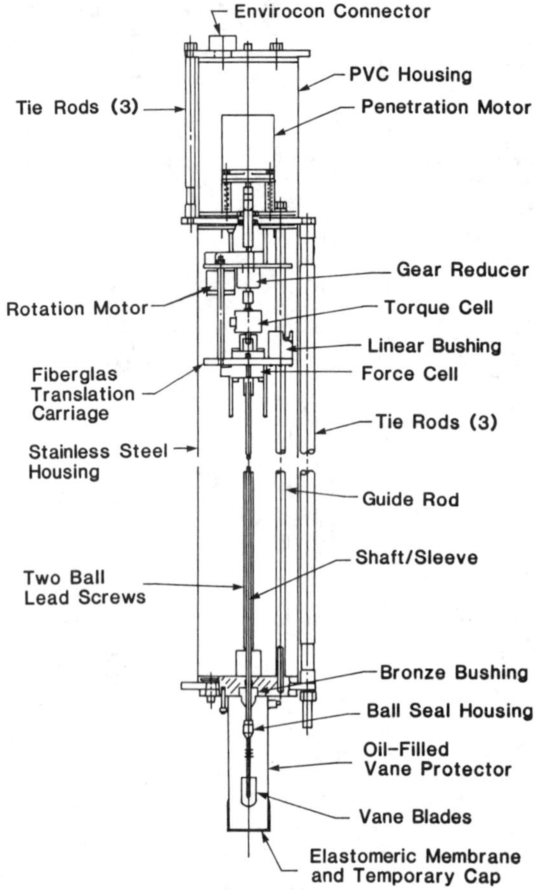

FIG. 2—*Assembly drawing of mechanical systems of ASSP Mode. B. Penetration depth is approximately 1.5 m.*

sette tape deck, and four Airpax stepper-motor drive cards (Fig. 3). Sensors of different ranges can be substituted to match the system's sensitivity to the sediment conditions.

Power for the system is normally provided from batteries housed in a 0.43-m-diameter Benthos glass sphere. At present there are two separate battery packs within the sphere; one 18-V pack with 10-Ah capacity for the vane controller electronics, and one 18-V pack with 40-Ah capacity for the two motors. The actual power requirements are 10 V at 800 mA for the controller and 12 V at approximately 3 A for the motor. The power supplies can be altered to meet special requirements; in shallow water, the system can be hard-wired to a surface vessel or floating power pack.

The system electronics are normally in a powered-down state, except for the computer interface board, which remains powered up (with 5 mA) to wait for commands from an external computer system or from a timed signal. The present system has a timer built into it to allow for ten different start-up time delays; five between 1 to 8 h and five between 12 to 13 months. After taking initial readings, the microprocessor computes and issues a num-

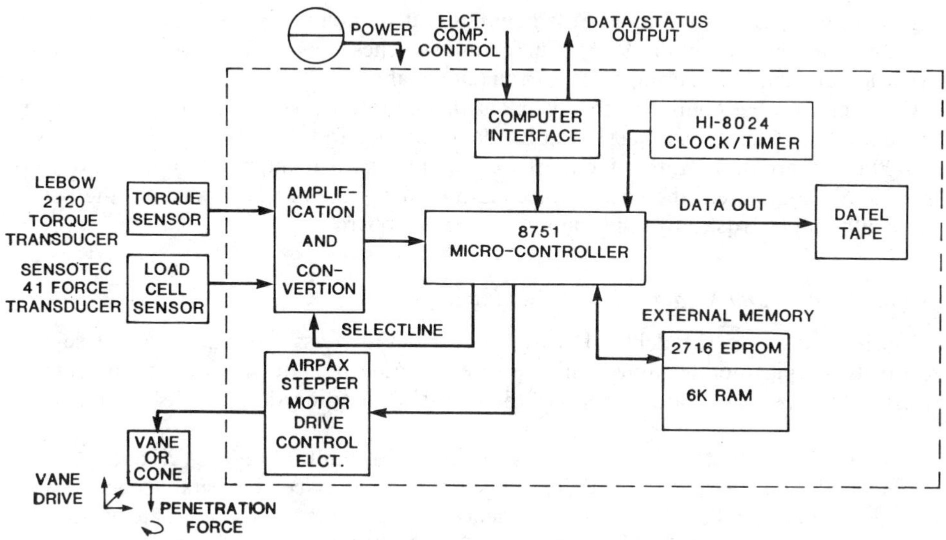

FIG. 3—*Controller block diagram for ASSP Model B.*

ber of individual commands to the stepper-motor drive card that controls the penetration motor. Each step of the penetration force is averaged to yield one force sample per 1.35 mm of penetration. The averaged data are stored on magnetic tape and placed in RAM for transmission to an external data acquisition system.

For operation with a vane, the penetration sequence is terminated at the appropriate depth for the next station, the microprocessor selects a rotation direction, and the vane is rotated at a rate of 0.0175 rad/s (60°/min). At present the total rotation angle is 1.5708 rad (90°) in a given direction. The direction of the vane is alternated at each successive station, keeping the vane blade rotation within the same quadrant. The output of the torque cell is read and the raw measurement stored on magnetic tape while the RAM receives averaged data. The 500 data points are averaged so that 100 data points are placed in memory for each series of torque measurements. The first 80 data points represent the first 0.7854 rad of rotation, and the last 20 points represent the final 0.7854 rad of rotation. This averaging technique was chosen to highlight the area where the sediment is expected to fail, usually within the first 0.1745 to 0.3491 rad (10° to 15°) of rotation. However, the full data set is recorded on the Datel tape. At the end of the rotation, just before the penetration to the next station, a final torque measurement is made to determine if a residual torque is being exerted upon the vane shaft. It is important to relieve the residual torque; otherwise the sediment at the next site will be subjected to an immediate torque. If a residual torque is sensed, the vane rotation direction is reversed to relieve the torque before penetration to the next station. However, a maximum reversal of 0.2618 rad (15°) is imposed as a limit, and the amount of reversal is subtracted from the next 1.5708 rad (90°) rotation sequence so as to keep the test rotation within the same quadrant. The microprocessor uses the same 8-bit analog to digital (A/D) converter to process the torque data. The torque measurement resolution is equal to the rated torque of the torque cell divided by 128. For a 1.412-N·m (200-in.·oz) torque cell this yields a sensitivity of 0.011 N·m (1.5625 in.·oz) per bit. Finer sensitivities than this can be achieved by changing torque cells.

Data from a pair of rotation limit switches mounted upon the motor shafts are collected and stored on the tape and in RAM. These limit switches allow verification of proper vane extension and rotation during post-mission data analysis.

Once the vane has completed its full extension and rotation sequence, it signals the master computer, which then requests that the data collected in the RAM be telemetered over a 2400-baud serial link. Either the master computer issues a command to retract, or as it is presently configured, the shaft retracts automatically. Upon full retraction, the power shuts off to all boards except the computer interface board.

Optional Modes and Modification to Existing System

The present system (Model B) of the ASSP provides for versatile capabilities, and several relatively simple modifications can be made to accommodate special requirements or enhance the ease of operation. A few of these possible modifications are discussed here.

1. Cone penetrometer: The system has a force sensor built into it (Fig. 2), and the vane probe could easily be replaced with a cone penetrometer device. A program change would be necessary to eliminate the rotation sequences and provide a constant a rate of penetration. The existing load cell has a capacity of 2.2 kN, but other cells can be substituted to increase the capacity. One limiting factor is the capacity of the penetration motor and drive system, but it is quite feasible to install a bigger motor in the fairly large space of the upper housing (Fig. 2).

2. General modification: Several components and aspects of the system can be changed to provide for increased sensitivity or capacity. These include

- vane size and configuration,
- cone size,
- force sensor,
- rotation motor,
- penetration motor, and
- programming changes: the controller system is quite versatile and almost any variable, such as rate of rotation, can be easily changed.

3. Power: The power can come from any suitable source, and it would be feasible to hard-wire the system to a surface vessel.

4. Weight reduction: For shallow water (less than approximately 2000 m), a PVC housing can be used in place of stainless steel for the mechanical system (Fig. 2). Because the system was designed for 6000-m water depth, it was necessary to use a thick-walled (0.4 kN in air) pressure housing for the controller. For shallow water operations a much lighter housing could be used with a very significant reduction in weight.

Field Trials and Laboratory Tests of the ASSP

General Comments

The present ASSP system (Model B) has been subjected to some rather severe environmental test situations. The system was mounted on two separate large platforms for deep-water tests and has been at a depth of over 5800 m on four separate occasions. During recovery of one of these platforms, a connection block failure caused the entire platform to free fall to the bottom. Upon recovery, the ASSP showed no discernible damage. Subsequent to the deep-water trials, the system has been mounted on a special platform (Fig. 4) and used in the 60 m of water. We feel that the design is sound and the construction of the system is very rugged. Following are brief descriptions of the test situations for which data are presented in the next section.

Atlas-84 Cruise, North Central Pacific

The main objective of the 1984 cruise was to test all the components of the in-situ heat transfer experiment near the deep-water site selected for the one-year experiment [*6*]. The main instruments are the following:

- thermal sensors, on heater and in sediment,
- thermal conductivity probes (line sources),
- piezometers for pore-water pressure monitoring,
- pore-water sampler for geochemical analysis,
- vane shear system for geotechnical analysis,

FIG. 4—*ASSP system mounted on platform for shallow water tests.*

- ion migration experiment with overcorer, and
- hydrostatically activated corers,
- cameras to monitor conditions at site and around heater.

For the 1984 cruise, the instrumentation was mounted on two separate platforms, and a heat source was not provided.

The vane shear measurements reported here were taken in a water depth of 5845 m in an abyssal hill region at 30°20 827′N, 157°50 921′W within a SDP study site designated as MPG-I. The sediments in this area have been studied quite extensively, and the upper few meters are generally characterized as being fine-grained illite-rich clays of medium sensitivity, low strength, low permeability, and high compressibility (Table 1).

There were two lowerings of the main platform. On the first lowering, a suite of ten vane shear measurements was made and then a piston core (HLC-1) was taken at a horizontal distance of 370 mm from the vane. On the second lowering only a few measurements were taken and another core (HLC-3) was taken. The horizontal distance between the two lowerings was 66 m. The core samples were inverted vertically and hydraulically extruded incrementally by using the corner piston and ram. A motorized laboratory miniature vane (Wykeham-Farrance) with a torque transducer, 12.5- by 12.5-mm vane, and 0.0175-rad/s (60°/min.) rotation rate was used. The vane apparatus was rigidly attached to the core barrel to minimize relative movement between the two. Samples were taken for water content determinations and several types of subsamples were obtained for detailed laboratory analysis.

Rhode Island Sound (RIS): Cruise MGL-3

After considerable reconnaissance work, a test site of relatively fine-grained sediment was located east of Block Island, RI, in 60-m water depth. Most of the surficial sediments on the continental shelf in this region consist of sands and gravels, but the area selected is a slightly deeper narrow basin that has been infilled with finer materials. Several large-diameter (104-mm) gravity cores were taken at the study site, and subbottom acoustic profiles were used to verify the areal extent of softer strata.

The ASSP system was mounted on an available steel platform, which was modified for use on the coastal research vessel R/V Schock (Fig. 4). A LORAN-C system was used to set two buoys at opposite ends of a 1.0-km-long test site. The instrument platform was lowered to the seafloor between these buoys with a 25.4 mm synthetic line. Two glass floats

TABLE 1—*Summary of geotechnical properties.*

Location	Water Depth, m	w, %[a]	e_0	w_L, %	I_p, %	Sand, %	Silt, %	Clay, %
MPG-1 (Pacific)	5845	110/(98 – 137)	3.05/(2.56 – 3.33)	90/(81 – 101)	50/(52 – 71)	TR	33	66
RIS	60	58/(44 – 74)	1.51/(1.15 – 1.93)	50/48.52	23/(22 – 27)	57	27	16

[a] Corrected for 35% salt.

NOTES: data format: average/(min – max). MPG-1: 30°20,827′N, 157°50,921′W. RIS = Rhode Island Sound: 41°08.03N, 71°17.61W.

were set at 20 m above the platform and the 90-m synthetic line was cast free with surface floats. Therefore, it was not necessary to hold station during the test sequence and the testing time was utilized to obtain gravity cores along the line designated by the two end buoys.

Before deployment, the pre-set timer (1 h) of the ASSP controller system was activated with an external signal. In this mode of operation, the ASSP goes through a short test sequence consisting of a short penetration, four vane rotations, and retraction to the original position; four vane rotations, and retraction to the original position; all within the probe protector housing (Fig. 4). The platform is then lowered to the seafloor as described above. After a 1-h delay, the ASSP goes through the full sequence, including 26 rotations (four within the protector housing and 22 within the sediment) and full retraction to the original probe position).

After an elapsed time of 123 min (8 min more than necessary), recovery of the platform was started. Upon retrieval it was found that the probe sleeve/shaft had not retracted fully and was in fact bent.

As will be shown later, only 15 usable stations were recorded within the upper 1.0 m of sediment. It appears that either the platform recovery operation was initiated during the penetration sequence or power loss occurred. The cause of this is not yet determined, but evidently there was either a human error, or a timing sequence malfunction. The vane blades and small piercing tip at the end were not damaged in any way, and therefore it appears that there was no obstruction in the sediment that would have prevented penetration. The sleeve-shaft assembly could not be straightened, and it was not possible to do the second planned deployment. A check of systems showed that no water leaked into any of the components, and the power pack was still up to a high level (total voltage loss was 6% for the motors and 5% for the logic electronics).

ASSP Test on Core Sample

Subsequent to the Rhode Island Sound test, the bent sleeve/shaft assembly was replaced with spare and all systems were checked. To test the repaired instrument the ASSP system was mounted above a secured gravity core retrieved from the site, and the full sequence was initiated into the core. However, the automatic timer was not used in this test. It was found that the system operated perfectly and a series of 19 vane tests were taken in the core. The core was then extruded, and samples were taken for water content determinations.

Results and Comparisons

General Comments

The results presented are intended to illustrate the more important features of the system and show comparisons between vane shear measurements taken with the ASSP and miniature vane measurements on nearby core samples. Many more operational tests have been conducted to observe performance of various components, determine calibration parameters for the sensors under differing environmental conditions, and ascertain some correction factors necessary for analysis of the raw data. Some of the general aspects are discussed here, and the specific details for each of the test situations are described with the data analysis.

Since the largest vane used in the ASSP tests (30 by 45 mm) was considerably larger than the one used on the cores (12.7 by 12.7 mm), a study was made using laboratory remolded illite from MPG-I to determine the effect of this on the results. The results from several measurements with each vane size showed no significant size effect, with the larger vane showing slightly lower values (within 2%). Therefore, it is assumed here that the results obtained with the ASSP vane can be compared directly with those of the laboratory miniature vane.

In the ASSP, a portion of the shaft above the vane is exposed to the sediment (96 mm length and 6.35 mm). The analysis should account for the torque generated by sediment traction on this section. A correction for the torque on this portion was made by assuming a triangular stress distribution on the ends. This correction (3.1%) was subtracted from the measured torque to yield the corrected torque used to calculate shear strength around the vane blades.

The equation for conversion of corrected torque T to strength S_u includes the vane dimensions: diameter D, height H, and radius R, where R is the radius of the circular arc at the vane bottom (see Fig. 2), and assumes a triangular stress distribution on the top and bottom surfaces. This stress distribution was chosen since it has been shown to closely represent the actual stress distribution [7]. The general equation is

$$S_u = \frac{T}{[(\pi/2)D^2H + (\pi/16)D^3 + (\pi^4/54)R^3]}$$

For the 30- by 45-mm vane this converts to

$$S_u = 0.0901(T)^{\text{oz-in.}}, \text{ kPa}$$

and for the 20- by 30-mm vane the equation is

$$S_u = 0.3609(T)^{\text{oz-in.}}, \text{ kPa}$$

The controller readings N are converted to voltage V, using the following relationship

$$V = \frac{5N}{128} - 5, \text{ V}$$

Analysis of Individual Vane Shear Results

The raw data obtained from the sensors are recorded on the Datel tape deck, which is then processed through a Datel reader into the computer. Plots of raw data versus rotation count (there are 500 data points through the full rotation angle of 90°) for the deep water test (north central Pacific, MPG-I) are shown in Fig. 5. These plots are essentially of raw data, where the Datel numbers are converted to voltage and multiplied by an approximate calibration factor for the torque sensor, and the angle of rotation is derived from the data formatting. Therefore the stresses are not corrected for such effects as friction and pressure. The two curves are for two different stations (depths) within the sediment; one is for clockwise and the other for counterclockwise rotation. The "zero torque" plateau region results from a small mechanical gap that was built into the coupling assembly between the shaft and the torque sensor. This assures that a true zero position can be ascertained when there is not torque applied to the vane blades during each succeeding vane rotation. Therefore the actual torque application and vane test begin at the end of this zero-torque position. The shape of the plots reveals an almost linear initial increase in stress, a very definite peak

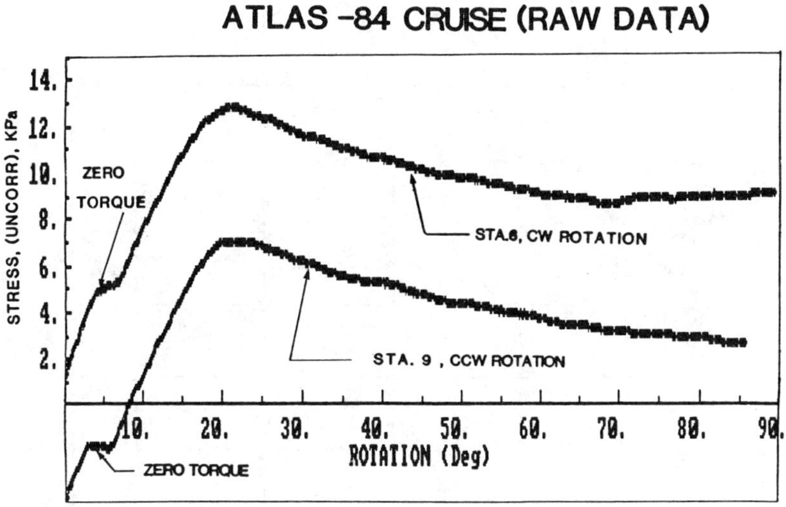

FIG. 5—*Vane shear analysis for Stations 6 and 9 of deep-water sediments, North Central Pacific.*

strength, and then a smooth transition to the residual strength. Ten vane rotations were accomplished with 1.5 m depth, and all the data plots were similarly of good quality.

Similar data for two vane tests in the shallow water sediment (Rhode Island Sound) are shown in Fig. 6. The main difference between this raw data (Fig. 6*a*) and Fig. 5 is that there is an added torque resistance caused by a new arrangement of the lead wires to the torque sensor. This resistance and other frictional resistance effects were evaluated with eight rotations before sediment penetration. After applying the necessary corrections, the torque sensor calibration factor, and the conversion equation, the shear stress can be plotted versus rotation angle as shown in Fig. 6*b*. The shapes of these curves are very similar to those obtained with the laboratory miniature vane apparatus. There is a very sharp peak strength followed by a rapid and large decrease in strength after failure. The ratio of residual to peak strength for these two tests is approximately 0.4. It should be noted that a correction for elastic rotation of the shaft has not been applied here. After this is done, it will be possible to obtain an estimate of shear modulus from the initial slopes of the stress-rotation curves.

Strength Profiles

Profiles of vane shear strength and penetration resistance are presented for three tests with the ASSP system: in deep-water sediments of the North Central Pacific, in shallow water sediments of Rhode Island Sound, and in a full-core sample obtained from the Rhode Island Sound study site. The ASSP results are compared with miniature vane results on nearby core samples.

Deep Water Tests—The hydrostatically actuated corers (HLC) used on the Sept. 1984 cruise (Atlas-84) use the ambient water pressure to drive the core tube into the sediment while preventing movement of the piston [*8*]. The core has an inside diameter of 102 mm, a smooth outside barrel, and tapered nose cone. This very controlled coring operation from a fixed platform usually recovers an excellent quality sample with minimal disturbance. However, at the site the manganese nodule cover is estimated to be 30 to 40%, and it is possible that nodules could be dragged down into the sediment.

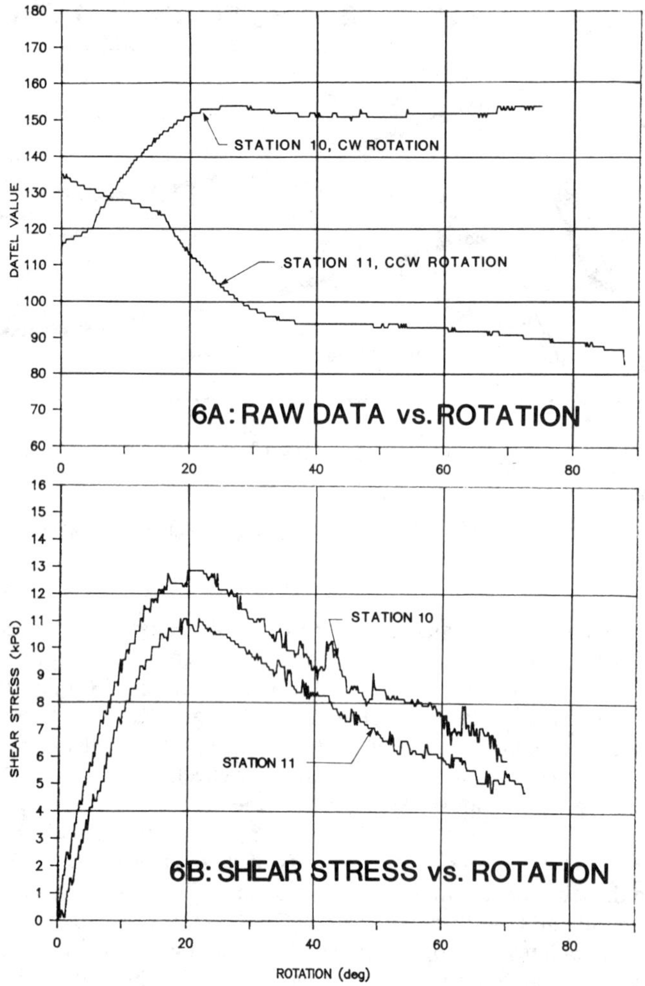

FIG. 6—*Vane shear analysis for Stations 10 and 11 of shallow water sediments, Rhode Island Sound, RI.*

Some typical physical property data for the upper 4 m of the illite-rich clay in MPG-I are shown in Table 1. Throughout the region the upper 3 to 4 m show high "apparent" overconsolidation with OCR values of more than 3 down to 1 m depth [*8*]. The water content profiles of two cores (Fig. 7) show that there are variations downcore. Core HLC-1 was taken on the same lowering as the ASSP test whereas HLC-3 was approximately 66 m away. Comparison with other cores in MPG-I indicate that this variability is fairly typical. In HLC-1 there is a rapid increase in water content from 135% at the surface to about 112% within the upper 0.2 m and a further decrease to about 105% at 0.4 m. The water content increases again below 0.5 m with an average value of 116% to the core bottom at 1.3 m. The average water contents for both cores show a gradual increase with depth, from 111% at 0.1 m to 118% at 1.4 m. The results of the ASSP vane measurements and miniature vane results on the two hydrostatic cores are also shown in Fig. 7. Overall, the in-situ

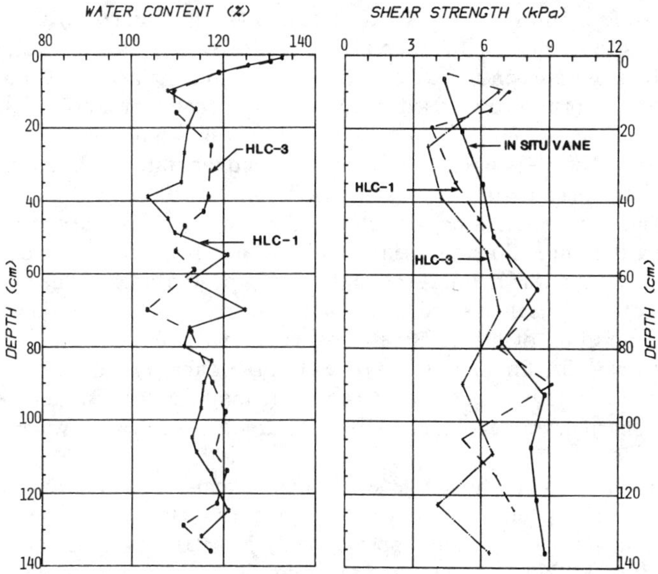

FIG. 7—*Profiles of in-situ results and laboratory results from two large-diameter piston cores. Cruise ATLAS-4 MPG-I, North Central Pacific. Water depth = 5845 m.*

results indicate a strength increase with depth, but at a decreasing rate. The intercept at the surface is slightly less than 4 kPa, and at 1.4 m the strength is about 8.8 kPa. However, between 0.5 m and 1.05 there is variability about the general trend.

In general, the strength trends for the two HLC cores are very similar. During extrusion the HLC-1 core appeared to be in very good condition, but a few points are worth noting. There was extreme variability in the upper 0.2 m with some points higher and one lower than the in-situ results. This may have been caused by compression during the extrusion process but is more likely to have resulted from relative motions caused by ship vibrations. Below 0.2 m the HLC-1 and in-situ trends are remarkably similar; although below 0.9 m the decrease shown in the core results is much greater. Visual observations of HLC-3 indicated some disturbance, possibly by intrusion of nodules, but overall the core seemed to be in good condition. Except for one point at 0.1 m, the shear strength of this core was considerably lower than both the in-situ results and the HLC-1 results. It should be noted that HLC-3 was hauled aboard manually and therefore may have been subjected to more disturbance than HLC-1. However, the mechanical disturbance imparted to these two cores was probably much less than that experienced by normal piston or gravity cores.

In order to quantify the differences between core and in-situ vane measurements, a numerical integration of strength between the depth intervals of 0.2 m to 1.15 m was made to determine the area under the respective curves. The most direct and reliable comparison is for core HLC-1, since this was taken on the same lowering. As shown in Fig. 7, the agreement between in-situ and core measurements between 0.5 and 0.9 m is excellent, but there is disparity above and below this zone. Based on core HLC-1 results, the average core strengths would need to be increased by 15% to obtain in-situ strengths. The corresponding value for HLC-3 is 33%.

Similar comparisons were made with two other nearby (within 1 km) large-diameter gravity cores with correction values of 15 and 44% [*4*]. Comparison of the ASSP vane profiles with all four cores (the two HLC cores and two gravity cores) showed that, as an average, the in-situ strengths are 26% greater than the core results. However, the cores used in these comparisons are considered to be of very good quality and smaller diameter cores, especially standard piston cores, used in deep water probably produce greater sample disturbance and the corrections suggested above may not be applicable.

Shallow Water Tests—The results of the in-situ and laboratory miniature vane tests are shown in Fig. 8. The core sample was obtained within about 110 m from the in-situ tests. The reason that the core data begins below the 20-cm depth is that the corer overpenetrated, causing the upper 20 cm of sediment to be lost into the corer weight stand. One other problem was that the bottom seal on the core pipe was cracked during transport, and therefore the core sample dewatered by an unknown amount. This could have a significant effect on the condition of this relatively coarse grained sediment (the sediment can be described as a gray silty fine sand with some clay, see Table 1). The water content profile indicates a zone of relatively high water content (average w = 66%) to a depth of 75 cm followed by a sharp decrease to about 57% and a trend of decreasing water content below 130 cm.

The in-situ vane test results are generally higher than the laboratory miniature vane results (Fig. 8). The upper 30 cm have low strength (4 kPa), and there is a very sharp gradient to a zone of high strength (approximately 13 kPa) between 30 and 70 cm. A low in-situ strength of 1.2 kPa was measured at 73 cm depth, which appears to correlate with a relatively low strength of 4.3 kPa at 66 cm in the core.

The penetration force data have some interesting characteristics. It must be remembered that the major axial force resistance occurs on the sleeve/shaft bushing at a distance of approximately 9 cm above the mid-plane of the vane. Therefore, for comparison with vane shear data, 9 cm should be subtracted from the penetration depth plot. When this is done, there is a good correlation between the two data sets. The most obvious and dramatic

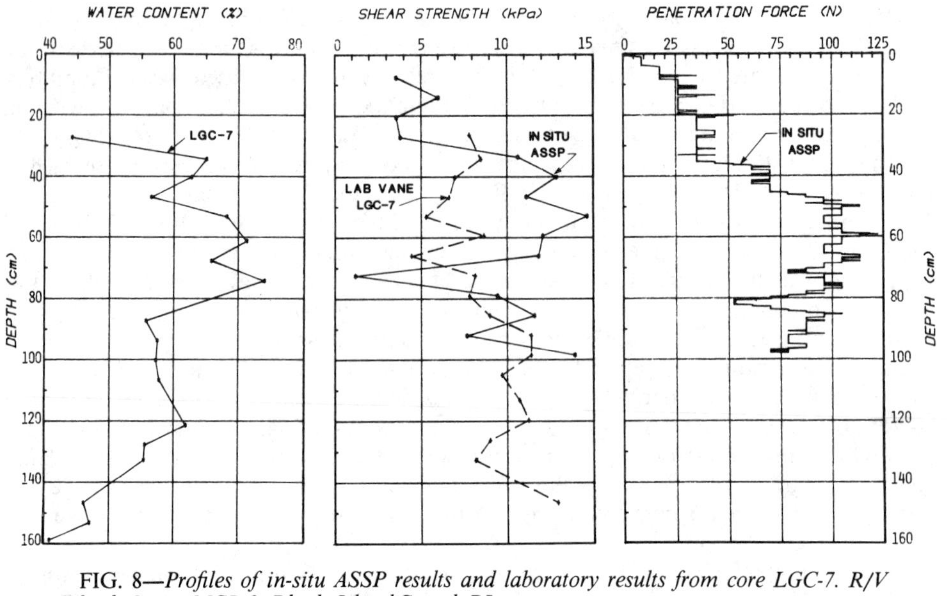

FIG. 8—*Profiles of in-situ ASSP results and laboratory results from core LGC-7. R/V Schock Cruise MGL-3, Rhode Island Sound, RI.*

correlation occurs at the 73 cm depth where there is a very large decrease in force (at 82 cm) corresponding to the low vane shear result. We have not attempted to determine a quantitative relationship to convert penetration resistance to shear strength. The analysis is complicated by the fact that the penetration force comes from a combination of the resistance on the vane, on the shaft/sleeve bushing, and from increasing traction on the buried shaft.

Tests on Full Core Sample—As mentioned earlier, the ASSP test was conducted with a new sleeve/shaft assembly after the Rhode Island Sound cruise. The two cores are approximately 130 m apart from each other. The water contents for LGC-5 (Fig. 9) are somewhat lower than for LGC-7 (Fig. 8). However, the vane shear strengths follow very similar trends, with the ASSP results generally a little lower than those from the laboratory miniature vane. In both cases there is a gradual linear increase of strength with depth and the well defined zones of higher, and lower strengths seen in the in-situ profile (Fig. 8) are not present. Except for a constant zone between 45 and 69 cm (corrected), the penetration resistance constantly increases with depth.

This core test shows that given the same conditions, that is, slightly disturbed cores from approximately the same location, the ASSP gives results that are consistent with other lab techniques. Thus the ASSP results should be a true representation of the sediment's strength. It should be noted that one possible complication with running the ASSP system into a core sample is that there may be significant boundary interactions between the probe and the walls of the core tube. This would affect penetration forces and may also be important in terms of drainage conditions and pore-pressure dissipation around the vane. The ASSP test in LGC-5 was conducted primarily to check operation of the system after the new sleeve/shaft assembly was installed. Based on the results, it appears that the system is operational.

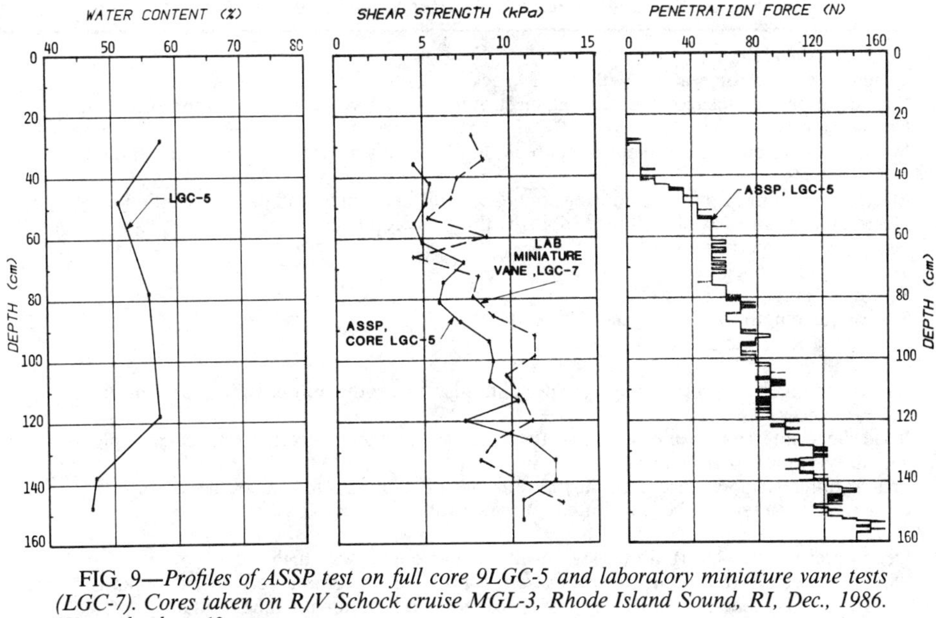

FIG. 9—*Profiles of ASSP test on full core 9LGC-5 and laboratory miniature vane tests (LGC-7). Cores taken on R/V Schock cruise MGL-3, Rhode Island Sound, RI, Dec., 1986. Water depth = 60 m.*

Summary

The autonomous seafloor strength profiler (ASSP) obtains a shear strength profile in the upper 1.5 m of seafloor sediments in water depths of up to 6000 m. A summary of capabilities, specifications, and results of field tests is shown in Table 2. The system is normally mounted on a bottom supported platform and consists of a mechanical system in an oil-filled housing at ambient pressure, an electronic controller/data acquisition package in a pressure housing, and a power package in a separate pressure housing (Figs. 1 and 4). The components are linked together with electrical cables in flexible oil-filled tubing.

With a vane probe, the system is presently programmed to take 24 shear strength tests within the 1.5 m depth. The penetration resistance record taken during each of the intervening penetration sequences affords an independent measure of strength variability. The controller can be reprogrammed for virtually any penetration/rotation sequence, and a cone penetrometer could be substituted for the vane blade. Special needs can be accommodated by using various sizes and capacities of force cell, torque sensor, and probe configurations.

The ASSP has been used successfully in both deep-sea (5845 m) clays and shallow water (60 m) clayey silty fine sand. The vane probe yields a continuous plot of stress versus angle of rotation (Fig. 6) showing a fairly linear initial slope, a definite peak strength, and the typical strain-softening behavior, which eventually reaches the residual strength. The plot of peak strength versus depth in the sediment column provides a profile of in-situ strength (Figs. 7, 8, and 9).

The in-situ vane strengths of the deep-water north central Pacific clay are 26% greater than miniature vane results on four nearby large-diameter cores. The trends of the in-situ profile are very similar to that of a core taken from the same platform lowering (within 370 mm).

TABLE 2—*Summary of ASSP capabilities and results.*

General Characteristics
System autonomy on seafloor with battery pack.
Four component packages: mechanical, electronics, power, volume/pressure compensator.
Can be hard-wired to surface.
Mechanical system is at ambient pressure in oil-filled housing; reduces sealing and friction problems.
Electronic system in pressure housing contains controller, timer, and data acquisition tape deck.
Electrical cables in flexible oil-filled tubing with Envirocon connectors.
Environmental Specifications
Water depth to 6000 m.
Sediment temperature to 300°C.
Sediment strength to 70 kPa (can be extended).
Time on seafloor: to 1 year.
Special Capabilities
Controller: can be reprogrammed for special needs to vary the number of test sites or other sequences.
Vane shear: interchangeable vanes and torque sensor to match special needs. Obtains plot of shear stress versus rotation angle.
Penetration resistance: measured during vane shear profiling, interchangeable sensor.
Cone penetrometer: can be substituted for vane probe.
Results
Deep water test; (5845 m): illitic clay, in-situ vane strengths 26% higher than core.
Shallow water test (60 m): in-situ vane strength profile is higher and very different than core measurements; penetration profile shows excellent correlation with vane results.

For the shallow water test, the in-situ vane strengths are, in general, considerably higher and show more variability than those of a nearby core and the profile are quite different (Fig. 8). The penetration resistance profile correlates very well with the vane profile. A thick zone of high strength and a narrow zone of much lower strength were not detected in the core sample.

Based on these preliminary tests it appears that: (1) in-situ strengths are usually considerably greater than those obtained from core samples and (2) some of the in-situ variability does not show up on core samples. These differences between in-situ and core obtained results are attributed to the changes in the state of stress of the sediment and the amount of disturbance incurred. It seems plausible that some soil types may be more susceptible to disturbance than others. This may be the case in the coarser (shallow water) material by showing a loss in variability from in-situ measurements. These results confirm a need for in-situ testing to a part of any sediment property investigation.

Acknowledgments

Sponsorship for most of the instrumentation development came from the U.S. Department of Energy by contract through Sandia National Laboratories (SNL), Subseabed Disposal Project (Contract Nos. 25-8667 and 95-3078). Personnel who contributed significantly to the early stages of the program were D. Butler, J. Babb, and L. Simoneau (University of Rhode Island [URI]), J. Lipkin (SNL), and P. Pietryka (Datasonics). Others who assisted with the recent work were H. Brandes, H. Mairs, M. Zizza, and F. Pease, all of URI. The assistance of these and others who are involved in the development and research is greatly appreciated.

References

[*1*] Babb, J. D., "Development of an In Situ Vane for Strength Measurement of Deep Sea Sediments," M.S. thesis, University of Rhode Island, Kingston, RI, 1982.

[*2*] Babb, J. D. and Silva, A. J., "An In Situ Vane System for Measuring Deep Sea Sediment Shear Strength," *IEEE/MTS Proceedings, Oceans '83,* Vol. 1, 1983, pp. 598–602.

[*3*] Silva, A. J., Babb, J. D., Lipkin, J., Pietryka, P. and Butler, D., "In Situ Vane System for Seafloor Strength Investigations," *IEEE Journal Oceanic Engineering,* Vol. OE-10, No. 1, 1985, pp. 23–31.

[*4*] Silva, A. J., "Comparison of In Situ and Ship-Board Vane Measurements on a Deep-Sea Clay," "Offshore Site Investigations 1985 Conference, London, England," in *Advances in Underwater Technology & Offshore Engineering,* Vol. 3, Offshore Site Investigation, 1985, pp. 219–230.

[*5*] Hollister, C. D., Anderson, D. R. and Heath, G. R., "Subseabed Disposal of Nuclear Waste?," *Science 213,* 1981, pp. 1321–1326.

[*6*] Percival, C. M., "1983 Subseabed Disposal Project Annual Report: Thermal Response Studies October 1982 Through September 1983," SAND85-1445, 1985, pp. 63–112.

[*7*] Wroth, C. P., "The Interpretation of In-Situ Soil Tests," *Geotechnique 34,* No. 4, 1984, pp. 449–489.

[*8*] Silva, A. J. and Jordan, S. A., "Consolidation Properties and Stress History of Some Deep Sea Sediments," *Proceedings of IUTAM Symposium,* University of New Castle Upon Tyne, United Kingdom, Seabed Mechanics, ISBN 0-86010-504-0, 1984, pp. 25–39.

Author Index

Subject Index

I–L

M

O

P

R

S

Y